The Sea of Lost Opportunity

*HANDBOOK OF PETROLEUM EXPLORATION
AND PRODUCTION*

7

Series Editor

JOHN CUBITT

The Sea of Lost Opportunity

North Sea Oil and Gas, British Industry and the Offshore Supplies Office

Norman J. Smith

ELSEVIER

Amsterdam • Boston • Heidelberg • London • New York • Oxford
Paris • San Diego • San Francisco • Sydney • Tokyo

Elsevier
The Boulevard, Langford Lane, Kidlington, Oxford, OX5 1GB, UK
Radarweg 29, PO Box 211, 1000 AE Amsterdam, The Netherlands

Notice
No responsibility is assumed by the publisher for any injury and/or damage to persons or
property as a matter of products liability, negligence or otherwise, or from any use or
operation of any methods, products, instructions or ideas contained in the material herein

British Library Cataloguing in Publication Data
A catalogue record for this book is available from the British Library

Library of Congress Cataloging-in-Publication Data
A catalog record for this book is available from the Library of Congress

ISBN: 978-0-444-53645-7

For information on all **Elsevier** publications
visit our web site at elsevierdirect.com

Printed and bound in Great Britain

11 12 13 14 10 9 8 7 6 5 4 3 2 1

Working together to grow
libraries in developing countries

www.elsevier.com | www.bookaid.org | www.sabre.org

ELSEVIER   BOOK AID International   Sabre Foundation

For my wife and family, who saw so little of me for so many years.

Contents

List of Tables

List of Charts

List of Figures

Acknowledgements

This book could never have been written without the help, assistance, and encouragement of many people. During the research phase, Professor Alex Kemp and Dr. Richard Perren of the University of Aberdeen gave unstinting help and advice, while Professor Roger Wootton of the City University was generous with ideas and sources. Mr. John Westwood of Douglas–Westwood Associates kindly read the draft.

I must also thank the libraries to which I paid so many visits, in particular, the Queen Mother at the University of Aberdeen, the Templeman at the University of Kent, the London Business School, the Energy Institute, and the British Library. In all cases, the staff gave freely of their time and expertise. The same is also true of the staff at The National Archives, the BP Archive, Lloyds Register, and UKOOA (now Oil and Gas UK) where much of my research was conducted. Many veterans of the North Sea oil and gas industry, mainly now in retirement, contributed to the work. Without their knowledge and opinions, so freely given, the content would have been very much the poorer.

Finally, I must thank my wife, Valerie, for her self-sacrifice in allowing this endeavour to over-shadow the early years of our retirement and my daughter, Gail, and her husband, Allan Graham, for helping when my IT skills proved inadequate for the task.

I was a member of the generation in Britain that came to maturity in the 1960s. Many of us were less concerned with the pleasures of 'the swinging sixties' than with what appeared to be a steady and irreversible decline in our country's economic and industrial fortunes. We felt that – unless 'something unforeseen' turned up – the country lacked an escape route from a future offering more of the depressing same. Many decided to emigrate. This was the era of the 'ten pound pom' when a cross-section of a million mainly young people moved to Australia, while the better qualified joined the 'brain-drain' to the United States, then excitingly engaged in a 'space race' with the USSR. My own new wife and I thought of joining them, but the 'something' unforeseen did turn up and we decided to stay.

That 'something' was North Sea oil and gas. Suddenly, the United Kingdom had become the possessor of a major new energy source, one that offered the prospect of injecting a massive new source of wealth into the economy at exactly the points where it was most needed – the balance of payments, government revenue, and as a regeneration opportunity for declining regions of the country and whole industrial sectors.

It was this last point which excited me, employed as I was then by a leading producer of capital goods, itself in need of new markets. As I saw it, with its strong oil companies, powerful shipping interests, and an unrivalled history of engineering innovation, the United Kingdom would surely not only be able to satisfy the new domestic demand for offshore goods and services but also go on to take a large share of the overseas offshore markets that were sure to emerge. The British government seemed bound to recognise and encourage this process. I determined to make my career in the offshore oil and gas industry, and for the next 35 or more years that is what I did.

My dreams were not to be fully realised. To be sure, the macroeconomic benefits of North Sea oil and gas duly arrived, facilitating the restructuring of the British economy; even now, they continue to contribute very substantially to the economy. Despite again being mired in an economic crisis, Britain is no longer singled out as 'the sick man of Europe'; after the 'credit crunch' crisis, others now are in much the 'same boat' as she is.

However, British-owned firms generally failed to emerge as leading players in the world market for offshore goods and services or even to succeed in unequivocally dominating the domestic market. Despite the fact that for many years the UK Continental Shelf represented the largest and sometimes also the most technically advanced segment of global demand for the goods and services required for the exploitation of offshore oil and gas, few indigenous offshore service and supply businesses of truly international scale developed, leaving the United Kingdom badly placed to gain a significant share of overseas markets.

To that extent, the United Kingdom has not enjoyed the full benefit of the North Sea discoveries, notwithstanding some 25 years of government support through its Offshore Supplies Office (OSO). The task of this book is to attempt to try to explain why. To the best of my knowledge, it is not something that has been attempted for 25 years, if really at all. In other words, it has not been a matter of concern except to those directly engaged in the industry.

Perhaps it should be, and for several reasons. Firstly, it is a contribution to the history of a vital stage of the UK technical and economic development, perhaps the most important since Second World War. Secondly, it shows, from an industrial viewpoint, how the British handled the exploitation of their most significant natural resource gain of the twentieth century. Thirdly, it may assist governments and industries faced with future instances of unforeseen, specialist, and large-scale new demand to manage their reactions more effectively. Fourthly, it throws light on how governments can pursue strategic industrial objectives whilst leaving market mechanisms to function with minimal interference, something some administrations – perhaps even the British – may wish to do now or in the future.

The book does not attempt to fully document the great scale, scope, and urgency of the effort required, mainly in the decade beginning in 1965, to come to terms with the unprecedented technical challenges posed by the exploitation of North Sea oil and gas. In addition to organisations individually identified, many other government agencies, professional bodies, academic and research institutions, and indeed individual companies also contributed to their resolution.

The main focus is upon British-owned businesses; deliberately so, though perhaps British-headquartered would have been a better criterion to adopt. This would recognize that some companies (although in reality only a very small minority) develop a shareholder register in which nationals of the country of origin become a minority, largely as a result of the activities of international institutional investors. However, the operational head office usually remains in the same place along with a decision-making process still rooted in the local culture and thus subject to local pressures.

Few major countries have had as open an attitude towards foreign inward investment or the foreign takeover of established domestic businesses as the United Kingdom, where virtually everything has been permissible, save possibly foreign takeovers of the major companies in the ultimate strategic industry, defence. While foreign direct investment is almost universally welcomed in developed countries, there is often much less enthusiasm for foreign takeovers of existing businesses, particularly where these are deemed to be of strategic importance, a term which itself can be interpreted in different ways. In France, it appears that it extends even to some consumer goods but many other countries take a more restrictive stance, confining controls to activities that affect national security. In most cases, these would include energy supply and its supporting activities where many would seek to ensure that foreign ownership did not come to predominate. Serious scope for disagreement exists on where the balance should correctly lie, making reciprocity difficult to achieve.

Foreign takeovers are usually justified (purely ideological free trade arguments apart) on the basis that they improve productivity and profitability through the introduction of new management techniques and inward technology transfer. As generalisations, there may be some truth in these assertions. However, many British companies in the energy support industry have been acquired, often before having reached maturity, because they offer opportunities for outward technology transfer as well as entry to new markets. Few attempts seem to have been made to assess the effect of foreign takeovers on the tax base (bearing in mind that most acquirers are multinationals with more opportunities for tax planning and transfer pricing than purely local firms) or on employment, particularly on how these extend to supply chains. Taking such factors into account over and above improved capital efficiency (itself primarily a benefit for the new owners), it is not obvious that foreign takeovers always increase the GNP, let alone the tax base.

Moreover, only the most extreme proponents of openness towards foreign-ownership will deny that it can sometimes bring disadvantages extending beyond the ownership of assets and income streams, though still falling short of threats to national security. For instance, major decisions relating to international investment, marketing, research, development, and design are almost always made in the country of control, usually also that of ownership. Employees who are non-nationals may also sometimes have additional obstacles to overcome to reach the most senior management positions in the controlling entity.

In the case of industries dependent on the exploitation of a non-renewable natural resource, the eventual decline of local activity is more likely to lead to the ultimate withdrawal of a foreign rather than a domestic owner, which will normally maintain its corporate functions and can seek additional overseas business without fear of intragroup conflict.

The period covered by the main chronological narrative begins with the drilling of the first well offshore the United Kingdom in 1963 and ends in 1993 the year of implementation of the European Single Market Act, which effectively ended government support for the British offshore service and supply industry. The main emphasis is on the industry's formative years, broadly 1965–1980. In addition to the introductory and chronological narratives, industry segment and corporate case studies are presented, giving some insight into the factors driving the decisions of individual managements and to the outcomes. A postscript deals very briefly with events since 1993.

Trying to assess OSO is a thankless task as it is impossible to know what would have happened in its absence. Inevitably, an attempt must be made to explain why the overall outcome was what it was and to suggest how British industrial performance might have been improved in the conditions then pertaining.

As far as the economic background to the period studied is concerned, the published sources employed were extensive. Official publications apart, the works of Robinson and Morgan (1976 and 1978) provided the most comprehensive coverage of the implications of the North Sea to the balance of payments. With respect to Britain's perceived economic decline, no parallel existed, with diverse opinions offered by many different authors. Explanations ranged from the very broad, such as the inherited institutional failings postulated by Elbaum and Lazonick (1986), to the very narrow, such as the social attitudes of the upper classes suggested by Wiener (1981).

North Sea oil and gas also generated a very large literature of its own, most in paper format but Internet and recorded speech resources were also involved. Only a small proportion of the material was of more than peripheral relevance to the issues of central concern to this work. A few of the general 'overview' works, particularly Arnold (1978) and Harvie (1994), did provide some useful material and there were additionally a small number of publications directly dealing with government industrial support policies, namely Jenkin (1981), Cook and Surrey (1983), and Cameron (1986). Hallwood (1986 and 1990) had published material bearing specifically on offshore supply business issues from both theoretical and practical standpoints, but his scope was narrow, focusing on service companies in the Aberdeen area. An unpublished PhD thesis (Pike 1991) took a rather broader view but still suffered from having a Scottish rather than a UK-wide focus.

Overall, a publications' review encouraged me to believe that his work would fill a gap in the North Sea literature by offering an integration of public policy concerns with 'hands-on' business issues, drawing in part on my own experience at a senior level both in OSO and the private sector. Nothing of this nature was identified in the literature.

Archive work was of more importance than the literature review and much of the previously unpublished content derives from the National and BP Archives and from United Kingdom Offshore Operators Association (UKOOA) records. The last had the added benefit of opening a 'window' to the records of bodies where UKOOA was a corporate member, such as minutes of the Oil Industry Liaison Committee (OILCO). In the first two cases, there was a 'thirty year' disclosure rule, but UKOOA granted access up to 1995. This obviated the need to make more than limited use of the Freedom of Information Act, a decision facilitated by the knowledge that relatively few of the files of the government department of most interest, OSO, had been preserved and that those that had seemed unrepresentative in character.

With respect to National Archive files, it was decided to try to identify the main themes that seemed to be present irrespective of the party in power. How to do this effectively presented a difficult problem given the large amount of material available from a variety of official sources. It is most unlikely that everything of relevance was unearthed. Parliamentary debates and political party policy statements were largely, though not entirely, ignored.

Over 30 participants in events were consulted directly, the majority by face-to-face interview and/or questionnaire, although there were also telephone and e-mail enquiries. A few people also provided me with private papers, in one case specially prepared. The respondents are characterised in the text by type of position rather than individually identified. Although not a statistical sample, it is believed that the informants were reasonably representative of decision takers and managers. They included ministers, senior civil servants both within and outside OSO, entrepreneurs and executives of oil and contracting companies, with US, French, and Norwegian nationals among them. My own recollections, as a participant, have been included, as far as possible without drawing attention to their origin. This last point raises the question of how objective I have been, particularly about OSO, which I had the privilege to direct for a short but exciting period. It is a fair question but one for the reader to answer. I can only say that I have done my best to avoid an autobiographical bias and to be as accurate as I can. Nevertheless, I must ask readers to accept that recollections stretching back 40 or more years may not always be entirely correct, an observation that applies to my kind respondents as well as me. If bias and errors arise from overreliance on my personal recollection, they are likely to be concentrated in Chapter 5.

Where considered necessary for comparative purposes, monetary values are shown in both 'money of the day' and constant (2008) price terms, the adjustment being made by use of the Gross Domestic Product (GDP) deflator calculators developed for the £ sterling by L. H. Officer and for the US $ by S. H. Williamson. I must thank the originators for their agreement to this. The oil price is always given in US $, whether at current or constant prices.

Finally, with reference to the book's illustrations, I was surprised to discover that even reputable image suppliers such as those I have used cannot always be absolutely certain that they are the copyright holders of a particular image. In such cases, I have made an effort to identify possible alternative copyright holders. However, if I have nonetheless been guilty of any inadvertent infringement, I can only apologise and invite the copyright holders to contact me.

Norman J. Smith
November, 2010

Chapter 1

In Europe's Sick Bay: Britain before North Sea Oil

Although it may seem strange to devote the opening pages of a book about the North Sea offshore oil and gas industry to broad economic and industrial issues, it is important to do so. Appreciating the circumstances and perceptions of the time offers the prospect of an insight into the mind-sets of those who made the policy decisions 50 or so years ago.

When the offshore industry reached the United Kingdom (UK) in the 1960s, the country already had a well-developed industrial base and much prior experience in the exploitation of oil and gas. It was the domicile of some of the world's largest exploration and production companies.

At the same time the British economy was facing considerable difficulties. It was inflexible in character – a result of government industrial policies, poor management, entrenched trade union power and a scarcity of venture capital. Governments became increasingly pre-occupied with a range of negative economic indicators such as balance of payment deficits, the public finances, poor productivity growth, strikes, increasing unemployment and rising inflation. There was a widespread perception that Britain was in relative economic decline. It became commonplace to speak of the combination of negative trends in terms of a 'British disease', which – unless cured – would condemn the UK to grow at a slower rate than its peers.

The long-term outlook appeared to be one of over-extension and continued relative decline, with short-term policy driven by the balance of payments and exchange rate considerations. Inevitably, once it became clear that this newly arrived industry was likely to add a significant increment to the nation's resources and to make its most substantial impact by easing the balance of payments constraint, it became the focus of political attention. This attention was to be heightened by the hope that substantial economic benefits would flow to the areas closest to the oil and gas fields, many characterised by declining heavy industries and rising unemployment. From the early 1970s, government concern over security of oil supply – shared by the oil companies, whose interests were otherwise purely commercial – added another powerful driver.

Norman J. Smith, The Sea of Lost Opportunity.

Whilst there are grounds to criticise British government policies towards the offshore supplies industry, such criticisms need to recognise that, in addition to immediate crises such as the 1972 and 1974 coal miners' strikes and the 1973 oil price hike, governments were heavily constrained by what were seen at the time as long-term economic problems of a structural nature. For investors in what was from the outset a costly and risky endeavour, having to address the high expectations of the government and the public whilst seeking to meet their own business objectives was challenging. It was not made easier by the fact that the very British industries to which it would be natural to turn to as suppliers, such as shipbuilding and major capital project construction, were clearly already facing great difficulties.

1.1 THE BRITISH BALANCE OF PAYMENTS PROBLEM

In the then world of fixed exchange rates and with the pound sterling still having the status of a reserve currency for many of the UK's former dependencies, the most pressing of the constraints faced by British governments was usually the balance of payments. The balance of payments was thus commonly perceived as the main 'driver' of short-term government economic policy. Thus, the importance of the shift from being a net oil importer to being a net exporter, which followed the development of the United Kingdom Continental Shelf (UKCS), should not be underestimated, particularly because it also brought security of oil supply and increased government revenues in its wake.

During the period from 1947 to 1976, Kirby (1991, p. 23) recognised no less than eight cycles of boom and slump, a pattern that became known as the 'stop–go' cycle. Two of these, 1964–1967 and 1973–1976, respectively, saw the genesis of the British offshore gas industry and the most active phase of British offshore oil development. Conditions in both these periods were 'extreme' in terms of factors other than their protracted length, with correspondingly 'extreme' implications for government policy. The first was characterised by a perceived speculative severity that led to sterling devaluation in November 1967, although the current account was actually close to balance (Thirwall and Gibson 1992, p. 238).

The second period, during which the pound sterling was already floating freely, combined a domestic crisis with the international one that followed the Yom Kippur War of 1973, the associated steep increase in oil prices and the Arab oil embargo. At home, a government lacking a clear Parliamentary majority faced industrial unrest, a depreciating currency and rising inflation. On this occasion, the British government was unable to contain the crisis by its own efforts. In June 1976, it was announced that a Group of Ten Nations had loaned it $5.3 billion (Thirwall and Gibson p. 249), roughly $16.2 billion in 2008 terms. This loan formed part of Britain's growing medium- and long-term foreign currency borrowings, which by November

1976 totalled about $18.5 billion (Arnold 1978, p. 327), over $56.5 billion in 2008 values.

Accumulated mainly in the previous 3 years, with the main repayments falling due in the 1980s when it was believed North Sea oil production would be at its peak, even this level of borrowing did not prove sufficient. At the end of the year Britain provided the International Monetary Fund (IMF) with a Letter of Intent containing commitments about the conduct of economic policy. In return, the IMF and the Group of Ten granted the country standby credits of $3.5 billion, or some $10.7 billion in 2008 terms, although these facilities never needed to be fully implemented (Wass 2008, p. 306).

Typically, balance of payment crises brought periods of economic expansion to a premature end through interest rate increases and associated fiscal and monetary tightening; sometimes more direct action was taken, such as the imposition of the import surcharge in October 1964. The resultant contraction in domestic demand reduced imports and took the pressure off the exchange rate without the need – except in 1949 and 1967 – to devalue the pound sterling. Even after the pound was floated in 1972, the fear of a deteriorating balance of payments being followed by the inflationary consequences of a weakening exchange rate was slow to disappear.

Although the improvement in the balance of payments permitted the reversal of the 'stop' measures and the resumption of 'go', the cycle reduced the long-term rate of economic growth and the productive potential of the UK because, in the view of Pollard (1984), the reduction of demand during 'stop' phases bore particularly hard on investment expenditure.

Pollard saw the origins of the 'stop–go' cycle in the legacy of Britain's position as a victorious power in 1945. The government itself was responsible, in his view, because of its slowness to adjust to Britain's reduced status in the post-war world when it ceased to be a global imperial power, leading to at least two pretensions.

The first was the maintenance during the period of fixed exchange rates of the Sterling Area, an arrangement through which countries – mainly former British dependencies and including a number of oil producers – used sterling as their international trading currency and held their foreign exchange reserves in London in the form of Sterling Balances. Although it need not always have worked to Britain's disadvantage, commentators such as Pollard mainly regarded the existence of the Sterling Area and sterling's status as an international trading currency as sources of weakness, amplifying cyclical balance of payment problems and constraining exchange rate policy by making the maintenance of a fixed parity a guiding principle of economic strategy.

The second was the government's own expenditure abroad on military expenses, foreign aid, and the servicing of overseas loans required to finance earlier deficits. In this analysis, the underlying cause of the balance of payments problem lay in the inability of the private sector to generate sufficiently large surpluses to finance the official sector's overseas deficits. Pollard (1992, p. 307) noted the total private balance was in surplus in every individual year between 1961 and 1970. Overall, its surplus for the decade totalled £6.45 billion. By contrast, the total current government balance was in deficit in every individual year and accumulated a total deficit for the decade of £6.14 billion. The position deteriorated in the next decade. Between 1971 and 1980 the total private balance was in deficit in three of the 10 years, although it returned a cumulative surplus of £17.44 billion. The total current government balance remained in deficit in every year, returning a cumulative deficit of £19.45 billion.

1.2 OIL AND THE BALANCE OF PAYMENTS

By far and away the worst deficit in the total private balance was that of nearly £2.08 billion (about £16.2 billion in 2008 terms) recorded for 1974, the year following the quadrupling of oil prices in late 1973 and to which in large measure it was due.

After the Second World War, the British economy had moved away from its dependence on domestically produced coal towards increasing reliance on imported oil, exposing the visible trade balance to events in world oil markets, as was dramatically illustrated in 1973–1974. Between 1968 and 1972 alone, oil increased its share in primary energy use in the UK from little more than 40% to approaching 50%. Although much of the increase reflected the growth of private motoring, there were also other factors, such as replacement of coal by refined petroleum products in the gas and railway industries and the construction of oil-fired power stations.

As long as oil was cheap (long the case prior to 1973), the growth in imports gave relatively little cause for concern – particularly as British-owned oil and shipping companies earned profits from the trade such that it could be estimated that even at 1974 prices, the foreign exchange cost of imported oil was only about 85% of the total cost (Baring Brothers 1974, p. 56). Moreover, Reddaway (1968) calculated that between 1956 and 1964 the annual post-tax profits of overseas subsidiaries of British oil companies averaged about £120 million, perhaps of the order of £2 billion a year in 2008 terms.

Oil imports grew steadily but did not increase their share of total visible imports much until the effects of the 1973 'oil shock' became apparent in 1974 (see Table 1.1).

In 1974, the visible deficit on the petroleum trade rose sharply to reach over £3.4 billion (about £26.5 billion in 2008 terms), more than three and a half times greater than the figure for the previous year. The deficit was almost entirely accounted for by crude imports, since trade in oil products was almost in balance.

It is worth noting that, by this time, substantial benefits were already accruing to the UK economy from the production of natural gas, which had begun to flow from the southern North Sea basin in 1967. However, Robinson and Morgan (1978, p. 148) concluded that this did not seem to have much

TABLE 1.1 UK Oil Trade 1964–1974 (Money of the Day)

Oil trade	1964	1965	1966	1967	1968	1969	1970	1971	1972	1973	1974
Imports (£m)	448	476	479	530	664	657	673	888	913	1,322	4,197
% Total visible imports	8.9	9.4	9.1	9.3	9.6	9.1	8.5	10.4	9.3	9.4	19.9
Oil trade balance (£m)	(306)	(338)	(339)	(390)	(483)	(480)	(450)	(642)	(674)	(952)	(3,429)
Oil price ($ per barrel)[a]	1.80	1.80	1.80	1.80	1.80	1.80	1.80	1.80	1.90	2.83	10.41

Sources: Treasury (1975), United Kingdom Balance of Payments. BP (2010), Statistical Review of World Energy.
[a]Arabian Light Crude.

effect on national economic performance, although by 1976 natural gas production should have been yielding a visible trade balance saving of around £1.7 billion (or over £9 billion in 2008 terms). This estimate was based on natural gas production substituting for about 34 million tonnes of oil at the then current oil price. An earlier official estimate – Treasury (1976) – had suggested slightly larger effects, but provided no details of its assumptions, a failing noted later in the same year in Robinson and Morgan (1976, p. 15).

Although the effects of southern North Sea gas production on the economy and on the balance of payments in particular were far from insignificant, they failed to excite the interest generated by North Sea oil, probably because they were not associated in the public mind with potential relief from the 'oil crisis' or with an awareness of great technological feats. Indeed, the main impact on the general population of southern North Sea gas was the progressive conversion of virtually all gas appliances to natural gas, itself a successful industrial project of considerable scale.

Even prior to the oil price increases of 1973, it had become apparent that North Sea oil reserves were significant and that there would certainly be production and hence import savings. After the price increases, interest escalated both for commercial and strategic reasons, for the effects were now clearly going to be magnified and almost certainly much larger than those emanating from the southern North Sea basin gas production. As Robinson and Morgan (1978, p. 147) pointed out, with the Saudi Arabian government revenues per barrel of light crude having risen from $0.90 in 1970 to over $11 by late 1975, striking oil at this particular represented *'remarkable good fortune'* for Britain.

Whilst the oil companies saw the potential of politically secure investments and improved supply flexibility, the government saw the prospect of an immediate improvement in credit-worthiness and a medium-term escape from the balance of payments constraint that had come to dominate its economic policies. The Bank of England (1980, p. 451) recognised that even if North Sea oil did not necessarily prove to be a large windfall gain, it would allow the UK to escape the big losses experienced by other countries.

Although the government necessarily had unpublished internal forecasts, much of the official financing activity undertaken during this period took place prior to the publication of official estimates of the potential effects of oil production on the balance of payments. In any case, with so many other more immediate issues also to consider, it would be wrong to over-estimate the role of prospective North Sea oil revenues in the thinking of those tasked with British economic management in the 1970s. That much is made clear by Wass (2008), a member of this small group. Nonetheless, in his account of the events of 1974–1977, he makes references to the anticipated effect of North Sea oil on the balance of payments, as well as one (p. 333) to its position as a *'security'* for borrowings.

Forecasts could only be attempted on the basis of comparing the position of a Britain with the North Sea with that of a 'non-North Sea' Britain. This required making assessments of a range of industry-specific issues, such as the extent to which foreign financing would offset the cost of imports required for development, or to which the anticipated repatriation of the associated interest, profits and dividends would eat into the visible trade benefits – see Treasury (1977a). Additionally, it demanded a continuation of traditional macro-economic forecasting. None of these could be predicted with much pretence of accuracy, a state of affairs that did not deter many from attempting the calculation.

Many projections of the potential balance of payments effects were produced. Seven, three official and four independent, are compared in Table 1.2 below for 1980, the year in which oil self-sufficiency was achieved. Although the 'interest rate effect' (the interest savings/earnings from reductions in official overseas liabilities/increases in overseas official assets) has been removed where explicitly present, other 'consistency' alterations could still be made to adjust for differences in approach. Even then, the time (and hence informational) differences would preclude any absolutely direct comparisons.

TABLE 1.2 Summary of Potential Balance of Payments Effects from North Sea Oil (or Oil and Gas) for 1980 (£ billion)

Source	Year Made	Coverage	Money	Forecast in Money of the Day	Forecast in 1980 Prices[a]
Baring Brothers	1974	Oil and gas	1973 Prices	2.03[b]	6.02[b]
Oppenheimer	1976	Oil	1974 Prices	2.05	5.08
Treasury	1976	Oil	Current prices	4.90	8.54
Hoare Govett	1976	Oil and gas (current account)	Current prices	5.83	10.16
Treasury	1977	Oil	1976 prices	4.30	7.50
Robinson and Morgan	1978	Oil	Current prices	4.00–7.00[b]	5.53–9.70[b]
Bank of England	1982	Oil and gas	1980 prices	7.50[b]	7.50[b]

[a]Adjusted by GDP deflator at market prices.
[b]Less interest rate effect.

It would have been asking too much of the forecasters to have expected them to predict the oil price and the sterling/dollar exchange rate with either accuracy or consistency. After the near quadrupling in 1973, the oil price rose slowly in nominal U.S. dollar terms for the rest of the decade before more than doubling in 1979 and rising somewhat further the following year. Until it strengthened sharply at the turn of the decade as the UK approached net self-sufficiency in oil, sterling progressively weakened against the U.S. dollar. Such movements favoured North Sea development and enhanced the size of the prospective balance of payments benefits.

Nevertheless, adjusted by the Gross Domestic Product (GDP) price deflator, only two of the other six forecasts remain outside the plausible range suggested in Robinson and Morgan (1978) and then only narrowly so. Such a 'consensus' would have no doubt reinforced the view among decision makers that great balance of payments gains were to be had from a rapid development of North Sea oil.

Space precludes more than a brief mention of the seven individual forecasts or of the differing definitions and assumptions employed. The earliest of the estimates was produced for circulation to clients of a merchant bank in early 1974 – Baring Brothers (1974, pp. 37–40, 51–59). While it made no attempt to consider the indirect balance of payments effects, such as resource displacement, it did address the full range of other issues considered by later forecasters, including both oil and gas effects – making some attempt to distinguish between them – and quantified one factor often ignored elsewhere, loss of earnings from producing oil overseas and shipping it to the UK. It was also highly specific in setting out the assumptions underlying its forecasts.

In March 1976, Oxford economist Peter Oppenheimer published a point estimate of a balance of payments benefit in 1980 based on 1974 prices (Oppenheimer 1976, pp. 18–25). His work was particularly interesting because it provided an early and very useful analytical framework. Although recognising the need to make assumptions about such matters as the oil price, UK production volumes and costs, which would ideally be used in a sensitivity analysis, he stressed the necessity to start with basic economic principles, showing that subject to the macro-economic assumptions he had employed, the net gain from North Sea oil production to the Gross National Product (GNP) would be equal to both the government's oil revenue and the gain to the balance of payments.

Treasury (1976) attempted to estimate of the potential balance of payments benefits of both oil and gas production. It was at pains to emphasise the complexity of the undertaking and the imprecision of the results. Given the particular caveats and assumptions employed, the resultant estimates showed the maximum balance of payments effect of North Sea oil.

Another estimate of the potential balance of payments impact of oil and gas combined produced in 1976 was one by stockbrokers Hoare Govett – cited by Arnold (1978, p. 328–330). Though differing in format and detail from the Treasury estimates and dealing only with the current account, it also forecast very large positive effects.

In the following year, Treasury (1977b) published an updated estimate. It again pointed out the lack of a clearly right way of making the calculation. It spelt out the assumption that the UK's oil demand would be unaffected by North Sea production.

Professor Colin Robinson and Dr Jon Morgan of the University of Surrey became particularly prominent academic commentators. Their approach was exposed in a highly developed form in Robinson and Morgan (1978, pp. 147–184). Although structured in the familiar 'bottom-up' form, it differed from earlier exercises in a number of respects. Robinson and Morgan were explicit in describing their conceptual framework (acknowledging a debt to Oppenheimer) and, most importantly, in specifying their assumptions in detail. For the two most sensitive assumptions, the level of oil production and the development of oil prices, they included three separate scenarios. With a 25-year period (1975–2000) to consider, the sensitivity calculations were computerised and dealt with a range of issues in addition to the balance of payments. It followed from this that Robinson and Morgan should produce their estimates for individual years in ranges rather than as point estimates. Having considered the probabilities, they were able to exclude some of the more extreme cases and produced what they considered to be plausible ranges for each of 1980, 1985 and 1990.

Net self-sufficiency in oil during 1980 did not bring to an end an interest in the balance of payments effects of the North Sea, and as time passed these would increasingly reflect history as well as forecasts. In early 1982, the Bank of England published a wide-ranging review of the implications of the development of the North Sea to the UK economy – see Bank of England (1982). As well as addressing the balance of payments, again pointing out that the results could be no more than indicative, it reinforced points already touched upon in some of the papers considered above. For instance, it stressed that the resource costs of developing North Sea oil were high and that the international oil price rises of the 1970s had cost the UK more than it had gained from the discovery and development of the North Sea, although that still left the UK better off than other industrialised countries forced to continue to import. This was because at the 1980 level of oil prices, the revenue per tonne of North Sea oil was in excess of £120 against an estimated real resource cost of £35, creating a 'rent', which the government sought to appropriate as far as possible through royalties and taxation.

Alone among the balance of payments forecasts mentioned in Table 1.2, this last contained mostly history, although the figures for 1981 were only an estimate and a projection was provided for 1985. While the issue of what exactly to cover remained, the need to make assumptions had become less. A discussion of the broader implications of North Sea revenues concluded they neither justified a spending 'bonanza' nor necessitated major structural change to the British economy.

From about 1980 onwards concerns about the balance of payments ceased to pre-occupy the government. Few could doubt that this was primarily a result of Britain's move in the space of a few years from importing virtually all its oil requirements to being a net exporter. Although other factors were probably also to some extent responsible for the accompanying strength of sterling, it was common to 'blame' North Sea oil for the resultant difficulties experienced by large swathes of British manufacturing industry and to claim the benefits of the North Sea had been 'wasted' on increased imports and in paying for unemployment. The resultant debate spawned an extensive economic literature, particularly notable contributions to which were Forsyth and Kay (1980) and Byatt et al (1982).

Great difficulties are involved in attempting to forecast the effect of any new industry on a national balance of payments, let alone on the entire economy. Most who attempted it for North Sea oil found it convenient to ignore some of the most intractable issues, such as the future course of oil prices and in particular the effect on the exchange rate and thus on activity in the 'non-oil economy'. Nevertheless, as the 1970s progressed, the growing number of optimistic projections gave the government good reason to believe that rapid development of the resources of the North Sea would offer release from what had been an over-riding economic policy constraint. They also hinted at the large increases in government tax revenues that lay ahead.

1.3 BRITISH ECONOMIC AND INDUSTRIAL DECLINE

If the balance of payments problem drove short-term economic policy, worries over Britain's relative economic and industrial decline dominated longer-term thinking. Concern with the country's international competitiveness went back at least as far as the middle of the nineteenth century (Dintenfass 1992). Interest in the subject increased as evidence of continued under-performance accumulated, peaking during the 1960–1980 period, the critical time for North Sea exploration and development. This is hardly surprising; between 1950 and 1973 the UK was overtaken by France, Germany and Italy in terms of real GDP per hour worked (Cairncross 1992 p. 6).

Britain's ability to pursue simultaneously the objectives of raising living standards, introducing a welfare state and remaining a world power became increasingly implausible as poor economic growth failed to deliver the increases in National Income required (Barnett 1986 and 2001) – leading Labour governments in particular to intervene.

The Labour government of 1964–1970 tried briefly to operate a National Plan and established an Industrial Reorganisation Corporation (IRC), whilst that of 1974–1979 tried indicative planning agreements and a National Enterprise Board (NEB), initiatives scrapped when the Conservatives returned to power. Ideological shifts led to alternate nationalisations by Labour governments and denationalisations by Conservative ones; in the case of the steel industry both occurred twice within a generation. The adverse effects of the resultant diversion of effort from routine tasks of industrial management, let alone more complex ones such as innovation and entrepreneurship, received little attention.

Governments of both persuasions tried to influence the behaviour of trade union and industrial leaders through dialogue and to modernise education and training for industry. Both parties devoted considerable effort to encouraging inward investment, particularly in regions of high unemployment. Although the original rationale was based on providing employment and sometimes import saving, as the evidence accumulated that Britain was lagging behind its foreign competitors, it came sometimes to be seen also as a means of replacing a failed British management model with a more dynamic overseas one. The model was usually American, since the United States of America (USA) was then the main source of inward investment.

The extensive literature on the possible causes of Britain's poor economic performance is essentially inconclusive. Many hypotheses were advanced but, as far as the author is aware, none became a consensus view. This is probably inevitable when the phenomenon operated over a period in excess of a century and almost certainly was the product of a number of factors that interacted in different ways at different times. Britain's historic legacy was without doubt an important influence. Elbaum and Lazonick (1986, p. 2) went so far as to say they believed the decline stemmed from institutional rigidities developed during the period, essentially the nineteenth century, when Britain was the world's strongest economic power.

The main theories circulating in the mid- to late-twentieth century can be grouped into three types. The first is the purely economic. For instance, the UK entered the twentieth century with a much smaller proportion of its labour force in agriculture than the other members of its peer group, offering the latter a 'built-in' productivity growth advantage as they had more scope to transfer workers from agriculture into higher productivity occupations as argued by Phelps Brown and Browne (1968, p. 63).

The relatively low rate of UK capital formation is given great importance by Pollard. (1992, pp. 295–299), not only for its direct impact on labour productivity, but also because new investment embodies the current technological 'state of the art'. Explanations for low investment include insufficient savings and the international orientation of UK capital markets, leading to a misalignment of domestic industrial investment needs and finance, and an increased cost of domestic capital. The 'stop–go' cycle initiated by balance of payments problems also disrupted investment programmes.

The 'crowding out' theory, associated with Bacon and Eltis (1978) became popular during the 1970s. It maintained that the public sector absorbed resources, which otherwise would have supported marketed output and productive investment, to a greater extent in Britain than elsewhere.

Crafts (2002) blamed the adoption of tariffs in the 1930s compounded by a lack of competition in the 1950–1970s when the nationalised industries functioned as monopolies and 'industrial policy' kept inefficient private companies in business. The population at large met the bill through high taxation and, others might have added, higher inflation.

Other economic explanations focussed on an export market structure orientated towards ex-imperial markets rather than the then faster growing industrialised countries where competition was often fiercer (Cairncross 1992, p. 19). This argument featured prominently in the debate over whether the UK should join the European 'Common Market.'

It was also sometimes claimed that Britain's research and development (R&D) expenditure was for long badly matched with the pattern of world trade, being heavily focussed on defence and aerospace (Dintenfass 1992, p. 49). Edgerton (1996, pp. 62–63) argued that this had not resulted in an under-spend on civil R&D and more generally showed the difficulty in attributing decline to failings in the fields of science and technology.

Two other classes of explanations might be described as socio-economic. The first concentrates on the attitudes, personal aspirations and educational backgrounds of the ruling and managerial elites. Some of the most radical views are those found in Wiener (1982), which sought to show that influential echelons of society developed an aversion for the capitalist ethos and an ensuing ambivalence towards industry, although less so towards finance. As summarised by Dintenfass (1992, pp. 12–13, 26, 39, 54–57), others found deficiencies including poor entrepreneurship, design and salesmanship, technological conservatism and adherence to outdated work practices. These could in part be laid at the door of inadequate education and training, particularly technical, at all levels. Cairncross (1992, p. 23) and others drew attention to a denial to qualified engineers in Britain of the prestige they enjoyed elsewhere.

A second socio-economic explanation, widely held when the North Sea oil and gas industry arrived, argued that poor industrial relations were the root cause of economic decline. Weak or ineffective management was perceived to be regularly making concessions to a recalcitrant labour force, whose leadership sometimes functioned 'officially' through the medium of an outdated trade union structure and sometimes 'unofficially' at the behest of 'left-wing militants'. This image was certainly exaggerated. Although the unionised proportion of the British labour force was higher than in most comparable countries, the UK was rarely near the top of the international strike league, although the validity of international comparisons is open to question.

The USA, which generally had a strike record if anything worse than that of the UK, was not normally thought of in terms of weak and ineffective management and some might even see a willingness to face up to strike action as an indication of strong management. Moreover, the industrial structure differed between the various countries and British strikes, at least, tended to be concentrated in a narrow range of industries. At various times motor vehicles, ports, coal mining, shipbuilding and construction all achieved notoriety.

The last two deserve a closer examination as important to the offshore supplies industry that came in the wake of North Sea oil and gas discoveries. Widespread demand for mobile drilling rigs and marine support vessels, which – or so it seemed – could be constructed in British yards was foreseen. The construction phase of an offshore oil or gas field development appeared to share many of the features of other major construction projects, apart from the fact that part of it took place offshore. Onshore facilities, such as terminals, seemed even more analogous.

The decline in British shipbuilding was easy to chart. Between the World Wars, Britain accounted on average for 40% of world output by tonnage, a proportion that had fallen by the mid-1960s to less than 10% (Lorenz and Wilkinson 1986, p. 116). Both export and domestic markets were lost. Governments from World War Two onwards were driven to intervene, culminating in nationalisation in 1977, a time when the global industry was engulfed in a collapse of demand following the 1973 oil crisis. During the 1980s, the Conservative government shut down most of the commercial yards of British Shipbuilders (BS), while the main naval yards were privatised. Heavy losses in the new Offshore Division accelerated this process (Johnman and Murphy 2002, pp. 212–214).

The reasons for the industry's decline were deep-seated and not confined to industrial relations. Failure to adjust to market changes hastened its demise. Over a relatively few years, demand for larger, standardised and generally simpler vessels, particularly tankers, bulk carriers and container ships, displaced that for 'bespoke' passenger and passenger-cargo liners, in which British yards had excelled. Other factors included a fragmented structure of mainly small private firms operating from long-established yards, often poorly laid out in cramped locations, reluctant to invest in capital equipment, technical training or modern management techniques and willing to allow the labour force largely to control the production process. An abundant supply of skilled labour, partly based on traditional shipbuilding crafts and partly on newer metal working trades, prepared to bear the costs of demand fluctuations in one yard by moving to another, was for long a major competitive advantage for the British industry.

The substitution of welding and prefabrication techniques for more traditional shipbuilding technology worked in favour of the generally larger and more modern overseas yards where there was greater reliance on up-to-date management techniques, semi-skilled labour and capital investment than semi-autonomous craftsmen. Yards that specialised early benefited from economies of scale and the learning curve.

Unlike the situation in competing countries, where trade unions tended to be industry-based and to encompass all trades, in Britain each trade (or craft) had its own union. As the British industry declined, the craft unions fought to protect their own 'patches', both against other crafts and against 'dilution' by semi-skilled labour. This proved a considerable barrier to the introduction of new techniques and equipment even when the employer was willing to introduce them. So-called 'demarcation' or 'who does what' disputes – many examples of which are given by Barnett (1986 and 2001) – became notorious, contributing not only to a growing reputation for missed delivery promises and cost over-runs but also to a picture of an industry dominated by a 'bloody-minded' self-destructive labour force. The very names of the crafts involved – shipwrights, boilermakers, platters, riveters, caulkers, braziers, drillers, woodworkers, joiners, etc. – were redolent of Victorian times and contributed to the industry's antiquated image. Dealing with them absorbed much management effort that might otherwise been deployed more productively elsewhere.

The oil and gas industry came to recognise that it would be unwise to rely on British shipbuilding too heavily as a source of capital equipment and that if the shipbuilding labour force, with its many relevant skills, was to be attracted into fabrication work for the offshore industry, there must be a risk that it would try to bring its 'bad practices' with it.

As for the construction industry, it was those parts that were most concerned with major capital projects which were both of most potential relevance to the development of oil and gas resources and also the most prone to time and cost over-runs. There was no single cause, with poor planning and project management, late deliveries to site, client design variations, contractual disputes and labour problems all being involved.

By the time the National Economic Development Office (NEDO) – an organisation where government, business and trade unions engaged directly with each other – had established a Working Party on large industrial sites in 1968, it had already been recognised that a national problem existed. From the late 1950s onwards there had been a steady growth in the number and scale of major site-based capital projects, particularly power stations, refining and chemical process plants and ferrous and non-ferrous metal plants. Whereas cost over-runs were serious issues in their own right for the companies concerned, delays in completion and the resultant lost production could have more extensive and far-reaching consequences, including adverse balance of payments effects.

When its report was published as NEDO (1970), the Working Party showed that in each sector concerned with the construction of major industrial projects, there had been lengthy delays and large cost over-runs, as illustrated in Table 1.3. Only in Oil & Chemicals, where performance was markedly better than Power Stations and Oil Gasification, did private ownership predominate at the time, suggesting that public sector involvement was detrimental to tight project control.

TABLE 1.3 Excess Costs and Programme Delays in Major UK Projects (Late 1950s to Late 1960s)

Industry Segment	Sector Control	No. of Projects	Median Original Cost Estimate in Money of the Day	Median Original Cost Estimate in £ 2008	Median Cost Over-Run	Maximum Cost Over-Run	Median Planned Time-scale	Median Time Over-Run	Maximum Time Over-Run
Power Stations	Public	13	£65.82 Million	£1.040 Billion	19%	50%	72 Months[a]	14 Months[b]	29 Months[b]
Oil & Chemicals	Private	16	£6.68 Million	£105.57 Million	3%	33%	20 Months	1.6 Months	9 Months
Oil Gasification	Public	7	£6.70 Million	£105.88 Million	11%	23%	Circa 30 Months	8 Months	23 Months

Adapted from NEDO (1970, pp. 125–126).
[a]Overall.
[b]First Generating Set.

TABLE 1.4 Attributed Causes of Delay in Construction of Large Industrial Sites

Attributed Cause of Delay	% of Companies Ranking as Most Important
Late design changes	27%
Late delivery of materials or plant	21%
Unexpectedly low labour productivity	10%
Labour disputes	8%
Delays in subcontractors	8%
Other contractors' performance	8%
Access	7%
Skilled labour shortages	5%
Other causes	6%
All causes	100%

Adapted from NEDO (1970, p. 86).

The most widely publicised source of disruption – labour disputes – rated only fourth equal in importance (see Table 1.4). A played only a minor role by comparison with late design changes and late deliveries from suppliers, respectively, first and second in importance.

Nearly all the issues mentioned in Table 1.4 were subsequently to feature in North Sea development and in much the same order. As the Working Party found, simply classifying reasons of delay did not advance matters much, since each was symptomatic of some underlying cause(s), usually of a 'cultural' nature. In all, the report made some 50 recommendations relating to management by the client and the contractor, project programming, craft training and labour relations, with the last also published separately in summary (NEDO 1971).

A follow-up report was written by a second Working Party (NEDO 1976), concluding that a single national site labour agreement was the *'primary instrument of reform'* of industrial relations. It also found that the provision of more stable employment would be the most important step the industry could take to improve employer/employee relations and productivity, that systematic training programmes for both craftsmen and management needed to be established and that clients and contractors both needed to reappraise their management of projects and their respective roles.

There were additionally a number of other significant observations such as the need to recognise that a project's eventual cost and date of completion more important than the amount of the original offer, the importance of high calibre project management, the advantages of a clear separation in time between the design and construction phases and the great significance of the form of contract, all lessons to be learnt by the North Sea oil and gas industry. It also concluded that inter-union demarcation disputes (a feature of large industrial sites as well as of shipyards) were less of a problem as a cause of labour disputes than as a reason for poor working practices resulting in low productivity.

The second Working Party had been established in 1975, with the remit of comparing engineering construction performance in the UK with that for similar projects elsewhere in Western Europe and the USA. Having studied seven British and eleven foreign projects active in the period 1968–1975, it found that for foreign projects in general both overall and construction times were shorter and less prone to delay. Projects were mostly executed with higher construction efficiency due to lower manning levels and higher productivity on specific tasks. Very importantly, delays incurred during projects were capable of retrieval by the end.

It was possible to have reservations in that exact comparisons were impossible, that British projects were more complex and British clients imposed a higher standard of engineering excellence than those overseas. The second Working Party did not believe these suggestions powerful enough to undermine its conclusions.

After examining both the pre-construction and construction phases, this Working Party found that 'site problems' leading to low employee morale were the major source of the UK's poor relative performance with capital projects. This manifested itself in many different ways including poor productivity, poor attendance records, a tendency to disregard procedures and striking without adequate cause. It also reported that site-negotiated bonuses were widely seen as a powerful source of labour disputes.

In addition to endorsing the initiatives of its predecessor, particularly the creation of a National Agreement for the Engineering Construction Industry, which eventually came about in 1981, the second Working Party made recommendations of its own aimed at improving 'on site' performance. These mainly called for further study of specific issues related to labour relations, attitudes and productivity.

NEDO continued to try to improve performance on large sites. In NEDO (1981), it showed that on typical projects it could be possible to reduce site man-hours by 10–15% by the use of vendor works packaged units, although at extra costs in areas such as design, management and transport. Such techniques were already widely employed for North Sea projects. There followed NEDO (1982), which sought to select key performance success factors from project definition via engineering, procurement and construction to commissioning, as well to address industry-wide issues like recruitment, training, motivation, supervision and experience sharing.

By the time of these two later reports, representatives of the upstream oil and gas industry and its contractors and suppliers dominated the composition of the Working Parties. Despite the unpromising background and a difficult start, they were eventually to show that major projects could be successfully executed in Britain.

The UK upstream oil and gas industry could have observed the risks of being a client of the British shipbuilding and construction industries from many perspectives. One of the most dramatic was from its involvement with Teesside, long geared to accommodating heavy industry – specifically iron and steel, engineering, shipbuilding and chemicals. In an attempt to constrain the growth of unemployment, from the mid-1960s until the end of the 1970s, both Conservative and Labour governments earmarked the area for large-scale industrial expansion, backed by major infrastructure investment. Industrial investment was heavily concentrated on two industries, chemicals (including oil refining and petrochemicals) and steel (Foord et al 1985).

Chemicals were the more important. Both British and overseas multinationals took considerable advantage of generous regional development subsidies, either to invest in new capacity or to replace outdated plant. They included Imperial Chemical Industries (ICI), Philips Petroleum, British Oxygen Company (BOC), Monsanto and Shell UK. For steel, the matter was entirely in the hands of the nationalised British Steel Corporation (BSC), which came into being in 1967.

During the critical years of North Sea oil development, Cleveland (the Teesside towns and surrounding area) became the UK's on-shore fixed capital investment 'hot-spot'. In every fiscal year between 1975/76 and 1979/80, it absorbed between 24.5% and 29.4% of regional development grant payments made in Great Britain (Foord et al, p. 32). Its proportion of the population for the Development Areas as a whole probably averaged less than 5%.

With investment on this scale and so heavily focused on a succession of large industrial sites, it was inevitable that the problems discussed above would manifest themselves on Teesside. Labour problems became almost endemic. Sound industrial relations on sites were made particularly difficult by the absence of a National Agreement. ICI had initially dominated the area's major construction sites and operated its site labour relations under the Engineering Employers Federation's Mechanical Construction Engineering Agreement. However, an influx of American oriented management contractors acting for in-coming firms eroded ICI's position by introducing a different site labour relations framework – that of the Oil and Chemical Plant Constructors Association. The resultant confusion enabled politically motivated militants to exploit the situation, spreading disruption from the Cleveland Potash Mine and Treatment Plant to the steel, chemical and infrastructure sites. Between 1979 and 1982, the problems were largely resolved thanks to the introduction of a new structured management approach by ICI, which was adopted by NEDO. Both the Engineering Employers Federation's and the Oil and Chemical Plant Constructors Association's agreements were superseded by the National Agreement for the Engineering Construction Industry and a newly instituted 'Teesside Understanding' involving clients, contractors and unions (Mullen 2002). By the time these improvements had occurred, the first and largest of the North Sea investment booms was already over.

Teesside's substantial involvement with offshore oil and gas began early. Of the 11 drilling rigs completed in British shipyards between 1961 and 1969, three were constructed on Teesside (Johnman and Murphy 2002, p. 211), only for the business to be abandoned. By 1975, oil was arriving at the Phillips terminal on Teesside from Norway's Ekofisk field. A year later, oil was being shipped by Shell to its Teesport refinery from its Auk field.

Teesside's fabrication yards in general fared rather better than its process plant construction sites and shipyards. The Laing Offshore platform yard at Graythorp, opened in 1972, had successfully completed three large oil production platform jackets by the time it closed in 1976 in the face of declining demand. Module yards established by Redpath Dorman Long (RDL), Cleveland Bridge and Whessoe functioned with relatively little labour disruption, it is said because the work force was drawn from different crafts and operated under different agreements from those on the construction sites. Local company, Wilson Walton, deserves mention as over 18 months in 1973–1975, it successfully completed the world's first conversion of a semi-submersible drilling rig (the *Transworld 58*) to a floating production facility to be utilised on the Argyll field, which produced the UK's first offshore oil. Significantly, a U.S. firm, Bechtel, which provided engineering, procurement and management services, supervised the conversion.

Oil and gas companies with exposure to Teesside in the 1970s suffered less than some other sectors from the local industrial turmoil, but saw first-hand the scale of disasters that could overtake major industrial projects.

1.4 AN INSUFFICIENT INHERITANCE: THE BRITISH OILFIELD SUPPLY INDUSTRY

Despite its early origins and long control of a large part of the world's known oil reserves, by the mid-twentieth century, the British petroleum industry had developed only a relatively small supply sector, necessarily based mainly on overseas markets. Though it contained elements that were petroleum specific (e.g. in oilfield tubulars), many of the products it sold to the upstream oil and gas industry were general mechanical and electrical engineering items. It was particularly weak in drilling and well services and in oilfield and offshore contracting. The British oilfield supply industry prior to the advent of the North Sea is discussed in more detail elsewhere (see pp. 41–48).

Part of the explanation of its weakness is that the major British companies traditionally provided in-house much of the requirements for their far-flung operations. By contrast, the U.S. petroleum industry, then heavily biased towards domestic activities and with many small oil and gas producers, had developed a large service and supply sector with accepted standards and specifications, with which British oil companies developed links around the world. Some American service and supply companies were already using the UK as their Eastern Hemisphere base before the initial North Sea discoveries. Others followed as their existing U.S. clients became active in the North Sea.

Crucially, in the decade and a half leading up to the start of North Sea exploration, a specialised offshore technology had evolved in the Gulf of Mexico (GoM), with offshoots in other benign marine areas. There was a small European component to this technology but it was mainly in American hands.

Shell and BP first became offshore operators in the early 1950s. Originally their contractors and suppliers were largely British, though specialist U.S. oilfield equipment suppliers were involved from the outset. As time went on, USA, French and Italian contractors were drawn in. State-sponsored offshore R&D had begun in France before the development of the North Sea and Italy also took the opportunity to enter the offshore industry early thanks to gas discoveries in the Adriatic.

The first phase of exploration and development of the southern North Sea was relatively modest in scale and of short duration, roughly 1965–1970, but did involve significant demand for items where there were no established UK suppliers such as offshore structures and pipelines and marine installation contracting. The government's main objective was, and always remained, the rapid exploitation of offshore resources. It limited industrial support to encouraging the oil and gas companies to place orders with British suppliers. The entrepreneurial response was weak and did not have the time to become well established. Most contracts went to American and Dutch suppliers, resulting in a widely cited estimate of UK content of about 30%.

The opening up of the southern basin had seen little increase in the still small indigenous capability or in the stock of trained labour to tackle the bigger technical problems and business opportunities in the northern North Sea. In the early stages of northern basin exploitation, the problems were more starkly defined than the opportunities, which were clouded by uncertainties over the size of the reserves base. For the American and Dutch firms involved, the development of the North Sea southern basin had shown the limitations of their existing benign environment technology, limitations which would be exacerbated by the much more severe conditions in the northern basin. Nonetheless, the southern basin experience allowed them to retain two vital attributes – the most relevant experience to date and the confidence of their customers, on which they were able to build as they entered the northern basin. Potential British competitors enjoyed no such advantages when they tried later first to enter North Sea business at a technically much more demanding point.

By late 1971, the government concluded that the expenditure involved in the development of the northern basin would be a multiple of that involved in the southern basin and that unless local content was dramatically increased, an unacceptable loss of jobs and foreign exchange would result. At the start of 1973, it set up the Offshore Supplies Office (OSO), with the remit of increasing the local content to 70%. This was too late to instigate the rapid expansion of domestic capacity necessary to prevent the huge surge of new orders then already being placed without a heavy reliance on imports and inward investment by foreign suppliers. As measured by expenditure, the all-time peak in UKCS development expenditure (then over 80% of total UKCS expenditure) occurred in 1976, largely reflecting orders placed in 1971–1974 (see Chart 4.1, p. 94).

The first published figures on the UK content in UKCS expenditure for a specific year related to 1974 (Department of Energy 1975) and showed it as only about 40%, illustrating how heavily UK industry must have 'lost-out' during the first great North Sea 'boom'. For it to have been otherwise would have required a government policy aimed at expanding the scope and capacity of the small export-oriented British oilfield supply industry several years earlier than was actually the case.

Chapter 2

The Genesis of the North Sea Oil and Gas Industry

Although few thought of Britain as an oil and gas producer prior to the North Sea, the domestic petroleum industry had a long history. Surface and coalmine petroleum and bituminous seepages were recorded at various locations from the late seventeenth to early nineteenth centuries. From the mid-nineteenth to mid-twentieth centuries, there was significant industrial activity, mainly in the West Lothian area of Scotland and also in other places such as North Somerset, based on the extraction of mineral oil products from oil shale. Less known was the small-scale production and use of natural gas at Heathfield in Sussex during part of the same period.

However, the origins of the British oil exploration and production (E&P) industry, a phenomenon of the late nineteenth and early twentieth centuries, were associated with more oil-prone overseas territories, particularly where they formed part of the formal British Empire or zones of British influence, rather than with the home islands. Territories such as Burma, Persia and Trinidad spring readily to mind. It was in overseas operations that the major British oil companies, British Petroleum (BP), Burmah Oil (Burmah) and Shell Transport and Trading (Shell) and their minor brethren had their roots.

Recognition of the growing importance of oil and of the difficulties of relying on imports under wartime conditions led in 1918 to the first serious exploration campaign in Britain, resulting in a number of small oil and gas discoveries and/or shows. None were commercial and interest declined (Woodward and Woodward 1973, p. 15). Following improvements in oil exploration techniques and the 1934 Petroleum Production Act, which vested ownership of petroleum deposits in the Crown and also allowed the granting of exploration licences subject to payment of a royalty on production, interest revived (Woodward and Woodward, p. 16). The exploration campaign mounted between 1936 and 1939 was much more successful than its predecessor. In 1937 the small Eskdale gas field in Yorkshire was found, followed in 1939 by Britain's first significant oil field – Eakring in Nottinghamshire (Arnold 1978, p. 33). Together with smaller adjacent fields and the Formby (Lancashire) discovery, Eakring made a modest but very

Norman J. Smith, *The Sea of Lost Opportunity*.

valuable contribution to wartime oil supply, while the Cousland gas discovery in Scotland was used to supply the town of Dalkeith.

After the Second World War, onshore exploration carried on more or less continuously, although the level of activity fluctuated. It was rewarded by the discovery of a number of small oil and gas fields, mostly in the South and Midlands of England, and one major oil field. This was Wytch Farm in Dorset, which – although entirely drained by onshore wells – has a large off-shore extension.

Despite the long history, onshore petroleum E&P activity has not been sufficient to have much more than a narrow local impact on economic and industrial issues. By contrast, exploitation of the oil and gas resources of the North Sea had impacts extending well beyond the adjacent coastal areas and directly or indirectly affected the UK as a whole.

2.1 THE MOVE TO THE NORTH SEA

For oil and gas exploration to take place in the North Sea, three conditions had to be satisfied. First, there had to be good reason to believe that reservoirs might be present. Second, an international legal framework allowing national jurisdictions over the continental shelf had to be in place. Third, economic extraction had to be feasible. The last condition is considered in more detail later and here it is sufficient to state that by the early 1960s it had been met.

The first condition was satisfied well before the others. Reference has already been made to the existence of oil and gas discoveries in Britain many years prior to the first North Sea discoveries. Small deposits had also been identified onshore in France, Germany and the Netherlands, suggesting that oil and gas reservoirs might lurk beneath the adjacent waters. It is not known when drilling in the North Sea was first proposed, but it was certainly no later than 1955 (Arnold 1978, p. 35).

In 1959, with the discovery of the huge Groningen gas field in the Netherlands, came the first indication that there might be large quantities of gas under the North Sea, given that the field's geology was analogous to the southern North Sea basin's geology (Upton 1996, p. 21).

As for the second condition, initial agreement on the principles for the division of continental shelf areas for exclusive mineral exploitation by littoral states had been reached in 1958 with the negotiation of the Geneva Convention on the Law of the Sea. This was ratified by the UK in early 1964, allowing enactment of the Continental Shelf Act, which, *inter alia*, extended offshore the provisions of the Petroleum Production Act and made the Gas Council – the precursor of the British Gas Corporation (BGC) – an effective monopsony buyer for North Sea gas. The same year saw the first Continental Shelf (Jurisdiction) Order creating English and Scottish areas of the UKCS. The award of exclusive exploration licences now became possible.

Whilst the Geneva Convention set out the general principles of dividing continental shelves among the littoral states, individual negotiations between neighbouring states were necessary to determine the exact marine boundaries. Agreements with the Netherlands, Belgium, Denmark and Germany were soon and easily reached, the last three having only short boundaries with the UK. Agreements proved more complex with Norway, France, Ireland and the Faeroes, the last not being achieved until the close of the twentieth century.

Agreement with Norway was also reached quickly (1965). The boundary adopted, which as with the other eastern littoral states followed the median line principle, nonetheless proved controversial in some circles in Britain. The cause was its disregard for a deep submarine trench close to the Norwegian coast, which in theory would have allowed the UK to have pressed for its sector to extend across the North Sea to the western edge of the 800-metre deep trench. As it turned out, much of Norway's extensive oil and gas reserves were subsequently discovered to lie in what could have been argued to be UK waters. With hindsight, there was, therefore, a huge penalty paid in economic and industrial terms by the UK for its failure to press for a boundary at the trench margin rather than the point of equidistance. Part of the penalty was a large reduction in the scale of the domestic market for the British-based service and supply industry.

Why this generous attitude was adopted towards the Norwegians, if such it was, remains unclear. The easiest explanation is that the Foreign and Commonwealth Office (FCO) was simply pursuing an established policy for using the median line to determine boundaries. Harvie (1994, p. 86) claimed it was because the British wanted to clear legal issues so that exploration could proceed unimpeded. This view was supported at an Institute of Contemporary History seminar by John Liverman (1999), a former senior Department of Energy (DEn) official. He believed that had the UK adopted a harder line, arbitration would have delayed drilling in the disputed area by 5–10 years.

Others at the seminar took a different view. John Guinness (another former senior civil servant) implied that the FCO did not pursue a strongly based legal case for the trench boundary. James Allcock (a former senior BGC executive) believed that the Norwegians were astonished by the British concession. Both suggested there was a desire to be friendly towards the Norwegians, a point with which the former managing director of the British National Oil Corporation (BNOC), Alastair Morton, concurred.

Settling the main North Sea boundaries by 1965 allowed rapid exploration of the UKCS, producing a succession of gas discoveries from 1965 in the southern North Sea and oil discoveries from 1969 in the central and northern North Sea. Gas production began in 1967 and oil production in 1975.

The First Licensing Round (1964) had been quickly followed by a Second (1965), with a lull before the Third (1970) and the Fourth (1971/72). This process started and ended under Conservative governments interrupted by the 1964–1970 Labour administration. However, throughout there was a single key figure in the licensing process, Angus Beckett, Undersecretary in the Petroleum Division of the Ministry of Power. His success in 'kick-starting' the exploration process in a new province can be judged from the number of blocks awarded – 348 in the First Round, 127 in the Second, 106 in the Third, and 282 in the Fourth. The Fourth Round took place after oil as well as gas discoveries and was more skewed towards the northern North Sea. Beckett was subsequently heavily criticised for the supposed 'leniency' of the licensing regime with its term of up to 46 years and alleged 'sell out' to multinational oil companies (Harvie 1996, p. 86).

There was always a large-scale involvement of foreign, particularly American, companies, but the authorities long used their powers to give preference in licence awards to British-owned companies, which received 30% of blocks awarded by 1980. The policy extended to existing independents and later to the new independents established by City institutions or by industrial and commercial organisations. The state-controlled Gas Council (later BGC) and National Coal Board (NCB) were encouraged to participate with American companies, respectively Amoco and Conoco. Under the 1974–1979 Labour Government, particular preference was given to the state-controlled BNOC. The British-operated proportion of reserves was far higher than it was of blocks. BP, Shell (also operating on behalf of Esso), and Burmah's group each obtained a licence found to contain a giant field, respectively Forties, Brent and Ninian.

2.2 THE TECHNOLOGICAL 'STATE OF THE ART'

Petroleum has been exploited from early times. However, the modern oil and gas industry is normally dated from 1859 when 'Colonel' Drake successfully 'bored' for oil in Pennsylvania in the United States. Discoveries quickly followed in several other states, of which California, Oklahoma and Texas soon came to dominate.

The USA remained the world's leading oil producer until well after the Second World War, although by then there were at least 20 other commercial producers. During the formative years of the early and middle twentieth century, its dominance was remarkable. Between 1910 and 1948 – a period in which world oil production expanded more than 10-fold – the American share fell only from 64% to 58% (Anglo-Iranian 1948, p. 7). This long period of supremacy ensured that specialised American oilfield equipment and service companies enjoyed a learning curve, economies of scale and opportunities for innovation unavailable to competitors elsewhere, though some of the more sophisticated technical advances originated in Europe. Assisted by overseas investment by U.S. oil companies and European investment in the USA, American oil field practice spread widely abroad. By 1943, the cumulative effect was such that American drilling crews in England's Eakring oil field were able to drill and complete a well in an average time of 1 week compared with a best time of 5 weeks for a British crew (Woodward and Woodward

p. 145). The difference lay partly in the more modern U.S. drilling equipment, but working practices were also superior.

U.S. oilfield technology originated in a predominantly agricultural society. It is no coincidence that an accumulation of oil or gas became known as a 'field' or that the fluid used in drilling as 'mud'. There are still occasional references to equipment as 'iron', reflecting former dependence on the rural blacksmith's craft, although increasing dependency on scientific principles and engineering technology can be observed as the industry developed.

Surface signs such as oil seeps originally dictated selection of drilling sites, as was the case with Drake's well (Yergin 1991, p. 27). The frequent association of oil deposits with anticlinal structures and, in particular, with salt domes became more widely recognised as the nineteenth century drew to a close; this led to the engagement of professional geologists to identify surface features related to such structures and to gather evidence from visible stratigraphy. Where surface evidence was unavailable, or to find more complicated types of oil traps, prospecting had to await the development of geophysical techniques, such as gravimetric and magnetic anomalies and in particular seismic surveys, which provided information on the subsurface.

Partly under the stimulus of the First World War, geophysical techniques were first developed in Europe, particularly by the Germans. They reached the USA in about 1923 or 1924 (Yergin pp. 218–219). Another European advance in the high technology field, this time French, was electric well logging. By the measurement of resistivity in the surrounding rock formation in an open well, this signals the presence of oil or water and allows correlation to be made between wells by matching zones of high and low resistance and the zones between them. Soon after its original discovery in about 1927, the technique crossed the Atlantic. When combined with the analysis of rock cores recovered from the well, this procedure permitted a great build- up in subsurface knowledge, helping to spark the emergence of the petroleum and reservoir engineering disciplines and thus rationally planned and managed field development.

If European scientists led the way in exploration techniques, the drilling of very large numbers of wells by practical American engineers, often untrained except 'on-the-job', allowed the USA to set the pace in drilling technology. Early wells were frequently sunk by animal-powered percussion equipment, with a weight repeatedly dropped from a derrick. A more sophisticated type of percussion equipment was the cable tool rig where a heavy bit (usually of a fishtail design) was lowered repeatedly into the hole with a spinning motion (Rundell 1977, p. 18), a method surviving in a few places into the 1940s. However, it was the rotary drilling rig (of ancient origin) that gradually became the standard method of drilling wells, which it still remains. Robert Beart had patented fluid circulating rotary well drilling in England in 1844 (Schemp. 2004). Developed in the 1870s for the drilling of water wells, it was introduced to the Texas oilfields in 1882 (Rundell, p. 25). The motive power for drilling also changed, with animals first giving way to steam and subsequently diesel and electric drive. Wooden drilling derricks yielded to steel structures, leading to development of the unitised compact mobile drilling rig with jack-knife masts.

Early wells used water as a lubricant and flush. The famous Texan Spindle-top discovery well of 1901 was drilled with a rotary rig and is believed to have been the first to employ 'mud' – produced by cattle milling around in a shallow pond. When pumped down the hollow drill stem, it proved sufficiently viscous to return the drill cuttings (Rundell, p. 37). The use of mud soon became standard drilling practice and was found to have valuable properties in addition to those of chippings carrier, lubricant and coolant. Most importantly, its weight was found to contain subsurface pressures; mud also restrained unwanted ingress of fluids and gases into the well by caking the sides. With time, a sophisticated mud technology developed, with muds made up of various minerals and chemical additives tailored to meet specific requirements.

Although the introduction of heavy drilling muds offered some means of downhole pressure containment, the prevalence of uncontrolled blow-outs (or 'gushers'), often leading to fires, remained a major problem until use of the blow-out preventer (BOP) became standard. It was first marketed in 1924 (Pike 2003). The initial frequency of blow-outs and fires in the USA fostered the development of well control and oil field fire-fighting expertise. Another practical skill developed early was 'fishing', the recovery of objects from downhole.

The first use of cement to seal and strengthen a well was probably in California in 1903; it was eventually adopted industry-wide (Franks and Lambert 1985, pp. 22, 36). In 1909, the conical roller cone bit was invented, allowing a straight round hole to be drilled quickly and efficiently (Rundell, p. 37). Early drillers were unable to measure unwanted angles in wells, although their presence was easy enough to recognise when attempting to set steel casing in a well. By 1924, angle measurement techniques had been introduced, at least in California (Franks and Lambert 1985, p. 108).

Whilst a high proportion of early discoveries flowed naturally, this was not always the case or sometimes it did not last very long, leading to a search for means to enhance or extend production. Surface drive suction pumps were certainly in use in California by about 1902 (Franks and Lambert pp. 16, 34). As in the case of drilling rigs, power sources evolved, with windmills and steam power giving way to mainly electric drives. By 1910, explosives were being employed downhole to fracture producing formations so that oil could move more easily to the well bore (Rundell p. 167). An alternative approach using hydrochloric acid (acidisation) was first used in Texas in 1932 (Rundell p. 174). By the 1940s, more sophisticated and capital-intensive techniques such as gas re-pressurisation and water injection were well understood (Anglo-Iranian 1948, p. 60).

By then, the practice of allowing oil and solution gas to separate naturally in open ponds had given way to the use of well-head separation plants and pipelines had become the preferred means of transporting oil and gas. It was no longer routinely acceptable to treat associated gas as an unwelcome waste product to be vented or flared regardless of volume. Increasingly, USA regulators were requiring it to be recovered as a by-product.

By the outbreak of the Second World War much of what was to become standard offshore drilling and production practice had been pioneered on land, but addressing purely marine petroleum issues remained in its infancy. This was despite the early origins of offshore petroleum activity, widely believed to have been in California in the late nineteenth century, where discoveries along the coast and oil slicks at sea led readily to the conclusion that there was oil offshore.

According to patent records, the notion of marine oil well drilling goes back at least to 1869 when a New Yorker designed an offshore drilling platform with hydraulically telescoping legs and an attendant service tender – concepts not to be employed until well into the following century – Kriedler (1997, p. 40).

Although Franks and Lambert (p. 46) report primitive attempts at offshore oil recovery as early as the 1880s, according to Rundell (p. 119), the first offshore well in the Santa Barbara Channel was not drilled until 1897 or 1898 – from a pier reaching out into the sea, almost the only method of offshore oil development until 1927. In that year, and also in California, the first directionally slanted (or 'whipstocked') well was drilled out to sea from the beach. This innovation – which was later to prove of great importance to offshore development – was accompanied in the late 1920s and early 1930s by drilling from artificial islands or from early offshore platforms. The latter were seen essentially as portions of piers detached from the shore as an economy measure or as a means of bypassing mineral claims (Franks and Lambert pp. 46–47). Simple adaptations of land-based systems otherwise sufficed.

The U.S. Gulf coast (GoM) rather than the Californian is generally regarded as the 'home' of the offshore oil and gas industry, but here the move offshore was later. According to Rundell (p. 119) the first well drilled offshore the Texas coast was in 1908 at Goose Creek. However, as the very name suggests, this was no more than the culmination of a drift towards the open sea via the swamps and bayous, which lie between dry land and open sea in Louisiana and east Texas. Kriedler (pp. 39–40) maintains that it was around Caddo Lake in northern Louisiana that commercial drilling for oil in a water environment began in the GoM states in 1906, although there was no production from the Caddo's waters until 1911. During same period, Russian engineers working at the edges of the Caspian Sea were developing similar techniques.

Drilling in calm inland waters was not confined to coastal states. In 1910, a gas well was completed a mile offshore in Lake Erie. In 1943, the same lake saw the first known underwater well completion in 35 feet of water (Goodfellow 1977, p. 197).

With the exception of oil slicks, there were no surface features to guide pioneer offshore oil producers and limits to the extent to which onshore geological knowledge could be extrapolated. As a consequence, there was very little swampland or offshore exploration as such prior to the development of geophysical techniques. From the early 1920s, the growing availability of geophysical techniques and the first state leasings of submerged exploration tracts set off a GoM exploration boom, initially through the swamps and bayous and gradually into the open shallow sea, with a variety of converted shallow-draft craft transporting seismographic equipment and crews. In 1928, a submarine undertook a gravity survey of parts of the GoM, whilst 1934 saw early attempts at underwater coring (Pratt et al 1997, p. 2).

In 1927 had come the first discovery of a salt dome field (Vermillion Bay, Louisiana) by geophysical survey of an underwater area, confirming earlier predictions that the salt dome geology of Texas and Louisiana extended offshore. Within a year, the same survey contractor, Geophysical Research Corporation (GRC), then a subsidiary of Amerada, had discovered no less than eight salt domes (Kriedler p. 46). Amerada, well known as the pioneer of geophysical, and especially seismic reflection, exploration, was British-controlled until the Second World War. As late as the First UKCS Licence Round, it appears that the British government with 9% remained Amerada's largest shareholder – The National Archives [TNA]: Public Record Office (PRO) POWE 29/388.

Drilling in the swamps presented considerable logistical difficulties and the initial use of fixed platforms for exploratory work soon gave way to the self-contained mobile light-duty barge. Such vessels were unable to handle the heavy rigging and large quantities of drill pipe and casing required to drill and complete deep wells. A technological advance was clearly needed.

This came from Venezuela's Lake Maracaibo, a shallow inland sea, lacking significant tidal currents and major wave action and enjoying a benign climate. Foreign companies, in particular Shell Oil, moved into the area shortly after the First World War. Mobile drilling barges and other equipment were originally imported from the USA, adapted to suit local conditions and joined by such innovations as crew and workboats and floating tenders. The latter, carrying personnel quarters, utilities and drilling equipment and supplies, allowed the inexpensive drilling of either exploration or production wells from small platforms between which rapid movement was possible.

In 1928, Italian Louis Giliasso devised the heavy-duty sunken (or submersible) drilling barge based on his perception of what was needed in Lake Maracaibo. Both single and twin hulled versions were developed to carry drilling equipment and supplies. While easily movable in a de-ballasted state, by flooding the hull so that it rested on the bottom, such vessels offered the support needed for the drilling of deep wells and dramatically reduced the cost of shallow water exploration. Although there were earlier prototypes, the innovation 'took off' after Giliasso granted an exclusive licence to the Texas Company (later Texaco), which had seven such barges at work by 1935 (Kriedler pp. 48–50).

The combination of state lease grants, geophysical prospecting and the submersible drilling barge maintained offshore exploration at the edge of the GoM into the late 1930s and early 1940s. The first oil well completed in open water was in 1937 in the Creole Field, one mile from the Louisiana coast. By the following year, Louisiana had seen the drilling of a total of 700 wells surrounded by water. Texas lagged a little behind, not seeing its first open-water well until 1941, at the end of which year offshore activity in the GoM came to an abrupt end following the entry of the USA to the Second World War, see Kriedler (pp. 53–55).

During that War, British scientists and engineers made what would much later prove to be a major innovation in offshore field development – flexible pipelines laid from a reel. Conceived as a means of securely supplying petrol across the English Channel to an allied invasion army, the first prototype was tested in 1942. Two types of flexible were developed, one a composite (derived from undersea telephone cables) and the other steel. Both are still employed. The first was the 'brainchild' of Anglo-Iranian's then chief engineer and the second of his opposite numbers in Burmah and Iraq Petroleum. Together with the installation and pumping equipment, the lines provided 'pipelines under the ocean', popularly known as Pluto.

GoM activity resumed with oil company seismographic surveys in 1944, soon supplemented by marine and airborne magnetic surveys by government agencies. This geophysical work confirmed the similarity of the onshore and offshore geology and the great oil and gas potential of the U.S. Continental Shelf, which sloped gently 50–140 miles from the shore to the 600-feet water depth line. It became recognised as early as 1946 that, if offshore development was to begin in earnest, other types of data were also needed. Oceanography (in particular the nature of the seabed, wave behaviour and foundation scour) and meteorology were especially important, so that platform design and naval architecture could be appropriate. Oil companies, drilling contractors and shipyards consequently launched R&D programmes. Although Shell Oil studied the geology offshore California and experimented with floating drilling methods, in the GoM the lead was taken initially by independents such as Kerr-McGee, Phillips Petroleum, Magnolia Petroleum, and Humble Oil (Kriedler pp. 57–65).

Independents needed to form multi-firm consortia to handle the risks of offshore development. It was a partnership led by Kerr-McGee and also involving Phillips Petroleum and Stanolind (later Amoco) that in 1947 discovered, and immediately went on to develop, with three piled steel platforms, the world's first offshore field out of sight of land. The discovery well was drilled from a small fixed platform supported by floating tender, a converted war surplus vessel – Franks and Lambert (p. 47). Brown & Root (B&R) undertook the platform construction. The company already had been involved in marine oil activity for more than 10 years, having in 1936 laid a short offshore pipeline for Humble Oil. Two years or so later it built a mile-long derrick ending with a drilling platform as well as the platform complex for the Creole field. These pre-war structures were constructed of timber (Pratt et al pp. 4–5, 7).

Using techniques and procedures developed prior to the Second World War, Kerr-McGee employed tender support and multiple completions of directionally drilled wells. It also took advantage of the cost savings derived from the use of converted war surplus vessels – air/sea rescue boats, landing craft, freighters and a tugboat (Kriedler pp. 66–67). Other innovations in the field included the use of a helicopter, although the first commercial helicopter service in the GoM did not start until 1958 (Schempf). The introduction of helicopters not only benefited operational crews but also facilitated the use offshore of temporary specialists.

The offshore industry flourished for a few years in the wake of Kerr-McGee's pioneering effort, despite having to come to terms with such new problems as structural integrity, metal corrosion and logistical difficulties. Notwithstanding the higher development capital costs of offshore production, the generally large discovery size and lower lifting costs enabled it to compete against post-war U.S. onshore production, which increasingly depended upon expensive artificial lift. However, a jurisdictional dispute between the state governments of California, Louisiana and Texas and the U.S. federal government seriously depressed activity between 1950 and 1953.

When activity revived from 1953, it did so against a background of an enhanced rate of technical change as the adaptation of land and swampland techniques and the conversion of war surplus vessels gave way to equipment specifically designed for open sea conditions, increasingly constructed in specialised yards. Nowhere was this more so than in the case of offshore drilling equipment. Exploration in waters between the 15 feet or so appropriate for the Giliasso type submersible drilling barge and about 100 feet had been carried out from small piled steel platforms supported by drilling tenders. Where production could be carried out from the same platform structure, the disadvantages of this method (slowness and expense) were mitigated, but when a dry hole was encountered, the need to remove the platform magnified them. Nevertheless, Conoco briefly employed a fixed platform in North Sea southern basin exploration in the mid-1960s when conventional exploration rigs were scarce (Walmsley 1995, p. 26).

Kriedler and others have charted the evolution of drilling barges specifically designed for open water exploration. The first, *Breton Rig 20,* entered service in 1949, brainchild of an English-born engineer, John Hayward, then employed by Barnsdall Oil. Hayward is widely regarded as the 'father' of the mobile offshore drilling unit. His new design concept represented a substantial advance, differing from earlier drilling barges by having an upper drilling deck separated from the submersible hull by pile supports, providing the wave transparency of a stationary platform. It was generally self-contained with drilling equipment, crew quarters and storage space. It could be moved from one location to another in a matter of days and drill in water depths of up to 40 feet.

Backed by Murphy Oil, improved versions of the Hayward design were supplied to what was probably the world's first specialist mobile offshore drilling contractor, Offshore Drilling and Exploration Company (ODECO), founded in 1953. By mid-1954, following successful exploration work for Shell Oil (by then with its own offshore technology research programme), ODECO had established that the mobile offshore drilling unit was the most cost-effective means of offshore exploration and relatively resistant in the water depths in which it operated to the GoM's most severe weather – hurricanes.

ODECO was soon imitated by oil companies such as Kerr-McGee, which established its own offshore drilling fleet (Transworld), and by rival independent contractors. With expanding demand and active interaction between oil companies and their suppliers, a widening range of GoM shipyards and equipment suppliers entered the drilling and construction markets, confirming the area's central role in the offshore industry and sustaining technical progress.

Table 2.1 illustrates some of the more visible, large-scale innovations leading to a steady increase in the water depths open to exploration drilling (with the 1000-feet benchmark being passed by the end of the 'sixties') and production (the 300-feet barrier being passed in the mid-'sixties').

There was also steady progress in important service areas, such as marine seismic, underwater services and structural design criteria. Seismic technology was a major beneficiary of general advances in computing and communications, themselves driven by military requirements.

In the underwater field, traditional shallow water diving techniques were used for a variety of purposes from the early days of the offshore industry. In geophysical surveys, for instance, prior to the development of specialist instruments, diving bells were used to allow gravity meters to be set up on the seabed, with the diver recording the readings (Cameron Bryce 1999, p. 194).

The 'Cold War' needs of the USA and other major navies drove the activity into greater water depths slightly in advance of the offshore industry's needs. Thus, mixed helium-oxygen gas diving was available by the 1950s, developing to the point where it allowed divers to work at 250 feet by the early 1960s and 600 feet by the end of the decade. The subsequent development of saturation diving (which greatly improved the productivity of deep diving) and the associated breathing apparatus, heated wet suits, transfer bells, underwater telephones, umbilical lines and decompression chambers was followed by the introduction of the underwater (hyperbaric) welding habitats and pipe-alignment techniques essential for the laying and repair of pipelines. The work of divers was extended far beyond their previous roles in observation/inspection, salvage and limited intervention. Increasingly, their work was supported from sophisticated diving support vessels (DSVs), often from the 1970s equipped with dynamic positioning (DP) systems.

With their origins in the defence sector, remotely operated underwater vehicles (ROVs) – first used for observation and then for a growing range of work tasks – were employed in the oil industry by the early 1960s, although

TABLE 2.1 Some Key Offshore Innovations 1949–1963

Innovation	Innovator	Year	Significance
Submersible hull with raised drilling platform	Barnsdall Oil	1949	Extended mobile offshore drilling capability by providing wave transparency in water depths up to 40 feet
Heavy-lift vessels	J Ray McDermott Co.	1949	Enabled installation of steel piled platforms and deck modules pre-fabricated onshore
Subsea blow-out controls	Koomey	1953	Facilitated safer offshore drilling
Pipelay vessels	B&R	1954	Extended ability to pipe oil and gas to shore from 'inshore' to 'offshore' waters
Jack-up drilling rigs	The Offshore Co.	1954	Extended drilling capability for bottom-supported mobile rigs to (initially) 100 feet
Supply boats	Tidewater	1955	Facilitated more efficient offshore operations
Drill-ships	Global Marine	1956	Allowed drilling in waters too deep for bottom-support
Single point mooring	Imodco and Shell Oil	1959	Essential for safe offshore loading
Subsea well completions	Shell Oil	1960	First step towards subsea production systems with unconstrained water depth capability
Diver transfer under pressure	Royal Navy	1961	Essential for development of offshore saturation diving operations
Semi-submersible drilling rigs	Shell Oil	1961	Provided stable mobile drilling unit in (initially) up to 600 feet
Dynamic positioning	Shell Oil	1961	Permitted drilling and other operations unconstrained by mooring requirements
Remotely operated vehicles (ROVs)	Shell Oil	1962	Opened the way to unmanned underwater operations
Turret moored drill-ships	The Offshore Co.	1963	Enabled vessel always to head into wind, reducing rolling motion and subsequently often employed in ship shaped floating production vessels.

Sources: Based on information from Kriedler, Ocean Star website and other sources.

more complex tasks beyond diver capability were also for a time undertaken by one-man atmospheric diving systems or multi-manned manned submersibles. During this brief period, UK manufacturers enjoyed a lead. A comprehensive account of the development of underwater services is to be found in Swann (2007).

Structural design criteria owed less to the military than to industry initiatives, particularly the systematic collection from the early 1950s onwards of oceanographic and meteorological data for the Texas, Louisiana and California offshore areas. This allowed the specification and prediction of current conditions, wind and wave forces, including the anticipated extremes, with the height of the platform above mean sea level necessary to provide wave clearance a key issue. Such data formed the basis for the development of models and formulae leading to the introduction of standard design criteria and recommended practices and codes for both fixed platforms and mobile rigs. The importance of seabed conditions at proposed drilling or construction sites also came to be recognised, promoting the use of sonar scanning techniques. Perhaps thanks to deeper near-shore waters and the greater presence of the U.S. Navy, Californian organisations were prominent in development work in the 1960s and 1970s relating to floating structures such as buoys and tension leg platforms (TLPs) and, at times, underwater services developed faster there than in the GoM.

From the mid-1960s, the application of computer-assisted design to offshore structures became practical, with B&R a prominent pioneer (Pratt et al p. 57). It dominated offshore design, fabrication and installation, having completed 315 platforms by early 1966, with J. Ray McDermott (McDermott) as its most serious rival.

From 1962 until its abandonment in 1966, B&R acted as manager for the U.S. government's Project Mohole (an attempt to drill down to the earth's mantle deep sea, where the crust is thinner), a role that allowed the company to advance semi-submersible drilling rig design, DP systems, well-head sonar re-entry systems, electric BOP controls and buoyant marine risers (Pratt et al pp. 77, 120–135).

An extensive technical literature began to emerge. An important early example dealing with wave forces on piles was Morison et al (1950). Statements of recommended practice for fixed platform design were not to be published for another quarter of century, appearing in the UK a year earlier than in the USA – Department of Energy (1974a). By then interest had focussed increasingly on the North Sea.

By the 1960s, offshore oil and gas technology was moving out of its pioneering phase. Centred on the GoM, where the world's largest land-based oil and gas support industry was close at hand for interaction; a secondary concentration was growing in California. Significant shallow water activity continued in Lake Maracaibo; new offshore areas (all enjoying a benign shallow water environment), such as Borneo, Trinidad and the Persian Gulf, had opened up from the 1950s. They offered export market opportunities for the U.S. offshore service and supply sector. Its opportunity to extend its expertise into more exacting environments would have to await the near-simultaneous opening of the southern North Sea in 1964, Alaska's Cook Inlet where oil was discovered in 1963 and Australia's Bass Strait where the first 'wildcat' well was drilled in 1965 and the first oil discovery made in 1967. In the same year, oil was found in Alaska's Prudhoe Bay. Important though these other harsh environment oil and gas provinces were, none was to offer a market comparable to the North Sea.

The new offshore 'know-how' was absorbed by the U.S. academic and training sectors, already far more developed than those of any other country in petroleum technology subjects. When combined with the wealth of practical experience entrenched in an expanding pool of specialist labour and equipment, this leading position in the field of education and training provided the U.S. offshore with a formidable competitive advantage.

2.3 THE UNITED KINGDOM POSITION

When the passage of the Continental Shelf Act opened the way for the exploitation of the UKCS in 1964, it would have probably appeared to informed outside observers that the UK was well positioned to go forward largely on the basis of its indigenous industrial resources. The country had long been the home of important international oil producing interests, had a sophisticated financial system, was a large producer of capital goods – including some specific to the oil and gas industry – and had strong marine and engineering traditions.

2.3.1 The British Oil Companies

British oil interests had had early exposure to the offshore scene. Shell Oil was prominent in the GoM and in Venezuela's Lake Maracaibo (where a British-built diesel electric drilling barge was introduced in 1952). As shown in Table 2.1 (see p. 34), it established itself as a leading offshore innovator, known for sharing its advances with the industry. By 1955, it had already drilled nearly 200 offshore wells in the GoM (Howarth 1997, p. 244). Unfortunately, from the UK point of view, Shell Oil was at the time a U.S. quoted company and operated largely autonomously of its controlling European shareholders, the Anglo-Dutch joint venture Royal Dutch/Shell.

Shell had already completed its first marine well a mile from the Borneo coast in 1952. In 1954, it began drilling five miles offshore Qatar and the first marine well in the Niger delta came in the following year, by the end of which the Shell group had nearly 300 offshore wells world wide. In 1955, Shell struck oil 55 miles offshore both Borneo and Qatar (Howarth pp. 244–245, 265).

Although Shell Brunei drilled its first true offshore well in 1957, it was not rewarded with a discovery for several years. The Shell executive mainly responsible remembered that Shell designed and constructed the initial production platforms itself. That the company had such a capability is corroborated by Higgins (1996, pp. 336–337), who refers to the first truly offshore Trinidadian field (in 40 feet of water) being developed between 1954 and 1957 by the local Shell affiliate using a platform design conceived for Brunei, and with Shell technicians constructing the platforms locally. However, early Shell activity offshore Brunei and elsewhere, offered British industry opportunities to supply, for instance, mobile drilling rigs (Jamieson 2003, p. 136).

From the 1950s, much of Shell's West European upstream interests (including the UK North Sea proper, but not other UKCS areas) were conducted through 50/50 joint ventures with Esso (now Exxon) of the USA. Although a Shell subsidiary, Shell UK Exploration and Production Limited (or Shell Expro) was the operating partner, Esso had an equal voice on major decisions.

BP's first involvement as an offshore Operator came in 1953, with Abu Dhabi Marine Areas (ADMA) in the Persian Gulf. It held a two-thirds interest, the balance being held by Compagnie Française du Pétroles (CFP, later Total). Following successful exploration and development, production began in 1958.

BP Archive files show that an offshore drilling study group was formed in late 1954. Its aims included investigating wave and weather conditions, the feasibility of a local supply base and, in particular, existing experience in offshore drilling, leading to visits to Shell in Qatar, the NCB in Scotland and to the USA. Notable among the findings of the first USA visit were:

i.   Forty-one drilling rigs supported by 500–600 vessels were operating in the GoM. The deepest production was in 67 feet of water. Most production wells were deviated drilled from multi-slot fixed platforms with tender support. Well productivities were low, typically only 100–200 barrels per day. Although most crew changes were by crew boat and crane-lift, as distances increased, the time and cost-saving advantages of using helicopters were becoming apparent;

ii. While submersible rigs were satisfactory in up to 40 feet water depth and fixed platforms sometimes used for exploratory drilling, the need to explore further offshore and in deeper water was stimulating the development of a new generation of fully mobile self-elevating drilling rigs;

iii. Steel jacketed platforms were the norm, commonly weighing about 200 tons without the deck. Installation was by crane barge, the maximum lift being about 250 tons. To reduce installation time and thus reduce weather exposure, it was desirable to use the biggest lift capacity available;

iv. Though loading oil into barges at the platform still took place, the growth of production favoured an increasing reliance on underwater pipelines, following primary separation on the platform;

v.   Overall, offshore costs were five times higher than onshore costs;
vi. There was a heavy reliance on contractors for a wide range of services,
     including metocean studies, drilling, offshore structure design, construc-
     tion and installation, casing and cementing services, support craft, and
     catering. Discussions were held with no less than 14 contractors (of which
     12 were subsequently to 'turn-up' in the North Sea).

The study group recognised important differences between conditions in
the GoM and ADMA. These included the latter's lack of metocean data, the
absence of local infrastructure and experienced labour and the higher individual well productivities anticipated. It considered that ADMA would require the
newer fully mobile rigs (subject to reliability), as well as a greater reliance on
helicopters.

After a second USA visit, it was decided to opt for a self-contained
rig based on a DeLong drilling barge design, a decision that BP based
upon DeLong's successful operational experience and to which it remained
loyal until the loss of *Sea Gem* in the North Sea a decade later. The rig,
the *ADMA Enterprise* was built in West Germany, as no British yard could
meet the required delivery date. The cost of the fully equipped barge was
£1.928 million (about £39 million in 2008 terms). Although the drilling
equipment itself (including logging and coring facilities) was USA designed,
most of the other equipment was European and indeed mainly of British
manufacture.

The BP Archive reveals only a modest British contractor role in the ADMA
exploration phase. Thus in 1953, Frenchman Jacques Cousteau with the
Calypso as the support vessel conducted an initial seabed survey; UK contractor George Wimpey (Wimpey) undertook subsequent seabed surveys. The
Calypso also supported a gravity survey team supplied by Geophysical
Prospecting Ltd. (Geoprosco), a British company, whose activities were followed by a marine seismic survey undertaken by a joint venture of Geophysical Services Inc. (GSI) and Geomarine Services Inc., both American, using
a new seismic survey vessel, the *Sonic*. The UK's Decca Navigator Company
provided survey control. A UK firm supplied weather forecasts, although wave
forecasts for platform design were sourced in the USA.

In 1956, Wimpey began the construction of the Das Island support base
(Figure 2.1) and by 1958 following a sequence of four successful exploration
wells on the Um Shaif structure, an initial development plan based on seven
to ten wells and production of 30,000–40,000 barrels per day was formulated.
Wimpey received the contract for installing the well-head platforms.

By 1971, there were over 60 platforms in water depths of up to 80 feet.
According to one of the BP engineers involved, the first platforms were
designed by BP Engineering and fabricated in the UK to their specification
by Tubewrights, with sections limited to an offshore lift of 20 tons (subsequently increased to 100 tons) for 'flange bolting' in location by divers; as

FIGURE 2.1   ADMA's Das Island Base. Reproduced by permission of the Energy Institute.

demand for platforms increased, BP switched sourcing to Bahrain where Wimpey and B&R jointly operated a yard.

Material from the BP Archive shows that jack-up gathering platforms for the development of the Zakum field required DeLong design input. They were built in France by Hersent for fitting out in Bahrain by Wimpey. Subsequently that joint venture was also involved in both gathering and well-head platforms in the Um Shaif field. A French company (Entrepose pour les Travaux Pétroliers et Maritimes or ETPM) supplied and installed another gathering platform in the Zakum field. Many other suppliers of various nationalities also played a part. Well services were sourced from Schlumberger (USA/French), Flopetrol (French) and Halliburton (USA), loading facilities from SBM (Dutch), process engineering from Tarmac (UK) and tanks from a variety of British, American and European suppliers. Of particular interest is the provision of contract drilling services from the Offshore Company (USA), a relationship originally entered into several years previously.

ADMA's capital expenditure budget for 1974 was disclosed as £361 million (over £2.8 billion in 2008 terms), showing the very great scale if the undertaking. The list of suppliers continued to grow. Among the new names appearing were Brush, Ferranti and Parsons (British suppliers of electrical equipment), Costain Process (British process engineer), Power Gas – Harris

(UK/USA process engineering joint venture), John Brown Engineering (British gas turbine supplier), AEG (West German supplier of electrical equipment), Dorman Long (South African platform supplier), Bechtel (U.S. project manager), Black, Sivalls & Bryson (BSB) and Air Products (U.S. process plant suppliers), and Saipem (Italian installer of risers and pipelines and a platform supplier). In addition, there were also new suppliers from Abu Dhabi, Kuwait, Japan, the Netherlands and Singapore.

BP Archive files record Wimpey was responsible for platform installation (Figure 2.2), hook-up and maintenance, for which it initially employed DeLong civil engineering jack-ups. A new purpose built vessel the *ADMA Constructor* came into service in 1971. Wimpey, by now in a platform installation joint venture with B&R, was still operating two other work barges in 1974 while B&R was mainly responsible for riser and pipeline installation, an activity to which the joint venture did not extend. Between 1962 and 1966, B&R installed a major pipeline gathering system as well as the Das Island loading line (Pratt et al p. 110). It was thus ADMA that introduced BP to the capabilities of specialist offshore contractors in general and B&R in particular.

ADMA allowed BP to innovate in the area of subsea production. The Zakum Subsea Production Scheme ran from August 1969 until April 1972 as recorded by Goodfellow (1977, pp. 114–123). It included the first use of a subsea oil and gas separator, something not attempted again for some 20 years. Its Archive confirms that BP concluded simple subsea production was viable. Most of the 'high-tech' components and services were supplied by USA and French firms. Among the British firms involved were Wimpey, which built, installed and serviced the manifold using divers supplied by the French company, Compagnie Maritime d'Expertises (Comex), Lucas Aerospace, J & S Pumps and Kent Instruments, which supplied communications and instrumentation. In 1969, BP and CFP both already belonged to the Comex-led consortium Subsea Equipment Associates Limited (SEAL), which sought to develop and market a 'dry' subsea production system (Swann p. 553).

In 1954, the BP/CFP partnership later joined by Conoco, constituted Dubai Marine Areas. Drilling began in 1964, with the giant Fateh discovery following in 1966. Besides Abu Dhabi and Dubai, BP was also involved offshore Trinidad, another benign environment.

Its corporate historian, Corley (1988), showed that Burmah too had offshore activities pre-dating the North Sea. In 1959, it entered into an agreement with Murphy Oil and its 52% owned subsidiary ODECO to jointly exploit leases offshore Louisiana where production was already established. In 1963, the company took an interest (originally one-third) in the Woodside offshore concession on the Australian North-West shelf and became Operator. This was destined to become a major offshore oil and gas province, though one from which Burmah itself would benefit little.

FIGURE 2.2   Early ADMA well-head platform installation. Reproduced by permission of the Energy Institute.

2.3.2 The British Supply Sector

A British oilfield supply industry was the natural corollary of the UK's long-established international oil industry. It was focused upon supporting the British oil companies and their land-based overseas operations, but also exported more widely. What was to become its most prominent trade association, the Council of British Manufacturers of Petroleum and Process Equipment (subsequently known as the CBMPE, and from 1981 the Energy Industries Council or EIC) was founded in 1943.

That the supply industry was well established before North Sea activity began (though very much smaller than its USA counter-part) can be judged from the fact that in the 5 years ending in 1964, Burmah, the smallest of the three UK majors, had placed export orders for £20 million (over £320 million in 2008 terms) with British manufactures, see Corley (p. 361). Exports of capital equipment were also specifically mentioned in a 1967 White Paper – Ministry of Power (1967, p. 25) – as an offset against oil import costs.

Steel was the indispensable material of the offshore industry. The UK had a large but operationally fragmented steel industry whose direction and investment levels had been disrupted by nationalisations and denationalisations. There was no reason for it to have paid early attention to the specific and demanding requirements of the offshore industry. Events were to show that even appropriately qualified welders to fabricate from such materials would to be in short supply in the UK during the early North Sea years.

Such issues were to prove to be particularly important since the steel involved could itself cost over half the total value of equipment and materials needed for an offshore development. It should, however, be noted that the expenditure on offshore services could exceed that on goods.

In 1959, the port of Churchill on Hudson's Bay was engaged in importing steel tubular goods for the Canadian oil industry and Stewarts & Lloyds, a leading UK tubular goods supplier, had an Alberta Manager (Cameron Bryce, p. 230,232). A few years earlier, the same company had supplied 40,000 tons of pipes for Burmah's Sui gas pipeline in Pakistan (Corley, p. 336). Searle (2004, p. 37) records that the company had also been the supplier of the piping employed in the steel variant of the wartime Pluto flexible pipeline system (see p. 31).

On several occasions in the Spring of 1963, there was Cabinet discussion of whether or not the UK should allow exports of large diameter line pipe to the Soviet Union as reported, for example, in TNA: PRO CAB 128/37 C (63) 23.5. Nevertheless, as early as June 1966, the FCO was expressing its concern at the inability of the British steel industry to manufacture the larger sizes of pipe needed for North Sea and export markets (TNA: PRO FO 371/187603).

The oil companies largely carried on the exploration function in-house, although independent geological consultants were available. Fairey Surveys and Hunting Surveys were involved in aeromagnetic work and Decca and Wimpey in hydrographic surveys. Seismograph Service Limited (SSL), set up by a USA parent in 1947, was for many years the only seismic contractor operating in the UK; although it maintained development and training programmes, it was hampered in its early days by lack of access to the foreign exchange needed to import American equipment. It was forced to design its own equipment and have it manufactured in Britain. Later USA arrivals found the UK a convenient base for Eastern Hemisphere marketing, recruitment and processing. They did not undertake research work in the UK, but by 1958 London had become a processing centre for seismic data from the Middle East (Anstey and Hempstead 1995, p. 34, 64). There were no local petroleum or reservoir engineering consultancies as such before Gaffney Cline Associates migrated from Trinidad after it became clear that the North Sea was a significant hydrocarbon province.

Domestic drilling was more concerned with coal than oil and gas, although established companies working for the NCB, did become involved in onshore oil and gas drilling to a limited extent. According to an ex-employee, drilling for coal and onshore oil and gas created sufficient demand for well-logging services for Schlumberger, the world leader in the field, to have established an operational base at Eakring in Nottinghamshire before the First Offshore Licensing Round. In the same Nottinghamshire area, from 1970 British Plasterboard Instruments (BPI) also offered logging services to the coal and non-hydrocarbon mineral industries. After a slow start, it eventually grew into a significant logging business within the oil and gas industry finally becoming known as Reeves Oilfield Services and passing into Canadian ownership. By contrast, an ICI logging unit did not survive.

The NCB also explored for coal reserves in the North Sea, employing various types of equipment. The BP Archive records that in early 1964, BP received an approach from one of the NCB's drilling contractors, Foraky, stating that the firm had drilled in the UK since 1925. It had been drilling up to five miles off its coasts for the previous 8 years in water depths in excess of 100 feet and in gale force winds. BP's response is not recorded.

Former employees have spoken of Wimpey's considerable early involvement with offshore drilling. The jack-up which drilled the UK's first offshore well two and a half miles off Lulworth Cove on the Dorset coast for BP in 1963, prior to the First Offshore Licensing Round, was provided by Wimpey. Like the ill-fated *Sea Gem*, also provided by Wimpey, it was a converted civil engineering unit with a DeLong pneumatic jacking mechanism. BP provided the drilling crews. In the mid-1960s, Wimpey, so it has been said, drilled a gas exploration well in the southern North Sea from the *Briarthorn*, a drill-ship that had worked for the NCB, suffering such a heavy loss that it became a 'one-off'. Later, the same company also owned a DP drill-ship (*Wimpey Sealab*) for use in marine geotechnical work.

The first dedicated UK oil and gas drilling contractor, KCA, was engaged in the Middle East at the beginning of North Sea operations. KCA had originated in 1957 as an Anglo-American joint venture known as Kier Calder Arrow Drilling. The BP Archive reveals that as early as mid-1964, KCA, in association with Global Marine, was offering to build jack-up or semi-submersible rigs for charter to BP for use in the North Sea and claiming to have financial backing of £5 million (over £76 million in 2008 terms) for this purpose. The offer was not accepted and KCA had to wait another decade before being awarded a major drilling contract for BP in the North Sea, the supply of labour and some services to operate BP owned equipment in the Forties field. Other '*mainly British*' companies interested in contracting rigs to BP in 1964 were recorded as Lithgows, Brown Brothers and Stewarts & Lloyds/Steel Company of Wales. There is nothing to show that any were seriously considered and none entered the business.

BP and Shell themselves owned and operated both onshore and offshore drilling rigs, but as demand increased were unable to meet 'in-house' the specialised labour requirements, leading to a growing reliance on contractors. For most of their sophisticated service requirements such as well logging, the oil companies also originally met their demands from within, though here too they slowly began to use contractors, mainly USA or French. 'Hybrid' cases occurred; there are a number of references in the BP Archive to decisions to purchase Schlumberger equipment that could be operated by either company or contractor personnel.

For equipment supply, the position was complex. A wide range of general mechanical and electrical engineering goods such as pumps, valves, compressors, diesel engines, gas turbines, electric and hydraulic motors and instrumentation were supplied from British factories, not always locally owned. Much of the more industry-specific items could also be supplied by domestic manufacturers, though choice was restricted, technology sometimes lagging and capacity limited. The strength of Stewarts & Lloyds in tubular goods has already been mentioned (see p. 42).

There was one long-established British manufacturer of drilling and related equipment, The Oil Well Engineering Company (OWECo). High -tech companies from outside the oil industry sometimes ventured into it. A notable example was the considerable effort put into unsuccessfully attempting to commercialise jet-engine pioneer Frank Whittle's sophisticated turbo-drilling motor between 1963 and 1971 by aero-engine manufacturers Bristol Siddeley and Rolls-Royce (RR). Some British manufacturers qualified as approved suppliers under the authority of U.S. institutions such as the American Petroleum Institute (API) and the American Society of Mechanical Engineers (ASME), whose standards were becoming increasingly important as the U.S. oil and gas industry expanded internationally.

During the period of the post-war dollar shortage, British governments were anxious that the British oil companies as far as possible sourced in sterling for their expanding overseas operations. This both stimulated indigenous firms and also encouraged investment from the USA (Smith 1985a, p. 268). Results included the establishment of British Oilfield Equipment Limited at Leeds to manufacture well-head equipment under licence from Cameron Iron Works. OWECo, with UK government and British oil company backing, entered into an arrangement with National Supply to manufacture its drilling and associated equipment under licence. The USA principals eventually absorbed both firms. Camco, a USA manufacturer of downhole items such as gas lift equipment and for a time British controlled, was established in Belfast prior to North Sea activity. Also in this period, Hughes Tool (drill bits) also established itself in Belfast while Gray Tool (well-head equipment) set up in Scotland.

Such businesses found the UK, with its then large stock of inexpensive engineers, a common language, similar legal system and good international communications, a welcoming and profitable base from which to service the Eastern Hemisphere, much of which still formed part of the Sterling Area trading bloc. They became large exporters. Other U.S. companies used the licensing route to access Sterling Area markets, but some products, for example large gas compressors and drill pipe, long remained unavailable from UK sources.

As far as process engineering and management contractors were concerned, prominent UK firms included William Press, an old company, long associated with the manufacture and distribution of towns' gas, and the more recently established Constructors John Brown (CJB), which had also worked in the 1960s on onshore natural gas pipeline construction in Algeria. Power Gas and Costain Process Engineering were involved in the exploitation of oilfields in the Middle East and elsewhere.

The attraction of the UK as an Eastern Hemisphere base was apparent to USA contractors. As a result, U.S. companies – some with upstream 'know-how' – were well established long before hydrocarbons were discovered in the North Sea. They were less linked with manufacturing than their British competitors, numbered British as well as American oil companies among their clients, and enjoyed a competitive advantage from the early USA lead in project management techniques, which Bechtel and others had originally developed for infrastructure and manufacturing projects. When adopted for space and defence programmes, project management acquired something of a celebrity status.

Bechtel had opened a permanent office in London in 1948 and Foster Wheeler far earlier. B&R opened its London office in 1959 (Pratt et al p. 204). It shrewdly appointed Sir Philip Southwell as its first British-based representative. After a long and distinguished career at home and abroad as an E&P executive, which included the development of the Eakring oil field, Southwell had gone on to become President of the Institute of Petroleum (IP).

Although independent consulting engineers (which specialise in advisory and supervisory work rather detailed design) were present in the USA, major projects were less overwhelmingly in their hands than was the case in the UK. By the mid-1960s, the offshore industry often made the same company responsible for design and project management, fabrication, other procurement, installation and commissioning during field development. Though popular with oil companies as means of speeding progress, it was alien to the prevailing British tradition.

In the British civil engineering sector (where again consulting engineers and contractors tended to operate separately), there were many firms with domestic and overseas experience of harbour and coastal works as well renowned international specialists in the field of bridges, notably Redpath Dorman Long (RDL) and Cleveland Bridge and Engineering. The latter was one of the firms responsible for the construction of the early Thames and Mersey estuary reinforced concrete sea forts designed by Guy Maunsell and installed during the Second World War (Turner 1996). Since Maunsell went on to design a movable marine drilling platform based on his tower fort concept for the NCB, from which wells were drilled in 17 locations in British coastal waters, the apparent failure of the oil industry to exploit his work was surprising (Turner 2004).

One established civil engineering consultancy that did have some success in the offshore industry was Sir William Halcrow and Partners. In the early 1970s, it was, a former employee recalled, engaged by BP for its ADMA project, first to analyse existing steel structures (probably in connection with newly introduced certification requirements) and later to undertake structural design for three and four legged jackets and flare stacks. It went on to supervise the design of a small platform for BP's West Sole gas field in the southern North Sea, reviewed other platform designs in the same field and carried out conceptual design work in connection with BP's large Magnus platform in the northern North Sea. However, it did not establish a long-term position in platform design, although it did continue in a certification of fitness role.

Civil engineering companies with operations in oil-prone areas subject to British political control had early exposure to simple shallow water offshore oil developments under benign weather conditions. Wimpey was a notable example (see pp. 38–40). The BP Archive makes it clear that the close relationship between Wimpey and BP continued into the early days of BP's UKCS operations, including the provision of jack-ups for conversion to drilling barges and co-operation in the fields of seabed survey and well-head platform design.

Wimpey became an attractive partner for B&R as the latter began to move into overseas markets. The two companies operated joint venture offshore construction facilities in Bahrain and Nigeria (and eventually in Scotland), with B&R handling engineering and Wimpey construction. The two firms also co-operated in the early UKCS supply base and supply boat fields.

British shipyards were quick to enter the market for mobile offshore drilling units which grew fast in the late 1950s, with a jack-up completed in an English yard in each of 1958 and 1959, although construction was by engineering 'consortia' with the shipyard assisting with fitting out (International Management and Engineering Group of Britain [IMEG] 1972, p. 89).

This initial activity was followed by a lengthy pause. No more rigs were completed in the UK till 1965, following a surge in orders associated with the First UKCS licensing round in 1964. According to an anonymous author from the offshore industry body, the United Kingdom Offshore Operators Association (UKOOA), writing in an undated document (probably written in about 1984), this had resulted in the construction in UK yards of five jack-ups and three semi-submersibles. He went on to write that for the next 10 years the yards showed no further interest bidding either for the construction of rigs or fixed structures.

Whether or not his statements were entirely accurate, broadly similar sentiments were expressed by IMEG (p. 89), which noted that only John Brown had shown persistence, with the other yards surrendering their expensively won experience. It postulated that this was either in response to painful losses or the ready availability of conventional shipbuilding contracts. Cook and Surrey (1983, p. 7), on the other hand, believed the early enthusiasm for mobile rig construction was probably mainly due to the availability of government subsidies.

As a leading maritime nation, at the time of the initial North Sea activities, the UK had a large merchant fleet, including many oil tankers operated by BP and Shell. Owners complained that operating under the British flag made them uncompetitive, especially relative to 'flags of convenience' whilst outsiders pointed to over-manning and poor labour relations. The large deep-sea fishing fleet went into steep decline as a result of the 'Cod Wars' with Iceland between 1958 and 1976, freeing both vessels and manpower for the offshore industry.

Of the 20 offshore support vessels operating in the UKCS by 1966, 80 were British built. British-controlled companies such as P&O Offshore Services, Offshore Marine and Wimpey (in association with B&R) were among the operators (Jamieson p. 145). Other British companies were involved in the provision of standby (safety) vessels (initially mostly ex-trawlers) and of vessels used in seismic and other surveys.

Turning to the underwater scene, the British made an early start as civil and marine engineers and salvage contractors. Indeed, both the diving helmet and the 'closed dress' diving suit were invented in England and for much of the nineteenth and early twentieth century the port of Whitstable, from which can be seen the Maunsell sea forts (see p. 45), was the base of an internationally important diver-based marine salvage industry (Evans and Bevan 1990). The British also pioneered atmospheric diving suit (ADS) technology and in 1935 a diver using such a device reached the wreck of the *Lusitania* in over 300 feet of water. By the late 1950s one small British diving company, Under-seas Ltd., was carrying out pipeline inspection and survey work for ADMA and Iranian Oil Services in the Persian Gulf.

However, post-World War Two experience of advanced saturation diving techniques was largely absent in the UK commercial sector. While the Royal Navy (RN) may not have lagged behind its USA and French counterparts in diving techniques and the transfer of saturation divers under pressure was an important RN innovation, it was not generally recognised as the leader, let alone closely meshed with British commercial interests. Like the British defence sector in general, it was often accused of being less willing than its USA and French counterparts to transfer technology to the private sector. Nevertheless, many former members of the RN diving branch went on to work as North Sea divers.

Some firms in the marine business services sector did become involved in the international offshore sector early. A notable example was underwriting in the Lloyd's insurance market, which in turn stimulated development of specialist advisory firms such as Noble Denton, which gained an early reputation in the long-distance movement of drilling rigs and offshore structures.

Lloyds Register of Shipping also became involved in the offshore industry relatively early, offering technical advisory, classification and survey services to designers, builders and owners of mobile drilling rigs and fixed structures, diving systems and submersibles. Its entry followed the decision of the BP–CFP ADMA joint venture to order the ADMA *Enterprise* (the first self-elevating drilling barge to be built in Europe) under Lloyd's Register survey. This was followed by work on ADMA production facilities. By 1966, Lloyds Register had completed or had on-going classification surveys for seven jack-up rigs, four semi-submersible rigs and 14 work barges. It was also '*well advanced in the preparation of rules for the classification, design and construction of drill rigs*' (Lloyds Register 1998, pp. 272–273).

In 1971, Lloyd's Register established its Ocean Engineering Department, which 3 years later became part of an Offshore Services Group. An early reputation was established in the use of computerised systems for the analysis of structural strength, fatigue and vibration in fixed and mobile structures. By 1972, IMEG could say that Lloyd's Register appeared '*to be scoring a notable success*' in a small but important area of the offshore business (IMEG 1972, p. 95).

While Lloyd's Register could examine the designs of others, it was not permitted to prepare its own designs. Its work always contained a large international component. Legislation, often a reaction to accidents, further widened the Register's markets both at home and abroad. When the UK required (from 1975) all offshore installations in British waters to have a valid certificate of fitness, Lloyd's Register inevitably became a major certifying authority because the development of the certification system the government required was an easy step from the classification of rigs and platforms (Lloyds Register p. 274).

Turning briefly to the academic scene, prior to the opening up of the North Sea geology – but not geophysics – was widely taught in British universities, as were most branches of engineering. Founded in 1908 by Professor John (later Lord) Cadman, who became Chairman of Anglo-Persian Oil (now BP), the Department of Petroleum Engineering and Refining at Birmingham University had enjoyed '*worldwide fame*' between the World Wars (Cameron Bryce, p. 32). However, subsequently, the university seen as most concerned with the petroleum industry was Imperial College, London (incorporating the Royal School of Mines) where petroleum and chemical engineering were taught alongside geology. There was no Chair of Petroleum Engineering or first degree in the subject (IMEG 1972, p. 66). There were no courses in offshore engineering, though naval architecture was available at the Universities of Glasgow, Newcastle and Southampton.

It was not until the early 1970s that the situation began to change, with encouragement from government, major companies and the IP. University College, London established an MSc course in Marine Earth Sciences in 1970 and another in Ocean Engineering in 1972 (IMEG, p. 65). Heriot Watt

University's Institute of Offshore Engineering followed in 1973. By the mid-1970s, there were Chairs of Petroleum Engineering (and students) at both Imperial and Heriot Watt. By then, the oil and gas industry had been active in the North Sea for more than 10 years.

Academically qualified manpower in specific offshore disciplines was thus initially almost non-existent. Fortunately, engineers from other disciplines such as civil and in particular aeronautical engineering could in part fill this deficiency. Britain's large and sophisticated aerospace industry fortuitously went into decline at much the same time as the development of the northern North Sea got underway, releasing manpower versed in advanced engineering techniques. Earlier, when the southern North Sea was undergoing its initial development, the industry was extremely busy. The same was true of the civil engineering sector, then actively engaged in motorways, major estuary cross-ings and nuclear power stations, all state financed. This demand stretched the supply of well-qualified engineers, making it seem commercially sensible to employ them on profitable low-risk public sector work in preference to the more problematic opportunities arising offshore. When the public sector work was defence 'classified', as it often was, possible collateral benefits that might have arisen from the work were denied to the offshore sector. It was rare for a company to make a successful transition from a defence to an oil industry supplier. Hunting (in the field of well-related precision hardware) was a notable exception.

Whereas the British oil companies had significant numbers of personnel with (predominantly) onshore upstream petroleum expertise, these were almost exclusively deployed on overseas operations. This applied at the technician and craft level as well as the professional. There was much emphasis on 'on the job' training, given as one reason why both BP and Shell operated own drilling rigs (IMEG p. 67).

The decline in the shipbuilding and heavy engineering industries that gath-ered pace from the mid-1960s meant that craft skills were to become readily available, whilst the existence of large numbers of ex-service technical person-nel was also to prove helpful. Whilst many such workers became employees of E&P companies, the majority of those who entered the industry from these and other backgrounds joined contracting companies of one sort or another. Although some companies, particularly in the drilling sector, did offer formal as well as 'on the job' training, most did not. Eventually, an impressive array of specialist training institutions offering courses in such varied subjects as diving, drilling, fire-fighting and offshore medicine survival were to come into being, but mostly long after the arrival of the North Sea industry. There was little service capability in place at the time of the IMEG study in 1972, beyond the ability of established firms in the aviation, marine and industrial equipment areas to service and repair their products employed in the offshore industry.

As far as industrial management and associated issues are concerned, serious problems existed, particularly in the shipbuilding and major projects sectors (see pp. 13–18). Given the scale and complexity of offshore projects, with the heavily negative economic consequences of delay and loss of production, such problems were of great potential concern to companies engaged in offshore oil and gas.

Formal management education was late to develop in the UK, with the first two business schools (London and Manchester) dating from the late 1960s. The emphasis was on general business skills. Project management, a discipline then fast developing in the USA and destined to become of critical importance to the offshore oil industry, was not taught.

2.4 POTENTIAL EUROPEAN COMPETITORS

Three other European countries could have been identified as likely to materialise as powerful forces in the European offshore oil and gas industry. They did not include Norway, which would eventually emerge as Britain's main European rival in North Sea markets.

Norway had no previous upstream petroleum industry expertise, its oil industry being effectively confined to tanker operations and small-scale refining. However, its strong marine sector and modern shipbuilding and marine engineering industries were all to prove successful at diversifying into offshore activities. A tradition of working in reinforced concrete, particularly hydroelectric dam construction, provided the base from which Norway came to dominate the North Sea market for concrete structures. Nor did they include Germany, home to significant onshore production (mainly of gas), an important international land drilling contractor (Deutag) and a number of other equipment and service suppliers. These rivals were in fact France, Italy, and the Netherlands, all combining existing upstream expertise with relevant marine activities.

The most prominent international oilfield service company, Schlumberger, was of French origin. By the 1960s, it was best regarded as a Franco-American enterprise, headquartered and quoted in the USA but with legal domicile in the Netherlands Antilles. Its starting point was the work the Schlumberger brothers carried out in small French onshore oil fields in the 1920s on electric logging (see p. 27). It went on to absorb other French oilfield service companies, such as Forex (drilling, 1952) and Flopetrol (well-testing, 1971), as well as companies in the USA and elsewhere. The Schlumberger family were also associated with the formation of the Compagnie Générale de Géophysique, a geophysical contractor. Another French family controlled company, Géoservices, became best known for mud logging. Such private firms remain prominent members of a strong trade association founded in 1953, le Groupement des Entreprises Parapétrolières et Paragazières (GEP), which has assisted the French oil and gas supply and service sector to speak with a unified voice.

However, in France more generally, the role of the state was critical. The original state sponsored oil company – CFP – was founded in 1920 to handle French interests in the Middle East GEP (2003–2010). Following the discovery of natural gas in southwest France in 1939, the French government set up the Régie Autonome des Pétroles (RAP) to develop it. In 1941, the Vichy regime established the Société National de Pétroles d'Aquitaine (SNPA) to seek domestic oil and gas resources.

Immediately after the Second World War, the French authorities made the development of oil production in territories they controlled a high priority, concentrating on French West Africa and Algeria in addition to metropolitan France (Yergin, pp. 525–527). This led to the establishment of new state companies under the aegis of the Bureau de Recherches Pétroliers (BRP), which also became the majority shareholder in SNPA. BRP and RAP were merged in 1966 to form the second French national champion in the E&P sector known as the Entreprise de Recherches et d'Activites Pétrolières (ERAP). ERAP controlled SNPA, with which its operational assets were merged in 1976 to form the business soon known as Elf.

Post-war, France recognised that a truly independent national oil industry required not only French E&P companies but also French supply and service companies, with their own proprietary technology. Accordingly, priority was given to developing a French technology base. Critical to this effort was the foundation in 1946 of the Institut Français du Pétrole (IFP), to act as a national R&D centre funded by a levy on hydrocarbon sales (Smith 1985a). Not only did IFP co-ordinate closely with E&P firms and their existing French suppliers, it also used its technical and financial resources to help launch new technology-based commercial ventures, in which the French oil companies were often co-shareholders. The first of such ventures was Technip, an engineering design and management company, founded in the 1950s. By 1981, there were about a dozen such 'spin-out' companies. Another important function of IFP was to organise technical training for the petroleum sector.

The departure in 1962 of France from Algeria – where substantial oil and gas production had been rapidly established – took place at a time of growing realisation of the potential scale of offshore resources. In response, in the early 1960s France established the Comité d'Études Pétrolières Marine (CEPM), which brought together IFP, the French oil companies and French industry. From its programmes emerged designs for semi-submersible drilling rigs, drill-ships and deepwater production systems, as well as enhanced expertise in areas such as pipelaying and repair, manned submersibles and ROVs, and structural fatigue (Cook and Surrey, p. 59).

Additionally, close co-operation developed between the French navy, the Centre National pour l'Exploitation des Océans (CNEXO) and underwater services companies, particularly Comex, which was soon to emerge as a leader in saturation diving (Smith 1985a). During the early 1970s Comex was a member of the SEAL subsea production system consortium (see p. 40). The French underwater pioneer Jacques Cousteau was involved in geological surveys of the seabed prior to the decision to drill on the ADMA concession in the 1950s (see p. 38).

By the time activity started on the UKCS, France was already engaged in a co-ordinated programme to create an offshore exploration and development capability ranging from seismic survey and drilling to platform construction (both steel and concrete), installation and underwater services. At that point, with state-funded R&D at its centre and the ability for contracts from the state oil companies to be steered in the desired directions, only the French offshore industry appeared to have the potential to mount a serious challenge to the American. This is well illustrated in the BP Archive which reveals that in 1970, when considering how to develop its Forties discovery, BP first considered an early production system based on an articulated tower that Elf was then testing in the Bay of Biscay. French effectiveness, it was often claimed, benefited from the existence of a cohesive national elite, moving easily between the public and private sectors.

Like that of France, Italy's oil and gas industry was primarily a state creation (Yergin, pp. 501–502). Of pre-war origin, the original core company Azienda Generali Italiana Petroli (AGIP) expanded on the back of post-Second World War natural gas discoveries in the Po valley extending offshore into the shallow waters of the Adriatic. Italy thus gained an early start in the installation of offshore structures and subsea pipelines. In 1953, AGIP became part of a state-controlled hydrocarbon sector conglomerate Ente Nationali Idrocarburi (ENI), which included companies supplying engineering products and services for oil and gas production, opening the way to industrial integration on the French model.

Italian private industry also supplied engineering goods and services for the oil and gas business. Its maritime sector included salvage and an underwater competence, originating in part to the Italian navy's wartime underwater expertise, which was to form part of the basis of an offshore construction capability.

Although its own offshore resources proved small, the Netherlands was well placed to develop an important offshore service and supply industry, essentially as a result of private initiatives and thus in marked contrast to France and Italy. Even before the discovery of the giant Groningen gas field in 1959, it had some domestic oil and gas production. More importantly, it had the headquarters of the Royal Dutch/Shell Group. Its unusual geography and history had given it extensive experience of subjects such as hydraulics and wave loading, matters of great significance to offshore construction. Its marine and shipbuilding background included salvage, towage, large barges, floating cranes and loading buoys. Towing and salvage became important Dutch marine service sectors and natural low-risk entry points to the offshore logistics and construction markets.

Dutch companies and entrepreneurs had early exposure to offshore requirements. Probably the best known was Pieter Heerema. After establishing his reputation with the use of pre-stressed concrete piles in Lake Maracaibo during the 1950s, he subsequently went back to the Netherlands to build a market-leading offshore construction business. His company was to play a major role in the development of the North Sea, particularly in the field of heavy lift barges. Its overseas work brought it into early contact with the emergent U.S. offshore contractors such as B&R, with which it co-operated in a joint venture between 1963 and 1966. In 1964 and 1965, this partnership was responsible for the construction and installation of '*11 fixed platforms and one that was re-installed*' (Pratt et al p. 207).

Thanks to the joint venture, de Groot Zwijndrecht (de Groot), a company originally involved in bridge and other civil engineering steelwork, received the first North Sea structural fabrication contract for fixed structures – platforms for the German sector. It went on to provide the early platforms for the UK southern North Sea.

Such early advantages, together with shipbuilders less constrained than the British by under-investment and labour problems, good technical universities and government-supported R&D made the Netherlands a strong contender to design and build equipment for the offshore drilling, construction and installation markets. Rotterdam in particular won a leading position in the construction, maintenance and repair of specialist offshore vessels, including drilling rigs, DP drill-ships and semi-submersible pipe lay and heavy lift vessels.

The key to this development was the founding in 1965 of Industrieele Handels Combinatie Holland (IHC) by a number of Dutch shipyards. The new company was to compete in both traditional fields and the new offshore market. In 1978, the design and related departments of IHC were placed into a separate subsidiary, Gusto Engineering (Novello and Araujo 2006).

IHC was not the only new Dutch offshore company formed through the collective action of industry members. A consortium of civil engineering contractors established the Netherlands Offshore, a second Dutch contender in the heavy lift field.

2.5 PERCEPTIONS OF THE UKCS HYDROCARBON RESOURCE BASE

Once it became clear that the natural gas reserves of the southern North Sea were sufficiently large to justify converting the distribution system and existing residential appliances, there was little further public debate of the extent of the gas resource base for many years. This necessary scale was achieved early and according to Arnold (p. 115) by 1970, 5 years after the initial West Sole discovery, proven reserves were already sufficient to maintain a production rate of 3.5–4 billion cubic feet for 25 years.

Substantial additional reserves were subsequently added, particularly associated gas in the oil fields and dry gas in Morecambe Bay. Nevertheless, by largely confining supply to the so-called 'premium' market (mainly UK households plus a few industrial users) and by entering into an import contract for Norwegian gas from the Frigg field in the mid-1970s, the behaviour of the BGC and its predecessor, the Gas Council, suggested to the outside observer that domestic reserves and thus the market available from their development were limited. At the Institute of Contemporary History Seminar in 1999 James Allcock, a former senior BGC executive, stated that the government did indeed believe that reserves were so restricted that the UK would never produce more than 3 billion cubic feet per day, which became the Gas Council's objective.

This modest perception of the natural gas potential appeared to be consistent with the short duration of the initial E&P 'boom' in the southern North Sea, largely confined to 1965–1970. However, this was not due to lack of prospects but was because companies had lost interest, after concluding that, under government pressure, the Gas Council/BGC was using its statutory privileges to acquire gas at cost plus a restricted margin, rather than at a price reflecting its competitive position in the energy market. The first managing director of Shell Expro – George Williams (1999) – made it very clear at the same seminar that exploration was stopped in the southern North Sea and efforts switched to oil prospects in the northern North Sea in response to the very low gas price.

The reasons for restraining the level of gas production and producer profits included a need to prevent an uncomfortably fast decline in the demand for coal, the desire to conserve gas supplies for the 'premium' market and a wish to limit the scale of profit repatriation by foreign-owned gas producers. In the event, when UK gas production peaked in 2000, during a brief period when the UK was a net gas exporter, it was at about three and a half times the Gas Council's early objective.

The position in respect of the scale and production profile of Britain's oil reserves was entirely different and there was a protracted and sometimes heated debate. To some extent, this was inevitable because of the much greater economic impact of the oil discoveries and their high political profile in terms of such issues as security of supply, the balance of payments, use of revenues, depletion policy and the rise of the Scottish National Party (SNP). Oil – unlike gas where international trade was in its infancy – was part of a large world market with producers, shippers and consumers of many different nationalities. Additionally, the debate became polarised between an academic (Peter Odell – for a short time advisor to the Labour government) and spokesmen for the oil majors. In 1975 Odell and his collaborator Rosing, published reserves estimates for the North Sea basin as a European whole, claiming that the much lower figures emanating from the companies reflected their acting in their own commercial rather than the national interest (Odell and Rosing 1975). This basic idea was subsequently elaborated at some length with respect to three major UK fields (Forties, Piper and Montrose) when they sought to calculate optimum oil recovery levels for both the companies and the British state, concluding that the former was 27.5% less than the latter (Odell and Rosing 1976).

The idea that the oil companies were either obtuse or devious in this matter had gained early credence when in April 1970 (6 months before BP's discovery of the giant Forties field and after the smaller Arbroath/Montrose discovery of the previous year) the then Chairman of BP, Sir Eric Drake, told reporters that he was pessimistic about the chances of finding commercial oil fields in the North Sea (Arnold 1978. p. 7). His scepticism was shared elsewhere; Williams (1999) recalled that at Shell board meetings a number of the senior geologists involved expressed the view that no oil would ever be produced in the northern North Sea.

As would still be the case today, any estimate of commercially recoverable reserves in a North Sea discovery had to start with a rough approximation of the original quantity of 'oil in place' and to judge how the reservoir would react under sustained production conditions, all subject to revision as additional wells and, in due course, production experience, generated new information. Reserves calculated for one purpose, such as prescriptive financial reporting, need not coincide with those calculated for another, such as field development planning, where the field Operator itself set the parameters. The growth of computing power helped improve reserves calculation by facilitating the use of probabilistic techniques. Reserves could be classified as 'proven', 'probable', and 'possible'. Nonetheless, the scope for differences in opinion, whether honest or contrived, was large.

It was hardly surprising therefore that the oil company statements about reserves remained cautious until the evidence was as unambiguous as it was ever likely to be. Even then their interest was largely confined to their own commercial discoveries, where – unlike Odell – they were constrained in various business-related ways such as rate of return objectives, financial disclosure regulations and banking requirements. Moreover, the concept of commerciality itself shifted with changes in price, tax, technology and the growth of infrastructure. The British government made its own reserves estimates from information supplied by the oil companies, to whose interests it could at times appear antipathetic, principally in respect of taxation.

A former Esso employee, Robinson (1999), suggested that this state of affairs created the atmosphere of a 'game', leading companies to formulate underestimates of oil reserves. If the oil companies were playing such a 'game' with government by downgrading reserves, an unintended consequence was the doubt sown in the minds of potential UK supply companies on the time-span and economic robustness of the North Sea market. Few people could appreciate that much of the difference between Odell's figures and those of the government and the companies lay in differing assumptions on oil recovery factors and investment levels. However, they did know the initial southern North Sea investment peak had been short-lived.

Odell and Rosing (1975) argued that the North Sea as a whole, mainly in UK and Norwegian waters, had (at a 90% probability) oil reserves of 78 billion barrels, assigning a 50% probability to reserves of 109 billion barrels. According to Arnold (1978, p. 70), contemporaneous Shell and BP estimates were 50–55 billion barrels. In April 1974 (before British offshore oil production began), DEn calculated UK reserves as between 8.5 and 14 billion barrels (cited by

Harvie 1994, p. 100). In 1993, when OSO ceased to intervene in the UKCS market, the official estimate of initially recoverable UK reserves (proven plus probable) was nearly 23 billion barrels (Department of Trade and Industry [DTI] 1994, p. 11). Since Norway alone held more than half the North Sea's initially recoverable reserves and with other littoral states holding small additional reserves, the BP and Shell estimates made about 20 years earlier appeared remarkably robust. However, by 2008, the North Sea's cumulative oil production had passed 50 billion barrels. Although production was already in decline by then, it was also clear that more remained to be produced. It was still too soon to write-off the Odell and Rosing prediction.

2.6 THE NEW TECHNICAL CHALLENGES OF THE NORTH SEA

Preliminary survey work began on the UKCS in 1962. Drilling began in the southern Sector of the UK North Sea in 1964, followed by the first discovery in December 1965 (BP's West Sole gas field), which commenced production only 16 months later in March 1967. Even this early, it had become clear that, although typical water depths for southern North Sea developments at around 100 feet were within the existing experience of developments in the GoM and elsewhere, straightforward direct transfer of existing technology would not suffice.

Although weather, wind, wave, current and sea bottom data relating to the North Sea did exist, it was for long inadequate, being focussed on the needs of shipping movements rather than the static location requirements of the offshore industry. Predictive models of wave and wind forces for structural design purposes of the type already developed for GoM conditions did not exist.

Southern North Sea temperatures were lower, currents stronger and seabed conditions more unstable than in the GoM. Most different of all were the wind and wave conditions, less because of greater maximum severity than because of the length of periods of consistent roughness. Not only did structural designs need to be strengthened and/or safety factors increased, but rig and platform re-supply also became more difficult and offshore installation opportunities scarcer and shorter, all leading to increased costs.

A number of early events in the southern North Sea, three of which follow by way of example, showed how much existing technology required both stretching and strengthening rather than simple extrapolation.

First, on a relatively calm day in late 1965 came the loss of BP's jack-up barge *Sea Gem* and 13 lives off Norfolk. This was attributed in part to metal fatigue and brittle fracture. The avoidance of brittle fracture – as through the use of tough fine grained steels enabling thinner sections to be employed – thereafter became an important criterion in platform and mobile rig design (Sentance 1991, p. 7).

Second, in March 1968 the 2-year-old Tees-built semi-submersible *Ocean Prince* was demolished, fortunately without loss of life, by 50-foot waves driven by 90-mile-per-hour winds. At the time it was resting on the sandy sea floor of the Dogger Bank about 125 miles from Scarborough, see Harvie (p. 30). The rig was operated and half owned by U.S. drilling contractor, ODECO, with Burmah and ICI sharing the balance.

A third early case of the unexpected were the problems that BP and its contractor B&R experienced with the West Sole gas export line (the first on the UKCS), showing that North Sea currents and seabed conditions would demand new pipeline installation techniques (IMEG p. 36).

Nor were early accidents confined to such dramatic, large-scale events. There were also small-scale accidents, often resulting in fatalities, across activities like drilling, diving, helicopter transport and crane operations. The latter can be most directly related to environmental differences between the GoM and the southern North Sea. Offshore cranes designed for the former area with its normally smooth sea proved inadequate for unloading supplies for rigs and platforms from supply boats functioning in the near-consistently choppy, or worse, waters of the North Sea. Slew-ring failure and resultant toppling were frequent, although operator inexperience may also have been a factor.

The move into the northern North Sea magnified problems to such an extent that an almost new technology had to be brought into existence. With water depths mainly in the range 300–600 feet, use of the jack-up rigs of the day was impossible. Wind speeds of up to 100 miles per hour with wave heights of up to 100 feet soon marginalised drill-ships, leaving semi-submersibles as the most effective exploration tools. Even their operations were heavily weather restricted, making it clear that with the then available construction vessels the installation of platforms and pipelines could only be undertaken for a few weeks in summer (the 'weather window').

To extend the 'weather window' required the development of a new generation of semi-submersible heavy lift and pipelay vessels. Larger and more powerful supply boats would be needed – not only because of the more difficult weather conditions but also because distances from shore had increased. Water depths would demand heavy investment in saturation diving equipment, surface support craft and a diver labour force capable of using them. In addition to their role in construction, divers became vital in on-going inspection, particularly after the problem of fatigue-induced cracking in steel structures came to be seen as important.

Platforms had to expand massively in size and weight. A large southern basin steel jacket would weigh 1000 tonnes whereas the first northern fixed platforms were about 12 times as heavy, with 'knock-on' effects on construction, transport to site and installation. The corresponding massive increase in structure costs meant the earlier practice of spreading topsides over several bridge-linked structures was no longer viable, requiring the juxtaposition of modules with different functions into a confined space (Archer 1991, p. 3).

Installation had to be fitted into a single 'weather window' lest a year's production be lost, with a serious 'knock-on' effect on project economics. The ability to float out a structure with its topside already in place increased interest in concrete structures, though other factors also played a part.

The northern North Sea environment was a *"quantum step up from anything else"* according to Denton (1999), a notable marine specialist, speaking at the Institute of Contemporary History Seminar. Design and construction was not the end of the matter because, as he pointed out, the ability to work, or even to survive, at sea had to be massively stepped-up.

Learning about the North Sea's physical environment and how to cope with it understandably became a high priority for all E&P companies. By the mid-1970s, their trade association, UKOOA (founded in 1964 as the North Sea Operators Committee or NSOC) had acquired more than 20 advisory committees and working groups, dealing mainly with the metocean environment, safety, employment, pollution, technical standards, and industry–government relations. Few of the committees or groups ever had budgets provided by UKOOA on behalf of the entire industry, largely because it was normal practice to refer funding requests direct to members, whose individual priorities differed.

A notable exception was the Oceanographic Committee. Originally founded in 1972, it was to operate for some 20 years, latterly under the auspices of UKOOA's Metocean Committee. Working in association with government agencies, its task was to broaden and accelerate data gathering. This involved weather ships, buoys and offshore installations. Its commitments did not stop there and, for example, it was soon bearing one-third of the cost of developing a numerical wind/wave model of the North Sea, one of the inputs to which would be the work of the Joint North Sea Wave Project (JONSWAP), which had begun in 1973.

As more data on North Sea conditions became available, it generated an extensive civil engineering literature. Significant early examples dealing with wave forces on fixed structures were papers from Paape (1969) and Hogben and Standing (1974), while Bjerrum (1974) addressed geotechnical problems with structural foundations. Structural design and appraisal became an increasingly complex process, much aided by the growth of electronic computational capacity and the development of specialist analytical and simulation programmes.

As the literature developed and the scale of both the problems and the opportunities emerged, a North Sea 'conference industry' came into being, assisting in the dissemination of information. A paper given at an early conference by the then General Manager of Shell Expro compared the physical conditions in the North Sea with those in other offshore provinces, showing how much worse they generally were and put forward the ideas the industry then had about how to counter them (Williams 1972).

Solutions had to be found and found very fast. First commercial discovery to first production in the UK southern North Sea basin took less than 2 years.

In the UK northern North Sea basin, where the scale and technical unknowns were much greater, the comparable timescale was about 5 years, with the move from discovery to initial budget and contract stage for Forties less than 2 years. This very speed, demanded by the circumstances of the time, required the mobilisation of global resources and thus helped prevent the UK turning the fact that the North Sea environment was fundamentally different from that of the GoM's into the source of national commercial strength that it might otherwise have been.

Chapter 3

Motivations and Constraints

Exploitation of the UKCS involved a complex interaction between several distinct 'constituencies'. The most influential of the 'constituencies' were the E&P companies and the British government, with their suppliers and the financial community playing a secondary role. This statement is, of course, a considerable simplification. In fact, none of the constituencies was entirely homogeneous. For instance, governments changed while the E&P sector included both British and foreign companies, varying widely in size and character. Each of the other constituencies also contained organisations of widely differing scale and national ownership, as well as individual entrepreneurs. All parties were also subject to a variety of external pressures, including organised labour, regional and special interest groups and international political and market forces.

Nevertheless, it serves an important purpose to generalise the circumstances, motivations and constraints, which characterised the four main sets of players, giving particular importance to the E&P companies and the government. It was these two that created the economic and business environment within which the other two had to operate.

3.1 THE EXPLORATION AND PRODUCTION COMPANIES

Exploitation of the UKCS was never likely to have been solely a British undertaking. Although all three major British oil companies – BP, Burmah and Shell – had already been engaged offshore prior to the advent of North Sea activity, their collective offshore experience was less than that of some U.S. companies.

The main objective of commercial businesses is to make a profit. Regardless of national origin, the commitment of their shareholders' resources to the UKCS made it clear that the E&P companies believed that there were reasonable prospects of earning a return there. An additional attraction of the area was the prospect of political and contractual security, which by the late 1960s was becoming increasingly important to an oil industry depending more and more on the volatile Middle East and North Africa. To be successful, companies needed to obtain and retain the goodwill of the host government. Finally, there was a common if unspoken interest among participants in furthering the development of offshore technology so that the areas available for exploitation could be progressively expanded into more difficult geographic zones.

Beyond these broad generalisations, the circumstances of individual licensees differed and it is worth noting that not all early North Sea participants were commercial companies. Two were owned by the British state. Since natural gas was the original target of North Sea exploration, the desire of the Gas Council to improve its raw material supply base and to at least participate in the E&P process both as a 'pupil' and as a 'policeman' needs no elaboration. Interestingly, the Gas Council (undoubtedly conscious of the fact that in the absence of access to natural gas, its main raw material would increasingly be oil-based naphtha) did not rule out the discovery of oil in its First Round License application. However, its expectations in this respect were modest. It was thinking in terms of perhaps finding a field producing about 20,000 barrels a day, which it considered '*a medium-sized oil field of European proportions*' (TNA: PRO POWE 29/388).

For the NCB, whose activities were already in managed decline by the mid-1960s, participation in the natural gas business offered a 'hedge' against acceleration in the rate of decline in UK coal demand, which was likely to follow large-scale introduction of natural gas into the market. It also offered a potential outlet for the considerable knowledge of the NCB of the geology and subsurface features of parts of the North Sea and possible alternative employment of some of its technical resources, such as core laboratories and seismic and offshore drilling expertise. Its entry came in the Second Licensing Round.

In addition to the British nationalised industries, state-controlled entities from France and Italy also came to participate. Here, potential diversification of oil and gas supplies was probably an important initial motivation, though an outlet for their national offshore expertise may also have been a factor.

Both British and foreign independents acquired licence interests from the First Round onwards. Nevertheless important though such participants may be in their own rights, they did not influence the initial business climate in the way the major British integrated companies did. Up until about the time North Sea exploration began, the history of the companies had been closely intertwined, through crossshareholdings and abortive merger attempts and through trading relationships reflecting BP's abundant Middle Eastern crude reserves, Shell's downstream strengths and Burmah's access to the markets of the Indian sub-continent.

BP started with the advantage of having already conducted nearly all prior onshore exploration in Britain. By 1964, it had also acquired shallow water offshore experience in the Middle East and Trinidad (see pp. 37–40) and had drilled the UK's first marine well in 1963. Its well-regarded exploration geologists were therefore favourably placed to interpret the results of the joint geophysical surveys (initially with Shell and Esso but subsequently with other partners) of the North Sea sedimentary basins undertaken in the early 1960s.

The attractions of southern basin gas exploration to BP, which commonly chose to operate as a sole license holder, are not difficult to see. Technical risks seemed modest and gas markets available. There was an element of product diversification (BP's natural gas interests were then small) as well as geographical

expansion into a politically secure home market at a time when the company's
activities were mainly in areas of high political risk.

Having selected its First Licence Round application blocks, BP secured a
jack-up barge from Wimpey. Converted into a drilling rig, it was given the
name *Sea Gem* (Figure 3.1). It made the first commercial gas discovery on
the UKCS (West Sole) in December 1965, only to collapse and sink 6 days later
(see p. 56). Nevertheless, West Sole was in production by May 1967, although
BP was less than happy with the prices it received from the Gas Council, the
sole buyer. With further gas exploration and development not seeming worthwhile, interest was switched to oil prospects in the northern North Sea basin,
though with an apparent lack of enthusiasm. However, there had been a decision as early as the summer of 1964 to order a drilling rig capable of working
in the northern North Sea, the large *Sea Quest* semi-submersible to be built at
Harland & Wolff (Bamberg 2000, pp. 200, 202).

FIGURE 3.1   The jack-up drilling rig *Sea Gem*. Reproduced from the BP Archive.

At this time, BP still had access to abundant low-cost Middle East oil reserves, in excess of its own downstream needs. Any North Sea oil would be far more expensive as well as highly capital-intensive. However, more politically secure sources of crude were becoming of increasing interest, even if not low-cost. Otherwise, BP would have been involved in neither the giant Prudhoe Bay discovery in Alaska in 1968–1969 nor the northern North Sea. As put by a former BP director, despite the many risks, it was the desire '... *to diversify away from the Middle East*' that brought BP and other companies into the northern North Sea (Butler 1999). Nevertheless, BP still did not rush to drill on its northern blocks, a 1967 internal study having concluded that the chances of finding a field big enough for economic viability were small (Bamberg, p. 203).

The disappointing results of the few exploration wells drilled in the British and Norwegian northern basin waters in the late 1960s appeared to support this view. All that was to change with the discovery by Phillips Petroleum – a USA independent – of the 'giant' Ekofisk oil field in Norwegian waters in December 1969.

Meanwhile, BP's semi-submersible *Sea Quest* had been completed late in 1966; when not working for BP, was chartered to other Operators. Working for Amoco, *Sea Quest* discovered the small Arbroath oil field in the same month as Ekofisk was found. In 1970 the rig was back working for BP on its northern basin acreage and in November of that year BP announced its discovery of the Forties oil field, the first British North Sea 'giant'. Together with the then recent Prudhoe Bay discovery, it offered BP the prospect of breaking free of its long running dependence on the Middle East.

By this time at least, it would appear that BP also subscribed to the view that oil prices would rise. Butler recalled that development of the Forties field was justified on an oil price of $6 per barrel, although the development loan (1972) was based on a figure of $2.50. In 1975 when Forties came into production, the price was more than four times as much as the latter figure.

Shell, it might be argued, was not really a British company, given its position as part of an Anglo-Dutch joint venture and its 50:50 North Sea partnership with a U.S. company (Esso). However, as its operating North Sea unit – Shell Exploration and Development (Shell Expro) – was a British registered company with a British registered parent (Shell Transport and Trading) and with a predominantly British management, the author considers it is reasonable to treat it as British, as most people did at the time.

Shell's original motive for entering North Sea exploration undoubtedly stemmed from its discovery in 1959 of the huge Groningen gas field in the Netherlands. This prompted its Dutch arm to drill four unsuccessful exploration wells in Dutch coastal waters before ratification of the North Sea Continental Shelf agreement in 1964 (Pratt et al. p. 202). It already had shallow water operating experience offshore Borneo, Brunei, Qatar, and Trinidad, quite apart from Shell Oil being one of companies leading GoM activities into deeper water (see pp. 33, 34–36). According to a very senior Shell Expro manager of the time, engineers with offshore experience were thus available to call upon.

With its Dutch experience, knowledge of English geology and the results of geophysical surveys of the North Sea undertaken between 1960 and 1963 in association with BP and Esso, Shell was keen to explore in the shallow waters of the UK southern basin, where the jack-up rigs, *Mr Cap* and *Orient Explorer*, drilled most of its exploration wells. *Orient Explorer*, originally employed offshore Borneo, was Shell owned and British built (Jamieson, p. 136). Its expectation was that there was money to be made with discoveries substantially smaller than Groningen. When, after discovery and development of the large Leman gas field (1966–1968) (Figure 3.2), Shell had to accept the low gas price on offer, its interest also shifted to oil prospects on its First Round acreage in the northern basin.

FIGURE 3.2   Shell's Leman Field Alpha Complex. Courtesy Shell International Limited

Oil prospects inevitably had an attraction. Since the Second World War, the Shell Group was considered 'crude short' relative to its downstream interests (Howarth, p. 315). Moreover, like BP, it was anxious to lessen its dependence on Middle Eastern supplies.

Northern North Sea weather and water depths were seen as extreme and the geological uncertainty was great and, it is said, as late 1970, the Shell Group's Chief Geologist remained very sceptical of northern basin prospects. However, from the mid-1960s, there was a belief in Shell that oil prices would rise making the exploitation of 'difficult' oil more feasible. Nevertheless, there was no concept of what northern basin oil economics would actually be (Williams 1999). At the Seminar where Williams spoke, others also offered explanations of why Shell was willing to explore in areas like the North Sea where at a then current reference price of $2.50 per barrel economic production would not be possible. Thus, Kassler (1999), a former Shell geologist, stated that management was fearful of the consequences of a resource scarcity scenario, which would inevitably increase the oil price. This view was supported by Odell (1999), who recollected that in the late 1960s the then Head of Shell's Economics Division had spoken of the dangers of a rise in the 'real' oil price to $7 per barrel by 1980.

Quite apart from oil price increases, Shell also knew that to justify North Sea oil development, it "... had to find big fields, with high productivity" (Williams 1999). Fortunately, its main early discoveries, particularly the Brent complex, met those criteria. The first discovery (Auk) came in 1971, after several dry 'obligation' wells on its First Round acreage. Most were drilled by Shell's British-built semi-submersible drilling rig, *Staflo* delivered at the end of 1967 (Jamieson p. 137).

Of the three companies, Burmah stood in the greatest need of the new sources of revenue potentially on offer in the North Sea. Historically, its activities had been heavily focused on Britain's old Indian Empire. The company's production and refining facilities in Burma had been destroyed in the Second Word War. In the company's view, compensation had been inadequate. The subsequent Burmese business environment was unattractive so that the company sold its remaining interests to the Burmese government in 1963. It was similarly to withdraw from India, though its presence in Pakistan was reinforced by the discovery and development from the early 1950s of the giant Sui gas field where commercial production started in 1955.

In the 1960s, Burmah sought a new strategic direction in other parts of the world. This was possible because it retained an able and innovatory technical work force and the financial resources necessary to deploy them in new areas. These resources included the collateral offered by its large shareholdings in both BP and Shell and a near debt-free balance sheet (Corley p. 349). The result was a series of corporate acquisitions, shareholding stakes and investments in E&P ventures. The company acquired operating and non-operating upstream interests (both on- and offshore) in Australia, Canada, Ecuador and the USA. It also acquired refining and marketing interests in the UK, as well as subsequently moving into the oil tanker and liquefied natural gas (LNG) carrier fields.

Additionally, as Corley records (p. 376), the company's geologists were well aware of the growing interest in the possibility of large hydrocarbon discoveries in north European waters. An added interest in such prospects arose with the introduction of Corporation Tax in 1965. This removed double taxation relief on Burmah's overseas income, then substantial relative to its UK income and dividend payments, giving it an additional motive to develop UK income streams, which North Sea success might offer.

In 1963 Burmah had carried out a geophysical survey of the North Sea in association with its GoM partner, Murphy Oil. This led to the formation of the Burmah North Sea Group (BNSG), comprising Burmah and ICI each with a 40% interest, with Murphy and its drilling contractor subsidiary, Offshore Drilling and Exploration Company (ODECO), sharing the remaining 20%. This structure helped Burmah husband its more limited capital resources compared to those of BP and Shell in what was likely to be an expensive undertaking.

The consortium was awarded thirty-four blocks in the First Licensing Round in 1964 and a further eight in the Second Round the following year. ODECO ordered a semi-submersible drilling rig from a Teesside yard – the ill-fated *Ocean Prince* – in which both Burmah and ICI held a quarter share. The *Lady Alison*, a purpose-built supply vessel, owned and operated by the Peninsular & Orient Steam Navigation Company (P&O), supported the rig. Most of the offshore personnel were British. It was this combination, which made the first oil discovery (though a not commercially viable one) in British waters off Cromer in October 1966 (Corley, pp. 376–378).

Despite the loss of the *Ocean Prince* (see p. 57), Burmah remained active in the North Sea until its near collapse in the financial crisis of 1974. During the subsequent reconstruction, most of its North Sea interests were sold to the state-owned British National Oil Corporation (BNOC), including shares in the major Thistle and Ninian discoveries, where Burmah had been Operator. A low level of North Sea exploration activity continued for some years after 1974. BP acquired the company in 2000.

Although the interest of the oil companies in the North Sea was always commercially driven, they knew that their activities would be constrained to a greater or lesser degree by the needs of others, most particularly those of the British government. As time went on, government imposed constraints became increasingly codified. With the notable exception of the original licensing rules, these tended to lag behind events. They were only occasionally coloured by political ideologies.

Some of the more important of the legislative enactments in the formative years are listed in Table 3.1 below. The 'target' was normally the E&P sector. Legislation could be subject to revision, with the revisions sometimes having a greater practical significance to those affected than new primary legislation.

In addition, the government also issued many regulations and guidance notes, covering a range of often-hazardous activities, such as diving and

TABLE 3.1 Important Offshore Legislation 1964–1978

Year	Legislation	Main Objective	Important Provisions
1964	Continental Shelf Act	Opening of the UKCS for exploration	Offshore licensing regime and natural gas purchasing privileges of Gas Council
1971	Mineral Workings (Offshore Installations) Act	Enabling legislation	Health, safety, and welfare regulations
1975	Oil Taxation Act	Increasing government revenue share	Introduced Petroleum Revenue Tax
1975	Petroleum and Submarine Pipelines Act	Extension of state control over E&P activities	Established BNOC, with rights to 51% of licence equity or of field production
1978	Participation Agreements Act	Enabling legislation	BNOC participation exempted from restrictive trade practices legislation

offshore installation construction, inspection and operation, backed by inspectorates and/or certification requirements. An early and important example dealt with offshore installations (Department of Energy 1974b). DEn for many years acted as agent for the Health and Safety Executive (HSE), functions that were returned to the HSE after the Cullen Enquiry into the Piper Alpha explosion of 1988 (Cullen 1990).

Taxation requires special mention. Partly as the result of the highly critical House of Commons Public Accounts Committee (PAC) report published in March 1973, see Arnold (1978, pp. 49–52), and partly from a wider political recognition of the need to capture for the state the large excess profit or 'rent' available from oil production under conditions of high oil prices, it became clear that a 'special North Sea tax' was needed in addition to the production royalty and corporation tax already levied.

The consequence was the Oil Taxation Act of 1975 and the introduction of Petroleum Revenue Tax (PRT) in the same year. PRT was a complex tax with high marginal rates. It added a 'ring-fence' around individual fields to the Corporation Tax (CT) 'ring-fence' around the UKCS as a whole. PRT changes were initially mainly upward but later modifications aimed at improving marginal field economics became a feature, creating an increasingly complex system, which interacted with Corporation Tax and royalties. On occasions tax was a determinant of both the level and nature of demand for parts of the service and supply sector.

BGC's capture of the gas 'rent' for the nation through contractual arrangements was for long seen as a constraint by the E&P companies. Its eventual removal and BGC's privatisation, the creation and dismantling of BNOC as well as the absorption within BP of its privatised part-successor, Britoil, also affected both the E&P and the service and supply sectors. BNOC, created by the 1974 Labour government, had had the potential through its nascent role as Operator, widespread representation on operating committees and preferential licence rights to provide OSO substantial support in developing the UK's service and supply sector.

As they played a role in defining the economic environment for both E&P and service and supply companies, two issues not involving legislation deserve mention. One was the Varley depletion 'assurances' of 1974, under which the Secretary of State (SoS) for Energy (Eric Varley) undertook that no investment or production constraints would be imposed on any field discovered by 1975 until 1982 or later. The other was the agreement in 1975 of a Memorandum of Understanding (MoU) and Code of Practice (CoP), defining the framework under which British industry was offered 'full and fair opportunity' (FFO) to win work from UKOOA members, which will be discussed further in Chapter 5.

3.2 THE BRITISH GOVERNMENT

The way the government exercised its powers over the UKCS was of critical importance to oil company operations and thus to the development of the offshore supplies market. Given changing circumstances over so long a period, it would be surprising if a totally consistent picture of government motivations and constraints emerged. There were periodic changes in the political party forming the government, There were also the differing perspectives of the various civil service departments involved, to say nothing of the different individuals holding government offices, to consider. Nevertheless, the early policy thrusts were revealed as surprisingly stable and bi-partisan.

So as not to rely entirely on what is contained in files held at the Public Record Office (PRO) or in published material, the author also undertook a limited number of interviews with surviving policy makers, a mixture of former ministers and senior public servants. While a range of PRO departmental files were also examined, considerable effort was devoted to reviewing the Cabinet Conclusions (or Minutes) and the associated memoranda – TNA: PRO CAB 128 and CAB 129 series – for the 15 years from the agreement of the international convention in 1958 allowing the division of the north-west European continental shelf into exclusive economic zones to the energy supply and price crisis of 1973 and its immediate aftermath.

In addition to a search for the origins of government policy towards it, an important aim was to try to determine when North Sea oil and gas first became a topic of interest at the highest levels of government. Unfortunately, for much of the period, specific references to both the North Sea and to the broader question of continental shelf mineral resources were conspicuous only by their absence.

Energy issues featured throughout the period, with a number of recurring themes. A powerful one was the need for diversity in electricity generating fuels, allowing nuclear and oil fired power stations to be constructed by the state-owned generators. Similarly, the gas industry was granted freedom to move away from coal as its sole primary raw material, initially in favour of petroleum derivatives – specifically naphtha – and imported LNG. Both these themes were related to the need to manage the inevitable decline of coal production as costs increased and cheaper substitutes became increasingly attractive; clean air also mattered.

The balance of payments was a regular pre-occupation. However, the rising volume of oil imports was for long not seen as a critical influence, given the generally modest cost of the commodity, the foreign exchange offsets offered by the profitable overseas production interests of British companies, the use of British tankers and the growing domestic refining capacity. However, security of supply was of concern.

In the early years, there were several occasions when the Geneva Convention on the Law of the Sea featured, but only in such contexts as the implications for fisheries or intelligence gathering of differing territorial water limits. The potential for oil and gas did not seem to be part of the discussions. Indeed, it appears that it was not until 1964 (the year of the First Licensing Round) that North Sea exploration finally figured in Cabinet meetings. At the beginning of April, in the run-up to the Third Reading of the Continental Shelf Bill, the Minister of Power circulated an information memorandum to his Cabinet colleagues – TNA: PRO CAB 129/117/C.P. (64) 82. In it, he presented what were to be core elements for licensing policy for many years into the future, specifically highlighting five:

i.   encouraging the rapid exploration and exploitation of UKCS petroleum resources;
ii. the need for licence applicants to be incorporated in the UK and for their profits to be subject to British tax;
iii. with respect to foreign applicants, the extents to which British oil companies were treated fairly in their home countries;
iv. the applicant's proposed work programme, together with its ability and resources to execute it;
v.   the applicant's past or current efforts to further UKCS resource developments and the UK's fuel economy.

It was made clear that a prominent contribution was to be expected from American companies given their greater experience of offshore operations, that arrangements must allow for British-owned participants as many opportunities as they were capable of properly using and that a discretionary licensing system should be adopted as it was more likely to produce a result acceptable to the government. References to support for the British supply industry were absent.

In September 1964, the Minister updated the Cabinet – stating that approximately one-third of the UK North Sea would be licensed with UK companies receiving about 30% overall and 40% in the most favoured area. Licensees had committed £80 million (about £1.22 billion in 2008 terms) to exploration and had already placed equipment orders worth £14 million (about £214 million in 2008 terms) with British suppliers – TNA: PRO CAB 128/38 CM (64) 47.7.

Shortly afterwards, the Conservative administration was replaced with a Labour one and fuel policy began to move more centre stage in Cabinet discussions. Ideas set out in October 1965 drew heavily on earlier pre-occupations but had to be adapted to place more emphasis on security of supply and the balance of payments following a hardening in attitudes towards Western companies among oil producing countries. Additionally, positive developments in the North Sea of potentially great economic benefit might have to be accommodated and would be energetically pursued – TNA: PRO CAB 129/122 C. (65) 130.

As early as the beginning of 1966, the Minister of Power, was able to advise the Cabinet that an initial 3-year supply gas deal for 100 million cubic feet per day at 5d (about £0.29 in 2008 terms) per therm had been negotiated between BP and the Gas Council, which was seen as an attractive price, with the possibility of price falls to come – TNA: PRO CAB 128/41 CC (66) 5.5. This factor had to be reflected in the delayed White Paper presented to Parliament in November 1967 – Ministry of Power (1967). The possibility of North Sea oil reserves was not mentioned.

By this time it was already apparent that the southern North Sea contained large natural gas reserves, which would provide a new source of primary energy and important cost-savings. Natural gas production equivalent to 15% of national energy demand was seen as possible by 1975. To enable natural gas to gain market share rapidly, construction of a nation-wide transmission system would be complete by 1970, with the mass conversion of appliances taking place over 10 years. Though coal would lose some custom, the main substitution would be for oil, then providing in the expensive form of naphtha, two-thirds of raw material used for gas supply. While no estimate of the potential balance of payments gain was presented, the implication was that it would be substantial, helping to justify the policy of rapid depletion.

While energy continued to feature on the Cabinet agenda over the next 3 years, there was little further reference to the North Sea. However, the return in 1970 of a Conservative government soon led to the call for a new Cabinet briefing paper from the Minister of Technology –TNA: PRO CAB 129/153 CP. (70) 80. This contained little that was new. However, it did provide an estimate of the foreign exchange savings anticipated from the introduction of North Sea gas (£100 million a year, about £1.4 billion in 2008 terms) and expressed a hope for a major oil discovery (probably the first such mention at this level) or further gas finds under British control. It was discussed and accepted –TNA: PRO CAB 128/47 CM 31 (70) 6.

From early 1971, Cabinet was devoting considerable attention to the attempts by members of the Organisation of Petroleum Exporting Countries (OPEC) to increase the price of crude oil and threaten its uninterrupted supply. TNA: PRO CAB 128/49 CM 10 (71) 4 recorded the Prime Minister (Heath) introducing to these discussions the most positive view yet of Britain's offshore potential by seeing the benefits in '... *increasing the pace of development of our own resources of power, including North Sea oil and gas*'.

As the year drew to an end, the Cabinet saw domestic energy concerns added to those with OPEC, culminating in the outbreak of a coal miners' strike in November followed in early February 1972 by the Declaration of a State of Emergency and a need to shed electricity load.

By the start of March the miners' strike was over, but energy crises were not. During early summer, the Cabinet discussed the nationalisation of the British-controlled Iraq Petroleum Company –TNA: PRO CAB 128/50 CM 29 (72) 2. Shortly afterwards, came the abolition of fixed exchange rates, effectively bringing the Sterling Area to an end. By December, coal had returned to centre stage amid growing concern about the security of oil supply, with the UK's good fortune in having recently discovered oil and gas fields being noted. The same month also saw discussions relating to a new Development Strategy for the British Steel Corporation (BSC), in which it was noted that pipes and structural sections for the North Sea represented its fastest growing market – TNA: PRO CAB 128/50 CM 59 (72) 3. The Cabinet does not appear to have discussed the wider industrial implications of North Sea development, although the government had already commissioned the *IMEG Report*.

As 1973 saw the Yom Kippur war and associated OPEC supply boycott, sharp increases in oil prices, another coal miners' strike, a further State of Emergency and a 'three-day week', inevitably energy supply and its international ramifications and economic repercussions continued to dominate Cabinet discussions. Evidence of friction between the government and the oil companies was clear.

References to North Sea oil remained sparse. However, with the prospect of advancing its supply to achieve national oil self-sufficiency by the early 1980s, its development was clearly now a priority at Cabinet level – TNA: PRO CAB 128/53 CC 48 (73) 2 and CAB 128/53 CC 53 (73) 2. Thus the case for exempting the construction of North Sea oil platforms from the three-day week power restrictions was to be examined –TNA: PRO CAB 128/53 CC 60 (73) – and the infrastructure programme for North Sea oil was to be maintained despite widespread cuts in public expenditure – TNA: PRO CAB 128/53 CC 61 (73) 6.

In March 1974, a Labour administration replaced a Conservative one. On 2nd July the new Cabinet discussed a statement to be made by the SoS for Energy, showing how the government proposed to implement its manifesto commitments to increase government control of offshore oil and gas – TNA: PRO CAB 128/54 CC 22 (74) 2. The main features were the introduction of Petroleum Revenue Tax (PRT), the 'ring-fencing' of the North Sea for corporation tax purposes, making it a condition of future licences for the government to have the option of a majority interest in fields discovered, beginning discussions with licensees aimed at government majority participation in commercial fields already found, establishing a state oil company to exercise government participation rights and extending DEn control over offshore operations through the Petroleum and Submarine Pipelines Act. The generality of files for the period examined suggest that any of the measures, with the possible exception of the state oil company, could also have been introduced had the Conservatives retained power.

Parliament received an account of the new strategy in a brief document – Secretary of State for Energy (1974). Although this statement and what followed from it represented the high point of British government intervention in the offshore oil and gas industry, it contained no promise of support for British industry beyond an intention to ensure maximum benefit for Scotland and other 'old' industrial areas. Establishment of the Scottish Development Agency (SDA) was announced at the same time.

The paucity until the 1970s of Cabinet discussion of North Sea issues should not be taken as showing disinterest by Ministers. Other papers examined from the Cabinet Office and the Prime Minister's Office, including some of Heath's Central Policy Review Staff (CPRS), show that senior Ministers and the Prime Minister at times took a close interest in North Sea issues, though this seemed only rarely to extend to the support industries, apart from pipe manufacture. There were clearly more important policy priorities.

The earliest prime ministerial file examined – TNA: PRO PREM 13/925 – began with the authorisation in May 1965 by the Cabinet's Ministerial Economic Development Committee of the Second Licensing Round, in which participation by the nationalised industries was to expand. The government made its overall objective clear in a statement to the House of Commons on 21st July 1965, re-emphasising that the rapid exploration and development of UKCS oil and gas resources remained its priority. As with the First Round, there was no explicit reference to supporting a British supplies industry though this might be inferred from indicating that in the choice of licensees the government would take into account an applicant's existing or planned contribution to the '... *strengthening of the United Kingdom balance of payments*' and to the development of the country's industry and employment, particularly in a regional context.

Following the discovery of gas by BP late in 1965, the contents of the file focus increasingly on the gas purchase price to be paid by the Gas Council – a focus that was to increase after the announcement of a 'temporary' price of 5d/therm in February 1966. A more immediate problem was explored in a memorandum from the Minister of Power to the Economic Development Committee dated 22nd June 1966 entitled 'Natural Gas: Supplies from the North Sea and Pipe Requirements'. This recommended that the inability of the sole British producer to supply pipe should not be allowed to delay the receipt of the first Shell-Esso gas supplies in 1967 and that recourse should accordingly be made to imported pipe. To act otherwise would delay benefits of payments from the substitution of natural gas for imported naphtha and oil. Thus was initiated a precedent that would that would normally serve as a policy guide in the years ahead, showing the importance of the balance of payments as a driver of government policy.

The twin themes of the need to increase pipe production capacity and to limit the price to be paid for North Sea gas also dominate another file –TNA: PRO PREM 13/1524 – which covers the period September 1966 to August 1967. The price issue was the overriding one, with concerns that if overseas producers were able to remit their profits, the balance of payments benefits of North Sea gas production would be dissipated and the coal industry possibly destabilised. Although there was discussion of the need to provide returns sufficient to maintain oil company interest, in the end the government laid down a maximum price that the Gas Council could offer Shell-Esso of 2.5d/therm (about £0.14 in 2008 terms). The need for the full economic implications of North Sea gas to be assessed was recognised and by 22nd March 1967, the Prime Minister was emphasising its urgency. Revealingly, in a record note dated 4th August 1967, the Chief Scientist responded to a query from the Prime Minister about a press report of a North Sea oil strike by following the BP and Shell line that that a commercial oil discovery was not likely.

By 1971, such a view had already been over taken by events. Interest in the oil potential of the UKCS was now very much on the government's agenda. It was to be heightened by the increased government awareness of the possibility of a crisis arising from the growing power of OPEC and the Middle East. In October of that year, the then Prime Minister received a warning from Shell's Chairman, that there was likely to be a major energy crisis and increased oil prices 'between ten and twenty-five years from now'. He added that since by 1980 the UK would be 30%–40% self-sufficient in oil, it would be better placed than most European countries. It appeared that the government had already formed similar views, although the Department of Trade and Industry (DTI)'s Working Group was accused of underestimating the extent to which prices could rise or that oil producers might start restricting supply, a key point in Shell's thinking – TNA: PRO PREM 15/1595.

Many important issues (including those of the critical period of August 1973 to October 1973) are covered in papers contained in files bearing the reference TNA: PRO CAB 184/61. In August 1971, the Treasury was suggesting to the Cabinet Office that the time to reappraise North Sea policy was approaching. It is clear that by then the Central Policy Review Staff (CPRS) at the Cabinet Office was now making a major contribution to how government thinking on the North Sea was evolving.

In December 1971, fearing that the government had 'inadequate appreciation', BP made a presentation to Ministers about the Forties Field development. The Prime Minister was not present but was fully briefed on what took place. Among the large number of points made in this briefing, four are of particular interest, namely:

i.   BP would probably be sourcing steel pipe from Japan on grounds of quality;
ii. BP was hopeful that at least some of the four large production platforms needed would be built in Scotland;
iii. employment in Scotland during Forties development would peak at 5,000;
iv. the Forties project financing would be novel, with the loan being amortised through a sale and repurchase arrangement with the banks.

Other issues, such as whether the export of Forties oil should be allowed, the level of government 'take', the possible impact of membership of the European Economic Community (EEC) and the implications for security of supply and the balance of payments generated most subsequent internal correspondence on the file. Nevertheless, 7th January 1972, the Cabinet Office, writing to the to DTI, made the strongest statement yet in support of the domestic offshore supplies industry, quoting the Prime Minister as saying that: '... *pipes and rigs and everything required for the recovery and exploitation of the oil in the Forties Field should be bought in this country, and that British industry should immediately be geared to what is required*'.

At last the offshore supplies issue had become a priority. The DTI responded weakly, saying that while it had encouraged British companies to provide equipment for BP's Forties field and would see what more could be done, to prevent project delays imports would be necessary, citing submarine pipe as an example. Positive suggestions on how things might be improved were lacking, apart from a reference to the possibility of British firms partnering with experienced U.S. companies.

When in March 1972, the DTI sent a draft paper outlining current oil policy to the CPRS, it made no mention of the British supply industry. Although not on that particular ground, the paper was heavily criticised by the CPRS and other Departments. It is evident that that the DTI was not regarded by its peers as an exemplar among policy departments.

In May 1972, the CPRS and the DTI jointly commissioned (from IMEG) a report on potential industrial benefits from the North Sea, with the CPRS likely to have been the driving force behind the action The report, despite its large scope, was received in September of the same year, suggesting considerable urgency, perhaps reflecting the Prime Minister's personal interest (see above). The content of the report and reaction to it are addressed in Chapter 4.

The sudden arrival on the scene of offshore supplies as a serious issue did not prevent continued discussion of other on-going North Sea issues, ranging from oil taxation, prices and depletion to balance of payments and the status of the UKCS under the Treaty of Rome. The question of the Norwegian median line also re-emerged, though it was by then 'water under the bridge'.

Only for depletion were the possible implications for the British supply industry considered. Thus, on 10th April 1973 Treasury made it clear to the DTI that it was generally against restricting production; it conceded that a conservative depletion policy would allow more time for British capacity to be developed but went on to argue that it might also discourage industry from making the large initial capital investments required to beak into the market. Essentially the same sentiments were expressed on 8th October 1973 in a draft DTI briefing paper, which also noted that the market was '*now expected to be £300 m per year*' (nearly £2.7 billion in 2008 terms). In both cases, the CPRS had been 'copied in'. It was certainly now well entrenched in the policy 'loop' on this particular issue.

If the North Sea had failed to attract much attention at the highest levels of government prior to the early 1970s (the gas price issue of 1966/1967 apart), the same cannot be said of the Whitehall departments directly involved, which were not confined simply to the Ministry of Power and its successors. Departmental activity was continuous at least from 1962 onwards. Although the question of how representative is the material chosen is always present, some useful insights of the early interactions over the North Sea between the civil service and the oil and gas industry are provided. The first evidence of a prospective licensee drawing to the government's attention its support for British industry came from Amoco U.K. Petroleum Limited in a letter dated 14th July 1964 covering its First Round application. In this it was at pains to make it clear that it had invited British yards to submit bids for the construction of a marine drilling rig –TNA: PRO POWE 29/388.

An internal Ministry of Power memorandum in the same file dated 2nd June 1964 points to one of the bidders probably being the North Sea Marine Engineering Construction Co. Ltd (NORSMEC). This was a recently formed consortium of five engineering firms and a shipping company, planning to offer construction and marine services, which came into being as a result of a personal acquaintanceship between its first managing director and an Amoco executive. NORSMEC went on to construct a jack-up at the Teesside facility of a consortium member.

Amoco's bid invitation to British yards could well have been in response to a specific request. This would have been consistent with the recollection by a senior Shell executive. During his company's First Round negotiations he was asked to undertake that any rigs required would be UK sourced. Whether such requests represented an official policy is not clear, but, if so, it was not one that had been formally announced.

With its responsibility for relations with overseas oil and gas producers, the FCO took an early and surprisingly detailed interest in North Sea affairs. Thus on 8th June 1966, the FCO voiced its concern to the Ministry of Power about the inability of the British steel industry to manufacture large diameter oil and gas line pipe for North Sea and overseas markets – TNA: PRO FO 371/187603. This may have been responsible for a statement in a Ministry of Power ministerial briefing dated 16th June 1966 (copied to the FCO) stressing the urgent need for an assessment of the demand for steel pipe. The same paper provided the FCO with a summary of the potential benefits of large-scale natural gas production, including an improvement in the balance of payments. An attempt to quantify this was made by the Ministry of Power in a paper dated 7th July 1966 entitled 'North Sea Gas – The Balance of Payments in the Short Term'. This produced a range of gains in 1970 varying between £15 million and £48 million a year, respectively, about £207 million and £664 million in 2008 terms. The figures were subject to many caveats, some relating to supply issues, such as whether foreign-owned drilling rigs and pipelay barges would be on-charter as well the ever-present question of the extent to which pipe demand could be met from domestic production. It was recognised that the investment necessary to expand pipe-making capacity would only take place if long-term demand could be foreseen. It had no forecasts to offer.

Some limitations on freedom of government action in the economic sphere resulted from the UK's international obligations the General Agreement on Trade and Tariffs (GATT), to say nothing of bilateral relations with individual foreign governments. However, in 1973, the UK joined the EEC, which increasingly restricted British government freedom of action in energy and industrial matters. Thereafter European concerns loomed large in FCO thinking. The main early issue was whether or not the UK could retain domestic oil production for its own use. However, other uncertainties extended into areas that could impact on broader matters of industrial policy. Thus a note entitled 'North Sea Oil and the EEC' dated 7th February 1974 pointed out that until determined by the European Court, it would remain unclear as to whether the Continental Shelf fell within EEC jurisdiction, an uncertainty the FCO preferred to stay unsettled –TNA: PRO FCO 30/243.

The same file contains a striking reference to the growing influence of events in Scotland. In the minute of an interdepartmental meeting on 17th March 1974, a Scottish Office official referred to his Ministers working in a '*highly charged atmosphere*' because five out of seven Scottish National Party (SNP) MPs represented areas associated with North Sea oil.

Even before this, North Sea oil had become a matter of Parliamentary concern in a way that North Sea gas never had. The PAC Session 1971–1972 is best remembered for its investigation of the North Sea tax regime, leading to conclusions which made it inevitable that government would have to take measures to increase the public 'take'. However, its enquiry, covered in TNA: PRO T 292/178, dealt with North Sea policy in general, looking at the whole range of issues that were increasingly being raised in Parliament and the press.

A critique of the licensing system was an important feature. On a number of occasions, the DTI (which had absorbed the Ministry of Power) was required to make written or verbal submissions explaining its policies. In a note dated 14th June 1972, referring to the recently completed Fourth Licensing Round, it repeated that its main objective remained the most rapid exploration and development of the UKCS as possible in order to improve the balance of payments and reduce dependence on oil imports, especially given what was happening within OPEC. Its second objective was to make sure '...*that the British interest was adequately represented*'. It went on to claim that that by early 1971, UKCS activity was in decline and required a stimulus to avoid resources being shifted elsewhere. Also, British equipment suppliers needed a large home market if they were to reduce reliance on imports and capture overseas markets. On 7th June 1972, it had been announced that a study of the industrial demand arising from North Sea development had been commissioned from consultants (see p. 75).

Shortly afterwards, on 20th July 1972, the Committee submitted a searching list of questions ranging over such matters as whether there were factors limiting British industry's ability to meet North Sea demand, the size of that demand, the proportion supplied from the UK and the extent to which the DTI had actively sought to support British industry. It specifically asked if the DTI a number of leading questions such as:

i.   whether it is was able to assist firms wishing to meet needs arising from North Sea activity;
ii. whether any steps to facilitate its ability to assist British industry had been taken;
iii. when it had become aware of the industrial and employment implications of North Sea activity and whether it had attempted to assess them in relation to the objective of rapid exploration and development;
iv. the extent, if any, to which the Department had sought to persuade licensees to seek British suppliers.

On the factual issues, it was perhaps unrealistic to expect answers before the recently commissioned report had been received. On the policy front, the Department's Permanent Undersecretary (PUS) had already admitted in verbal evidence given on 15th May 1972 that, unlike those of Norway, UK licence terms did not require preference to be given to national suppliers. He spoke of the difficulty in enforcing of such a policy and admitted that departmental policy had been limited to encouraging North Sea Operators to use British equipment. By the time of this exchange, however, it must have already been apparent to the DTI that it was inevitable that greater participation by British industry in the North Sea programme was about to join its policy objectives. The work of the PAC should also have warned that the public finances would be another.

To assess the relative importance of the lengthening list of government objectives, the author undertook structured interviews of a small number of people involved at policy level, although the passage of time alone prevented any attempt to create a 'representative' sample. The interviewees were five in number, two former ministers, two former very senior career civil servants and the then most senior surviving official of the short-lived state oil company, BNOC. All were caught up in the events of the 1970s, whilst two had also high-level experience from the mid-1960s.

Contributors were asked to respond to specific questions, volunteer their own opinions and attempt to quantify the relative importance of ten factors affecting policy, having 100 points to distribute across the factors. While all were prepared to discuss the factors, only three were prepared to attempt quantification. Two of these three each added an additional factor, respectively, relating management of oil and gas reserves and the creation of an independent technology base. These were not introduced into the analysis beyond their treatment as 'Other Factors'. Table 3.2 summarises the views of the 'quantifiers'.

'Security of Supply' clearly emerges as the most important factor influencing government policy, followed by 'Improving the Balance of Payments'. The next three factors, 'Supporting British E&P companies', 'Encouraging the British supply industry', and 'Strengthening Public Finances' were considered to be of roughly equal importance. None of the other individual factors were given much importance.

Nothing said by the two 'non-quantifiers' would have seriously altered the ratings, except possibly in two respects. One placed a greater emphasis on Scottish job creation and adherence to high standards of worker and environmental protection. The other went even further stating that it was wrong to have introduced worker and environmental protection into the rating system at all, as the need for legal compliance was absolute; he also emphasised the high political profile of Scottish issues up to Cabinet level.

TABLE 3.2 Relative Weight of Factors Influencing Government Policy

Factor	Average Score	Range of Scores	Rank
Security of Supply	21	17–26	1
Improving Balance of Payments	15.7	7–20	2
Supporting British E&P Companies	14.7	6–20	3
Encouraging British supply industry	14.3	5–20	4
Strengthening Public Finances	14	7–20	5
Employment in Scotland	5.7	5–7	7
Employment in non-Scottish Development Areas	3.7	2–5	8
Protecting Environment and Workers	2.7	2–4	9
Reviving Steel and Shipbuilding industries	1.7	1–2	10
Other Factors[a]	6.7	0–15	6

[a]Added by an interviewee.

None of the four who expressed an opinion on the matter thought that there was much difference between Labour and Conservative administrations in terms of the relative importance attached to the various factors, although it was pointed out that under the Thatcher administration there was an increased emphasis on 'Strengthening the Public Finances'.

Although it would be unwise to place undue emphasis on the opinions of so small a 'sample' of policy makers, the conclusions reached from their personal recollections appear clear-cut and consistent with the results of archival research. Both approaches clearly show the built-in momentum towards as rapid a speed of resource exploitation as possible, regardless as to whether or not it was in the interest of British suppliers. Economic necessity provided the driver of government policy, with other considerations acting more as constraints.

3.3 BRITISH INDUSTRY

As to the suppliers themselves, during the early years of North Sea exploration and development, most of British industry was a passive observer of events. Its position was thus totally different from those of the oil companies and government in that it could only react to demands initiated by the actions of the other two. In other words, its demand was a derived demand. Moreover, the potential direct exposure of individual firms to the events that were to occur varied from nil to very large.

The latter category included members of the UK's existing small oil supply and service industry. These companies had essentially been export oriented, mainly meeting the needs of BP, Burmah and Shell. With odd exceptions like Wimpey, Noble Denton and Lloyd's Register, they had no offshore experience. Although the potential benefit offered by a large domestic market would have been apparent to them, they were not bound immediately to make it their top priority, particularly if their existing resources were already committed to overseas markets.

Another group, whose exposure was potentially very large and in some cases would become virtually absolute, were businesses that benefited by virtue of their location. Initially these were ports and port related concerns along the east coast from Lowestoft in the south to Lerwick in the north, which supported the E&P campaigns. To these would be added many specifically created branches, subsidiaries, affiliates and newly established businesses, in some cases requiring investment in totally new facilities.

Widely scattered throughout the established industrial landscape were a great range of aviation, ship owning, ship building and civil, chemical, structural, marine, mechanical and electrical engineering concerns representing a high proportion of the nation's technical resources and its capital goods producing capacity. Many of these businesses already sold some part of their output to the international oil industry. However, their markets were generally diverse and they could not be considered as part of the oil supply industry as such. The arrival of the offshore industry in the UK offered an incremental market, with the requirement to address new customers and, in some cases, to adapt products to cope with offshore conditions. This could involve reducing product size and weight and improving resistance to a damp salty atmosphere, sometimes referred to as 'marinising', something that the more dedicated of oil industry suppliers would also have to address. It would not always be self-evident that the opportunities should be taken up, given the costs involved and uncertainties over the level and the timeframe of demand.

With such a varied business population, consideration of motives and aspirations should ideally be from the 'bottom-up', that is on a firm-by-firm basis to reflect the 'drivers' of the individual companies potentially involved. However, this is impractical in a work such as this, so that a 'top-down', or highly generalised approach, is all that can be attempted, based upon no more than a recognition that the underlying purpose of the private firm is to earn a profit for its owners through the provision of goods and services to customers.

Even this motivation can be over-taken in the short-term by a greater imperative – the need to survive. Given that the problems of British economic decline (see pp. 10–19) were particularly severe in respect of sectors of the investment goods industry upon which new and expanded demands were likely to be placed, this was a far from academic point. As shown in the report *Market – The World* (National Economic Development Office 1968), among the manifestations of a weakening British capital goods sector were a decline in its mechanical engineering exports from 21.1% of the world total in 1955 to 15% in 1966 accompanied over the same period by a rise in import penetration from 9% to 19%. Another unfavourable indicator was the declining profitability of British industry.

Some of the concerns interested in entering offshore oil supply would, it can thus be inferred, be doing so from a position of weakness and be driven by a desire as far as possible to utilise existing facilities and labour rather by than new investment in plant and training. The more specific the requirement, the less appropriate such a strategy would be. As requirements became increasingly specific, the greater the investment in specialised equipment, skilled labour and 'know-how' would be needed. Since these expenses would have to amortised over a number of years, this would have immediately raised the question of the scale and duration of domestic demand – something, as already been shown (see pp. 53–56), which was to remain contentious for a long time.

Especially in highly specific activities where foreign, mainly American, suppliers were already well established, there were also subtler barriers to entry to be addressed. There is an extensive economic literature dealing with barriers to entry, which it is not intended to address except to the extent where it seems to bear directly on the offshore industry.

Barriers to entry do not depend simply on access to economies of scale, patents or official regulation, etc. but embrace all the various investments, activities, 'know-how', and experience involved in developing a dependable brand with a respected history. Loyalty to firms with this attribute arises if customers can associate it with such benefits as lower cost or greater reliability (Demsetz 1982). With all the risks and complexities inherent in the exploration and development of North Sea oil and gas, it became easy to see that a new entrant to a 'core' activity lacking such attributes would, on purely commercial criteria, face an uphill task. Only an exceptionally determined management would be tempted to try.

Academic studies of barriers to entry faced by new entrants in the Aberdeen area were carried out in the mid- to late-1980s. They suffer from being narrowly focussed geographically and upon service activities where close client/supplier proximity at operational level is essential – activities at the heart of Aberdeen's oil economy but much less so more generally. Another drawback is that they address a period more than 20 years after the initial arrival of the offshore oil and gas industry in Britain. Nevertheless, it is useful to make some reference to them.

Cairns et al (1987) pointed out that supplying the offshore industry involved the services and products (often unspecialised) of many types of firms, resulting in a complex and difficult to define market structure. In this respect, they noted that many small specialist suppliers could come into existence in peripheral areas while only a few large general suppliers could be supported, their large market shares and initially superior knowledge alone being barriers to entry for newcomers.

They found no strong evidence that economies of scale or perception of relative risk provided barriers to entry but nonetheless concluded that these existed. Local, and other UK firms were faced with the need – and associated cost – of acquiring information, technology, skilled labour and expertise to compete with 'core' function U.S. incomers and thus tended to compete in the less specialised sectors. Those UK firms that did otherwise were judging that the time and cost taken to overcome entry barriers would be compatible with a reasonable return on investment.

The study – undertaken in the summer of 1984 – was based on a random sample of 185 wholly oil-related concerns, 69% of which were UK firms and most of the remainder USA controlled. There were over 1000 oil-related firms in the Aberdeen area at roughly that time according to Hallwood (1986).

Hallwood's own study was based on 1984 data but used a larger sample of 241 firms, 187 100% oil-related and 54 less than 100% oil-related. Of the first group 133 were affiliates (i.e. controlled outside the Aberdeen area) as were 35 of the second group (giving 70% 'absentee' control in total). The paper concentrated on the characteristics of affiliates and differences between the USA and the British controlled. It did not specifically address barriers to entry.

It is not considered further because the data involved were used as part of the source material for an important subsequent book (Hallwood 1990). In this, he added an investigation of procurement patterns of nine of the twenty-three

E&P companies active in summer 1984. On the basis of purchasing within thirteen 'core', service and supply sub sectors, he concluded the level of vertical disintegration in offshore oil and gas production was very high – with more than 90% of intermediate inputs purchased rather than supplied in-house.

Hallwood analysed how this structure came about, noting an important influence was the shifting and sometimes transient nature of demands for the specific inputs required for offshore E&P for a particular user and location. This could result in periodic over-capacity, or the converse. Since circumstances restricted the ability of an oil company to aggregate such demands and sell to their competitors, demand aggregation naturally fell into the hands of specialist service firms, willing to follow their client group to new locations. Such firms created trusted brands, presenting severe barriers to entry for potential local competitors. Hallwood argued that this applied even where patent ownership was unimportant and was enhanced by the fact that established firms enjoyed economies of scale based on the exploitation of global markets, showing how U.S. firms dominated the world market in oilfield machinery. He also singled out Schlumberger for its near monopoly in some well logging and data analysis market segments.

For the development of a specialist oilfield service and supply sector to take place, there had to be an initial concentration of activity. The classic examples were in the oil and gas prone areas of the USA. Nevertheless, even in a small province like Trinidad, where E&P was fragmented, it proved possible for a local company – Trinidad Oilfield Services – to come into existence in the 'core' well services area, to innovate and to operate overseas. The company functioned independently from 1938 to 1960, eventually being acquired by a U.S. competitor. Other locally controlled service companies succeeded it, such as Tri-Can Services, founded 1960, and Trinidad Oilwell Services, founded in 1966 (Higgins 1996).

For supplies specifically to offshore E&P prior to the development of the North Sea province, it could be argued that this condition of fragmented demand had been best met in the GoM. Many of the oil companies involved there would have been 'independents' with limited resources whose very presence in the business depended on access to a network of competent contractors. Hallwood (1990) himself recognised that where an oil company was operating on its own, it had to greatly increase in-house service provision, since suppliers would be reluctant to risk the commitment of fixed assets (and he could have added, scarce personnel) to the needs of a single buyer. Hallwood (1990, p. 47) quoted the (onshore) operations of Burmah in Pakistan as a case in point.

He might reasonably have gone on to consider the pre-North Sea operations of the British oil companies in general. He would have concluded that, a few small independents apart, it consisted of three large companies operating largely on their own in locations far from their U.S. peer group, compared to which their level of vertical disintegration would necessarily have been much lower. The major role played by BP, Burmah and Shell in the ownership and operation of offshore drilling rigs in the early stages of North Sea exploration together with their large in-house engineering operations suggests that this pattern persisted initially. It was the influx of U.S. E&P companies, bringing their U.S. suppliers in their wake that enabled the British firms to benefit more fully from vertical disintegration, thus leading to the imposition of U.S. market practice, see Hallwood (1990, p. 87). Many of Hallwood's 'core' suppliers were suppliers to upstream oil and gas companies generally, with offshore operations representing only an incremental demand, which must have been relevant to issues such as economies of scale. He paid little or no attention to that small group of suppliers whose operations, though 'core' to offshore E&P companies, did not extend throughout the oil and gas industry, a clear example being providers of underwater services.

Hallwood's observations on vertical disintegration and barriers to entry were not his main concern. He also addressed host country strategies towards newly developing offshore supply industries. However, as the title of his book indicated, he was primarily concerned with the question of transaction costs and trade between two groups of multinational corporations – the E&P companies and their suppliers of 'core' products and services. He concluded that the prevailing practice of procurement through the bid-tender auction system, where a small number of pre-qualified suppliers were invited to submit offers created price competition in imperfect markets at tolerable 'policing' cost. It also allowed for technological progress, including that based on supplier/client collaboration, and for the stimulation of additional competition should the clients deem it required. This was to be the market structure to which British industry would have to conform.

There is little to show that government took much interest in such matters prior to the early 1970s. The inadequate supply of British subsea pipes apart, the period 1964–1970, during which the initial southern gas fields were discovered and developed, apparently aroused little interest in the offshore supplies sector in senior government circles. Jenkin (1981) showed that public statements generally about marine science and technology in this period paid little regard for the potential for British suppliers to the offshore oil and gas industry, the meeting of whose needs was seen very much as an American fiefdom. He also pointed out that the simplicity and speed of the initial development of the southern basin allowed little time for new entrants to develop a business, that the expenditure involved was not especially large and that future prospects, at least at home, did not look encouraging. Indeed, Jenkin (p. 45) went so far as to say '... *as of 1969, the Government had more or less written off the offshore supply industry*'. He additionally noted a consensus among Ministers, whether Labour or Conservative, and senior civil servants that the involvement of foreign multinationals, including those in the oil industry, was beneficial to the national economy, resulting in a disinclination to seek to control their behaviour.

Such a position was evidently consistent with the failure by the government to take any action after a report requested in 1967 by the Ministry of Power from '... *a prominent American oil executive*' on the involvement of British firms in gas developments in the southern North Sea and any difficulties they had encountered. Jenkin (p. 43) claimed that the report revealed that that British firms were often reluctant to enter the offshore market and that many of the problems identified in 1972 during work on the *IMEG Report* could already be recognised. The author could find no other reference to the report.

It has to be said that Jenkin's version of these early events is at variance with the recollections of an anonymous UKOOA official. He stated that the Ministry made a verbal approach in September 1968 to UKOOA's precursor asking it to establish the extent to which British resources were being used in UKCS operations. When submitted to the Ministry in December 1968 (and later to the Cabinet), it evinced general surprise because '... *the percentage of UK participation was well above expectations and government voices were quieted*'. Unfortunately, nothing more is known of this report either, but its apparent conclusion sits ill not only with Jenkin but also with other evidence from the period, as will become apparent in Chapter 4.

With industry subject to commercial drivers, uncertainties over demand, novel technical requirements, barriers to entry arising from established foreign suppliers and lack positive government support, it would have been difficult for most potential entrants to make a convincing commercial case for early entry to the North Seas supplies market. There were exceptions of course, ranging from businesses such as Wimpey already involved abroad to east coast ports such as Yarmouth as well as a multiplicity of suppliers of goods and services not specific to the E&P industry. In addition, a small number dedicated managements took the decision that its potential did justify entering the new market. More were to hold back until the market exploded in size, with many others also waiting for evidence of government support.

3.4 FINANCE

Finance was fundamental for all commercial participants in North Sea development. The main concern here is with finance for the support sector. However, had financial resources not been available for the E&P sector, there would have been no role for the support sector so that it is worth briefly considering how the oil companies financed their North Sea activities.

Traditionally, the highest risk activity – exploration – had always been financed through the use of internal funds or by newly raised equity. To accommodate the much higher levels of cost associated with North Sea exploration, particularly in the northern basin, sole risk drilling became the exception. Typically, North Sea exploration was undertaken by consortia of two or more members, one of which (usually the largest interest holder) acted as 'Operator', managing the venture on behalf of the partners. The consortium arrangement also allowed companies to spread their exploration risk through participation in more than one consortium. Whereas the larger companies used internal funds to meet exploration expenditures, the smaller ones used equity funds raised from financial institutions and public issues. Up to 1977, some £88 million (about £412 million in 2008 terms) was raised from these sources in the UK (Wilson Committee 1978, p. 6). The authors noted the involvement in this of long-term investment institutions, including investment trusts, insurance companies and pension funds, though some banks also participated. The prospect of capital gain from a successful drilling programme, often followed by a trade sale, was attractive to such parties.

Onshore, development financing had not normally presented a problem for the main British oil companies, which were able to use their own funds or borrowings against their balance sheets. This approach was facilitated by fact that the large onshore fields where these companies typically operated could often generate revenue soon after a discovery well and could be developed incrementally, with cash flow from early wells financing subsequent phases. That said, in the USA with its many small independent producers, the banks evolved various forms of project finance.

No particular financial innovations were needed to finance the early southern basin gas fields. The scale of the projects was modest in comparison with the financial strength of most of the companies involved and where the time from discovery to production could be as little as 2 years. The northern North Sea was very different. The capital costs of individual fields were large, even relative to majors such as BP and Shell, which could find themselves participating in more than one development at the same time. The time from discovery to first production was long – typically 5–8 years – with up to four or more being required to achieve full payback. Initially, cost and time proved difficult to predict accurately, for reasons including technological uncertainties and design changes, unforeseen increases in inflation, labour shortages and disputes and loss of 'weather windows' for installation. Moreover, some fields were in the hands of consortia, the members of which had widely differing financial strengths. Such groupings could only move at the speed of their slowest member.

Nevertheless, the UK banking industry, with inputs of lending capacity and technical expertise from American banks, was able to ensure that funds were sufficient to support development. Its efforts to assist smaller companies were supported by larger oil industry participants and in one case (Tricentrol), the government, guaranteeing the obligations of financially weaker partners or by recourse to capital markets (e.g. LASMO). Putting these arrangements in place could lead to some delay, as could uncertainties arising from government policies.

The Wilson Committee (1978) found that in August 1977 bank loans outstanding (£1.65 billion, or over £7.7 billion in 2008 terms) amounted to about a third of estimated development expenditure up to that time and that the UK clearing banks had played an important part in innovative and high profile field financings, such as those for the Forties and Piper fields. The Bank of England, which monitored North Sea finance, subsequently endorsed its generally positive view of oil field finance (Tempest 1979).

In short, finance of oil company offshore activity was a special case to which the financial system quickly adapted, offering innovative solutions when required. The number of clients at one time was small, probably less than one hundred.

By contrast, the companies involved in the support sector numbered thousands, mostly already in existence as part of the established industrial structure so that their finance was not considered to be a special case. Most relied on retained earnings and bank finance, with very few having access to the London Stock Market, development capital or overseas funding.

The Wilson Committee (1978) also briefly considered the question of finance for the offshore supplies industry. It found no evidence of any general hindrance from a shortage of finance, indeed suggesting that some potential investors had been unable to find suitable outlets for their funds, which was attributed to the dominance in the industry of more broadly established businesses. Nevertheless, it noted that the minimum scale for a viable enterprise in many offshore supply sectors was large and that some large high-risk investments had not found funding. According to a study for Shell, this more pessimistic assessment should have been extended also to smaller innovatory firms, often denied access to capital by the lack of the collateral demanded by financiers (The Economist Intelligence Unit Limited 1984, p. 52).

Addressing the financial environment into which the North Sea support industry emerged is a formidable task and cannot be effectively separated from addressing that of British industry more generally. Estimates of the number of companies in the UK involved in North Sea exploitation varied between 2000 and 4000. A small proportion of these were newly established, whether as foreign implants, affiliates of established businesses or as start-ups. The vast majority, however, were existing firms to which the offshore market represented only a part, often small, of its overall activity, which makes it necessary to consider the general financial state of industrial and commercial companies.

The arrival of North Sea oil and gas coincided with a period of a general decline in profitability. After adjusting for the effects of inflation, the average return on the tangible assets of industrial and commercial companies fell from over 10% in the 1960s, to about 8% in the early 1970s and to under 3.5% in 1975 and 1976 (Wilson Committee 1977a). Manufacturing was particularly badly affected. Net of capital consumption and stock appreciation, its gross profits fell from 20.8% of net output in 1966 to 3.8% in 1976. By comparison, the decline in services was only from 33.2% to 31.6% (Kirby 1981, Table 22, p. 152).

The period also saw a continuation of the then established economic cycle of alternating periods of expansion and contraction against a background trend of rising inflation. There was growing direct interference in business by government, only halted by the return in 1979 of a Conservative government committed to the 'free market'. In addition to prices and incomes policies, the intervention ranged from outright nationalisations of entire industries to subsidisation of fixed investment through investment grants, Industry Act selective funding, regional policy and specific sector support/rationalisation schemes such as those for shipbuilding and shipping. Governments of both parties became involved in individual companies as an *ad hoc* response to a crisis, examples being RR, Upper Clyde Shipbuilders and Beagle Aircraft. Repercussions from the last case forced ministers and senior civil servants thereafter to exercise greater caution in statements or actions suggesting the possibility of public sector support for businesses in difficulty.

Long-term funds flowed into the stock market to be absorbed by government borrowing, company new issues or, overwhelmingly, secondary transactions. Quoted shares in particular were attractive during a period of generally rising inflation and a falling exchange rate because their income and capital was thought likely to be preserved in 'real' terms. At times, this so-called 'cult of the equity' had to compete with the attractions of property and commodities, particularly gold. The volatility of the economy was inimical to long-term thinking in both finance and industry. With portfolio distributions easier to switch than industrial operations, City investment horizons by the early 1970s were typically 3 years or less and thus out of step with industrial investment time-scales. Commentators at this time were frequently critical of City 'short-termism'.

As far as listed companies in general were concerned, the main source of funds was consistently their retained income, although this fell from 69% in 1966 to 49% in 1970 and stayed below 60% until 1975. New UK capital market issues declined even more sharply as a source of funds. In the late 1960s, they provided between 9% and 15% annually. In 1970, this fell to less than 4%, remaining below 6% in all years to 1976 apart from 1975 (over 8%). In 1974, the year of a great London Stock Market 'crash', the domestic capital market made no contribution to funding. The shortfall in resources created was met by an increase in bank lending. This provided close to 30% in each of the 3 years 1972–1974 (a period initiated by a 'boom' in response to lax monetary conditions and ending in economic crisis) as compared to a median of about 12% for the entire period (Central Statistical Office 1971, Table 77 and 1977a, Table 9.2).

The 'lax monetary conditions' resulted in part from the introduction of a new Credit Control and Competition Policy by the Bank of England in late 1971, allowing a rapid expansion of the so-called 'secondary banks' – small deposit-taking and credit-granting institutions. After 1971, the secondary banks diversified rapidly beyond conventional banking and hire purchase into areas previously confined to established merchant banks, stockbrokers, insurance companies and fund managers, some becoming effectively financial conglomerates and others industrial holding companies attracted by the prospects for 'asset stripping' long-established concerns.

The other factor involved was expansionist macro-economic policy pursued by the government from 1970 until the 1973 energy crisis and subsequent coal miners' strikes, which forced first retrenchment and then a change of administration. The ready availability of credit and a failure adequately to increase interest rates in line with rising inflationary pressures were central to this so-called 'dash for growth' or 'Barber boom' (after the Chancellor of the day). An unintended side effect was the recognition by entrepreneurs of the opportunities offered for capital gains to be made through highly leveraged property deals. The secondary banks often funded them, although the clearing banks and long-term investors such as pension funds and insurance companies also became involved, either directly or as financiers of secondary banks and property companies.

The resultant boom in property values became for a time self-reinforcing. Between 1970 and 1974, the proportion of total bank advances made to property companies rose from less than 4% to over 9%, representing an increase from £341 million to £2,820 million – respectively, about £3.88 billion and £22 billion in 2008 terms. This rate of increase in lending was more than three times than that to manufacturing, then the cornerstone of the economy (Central Statistical Office 1977b).

An increase in the Bank of England's Minimum Lending Rate to 13% in late 1973 and an emergency budget set against the background of a deepening international economic crisis punctured the boom. The secondary banks saw their deposits dry up and their collateral and share prices tumble in value. To prevent a possible collapse of the entire British financial system, the Bank of England led a large-scale rescue operation. Until the so-called 'credit crunch' of 2008, this was without doubt the severest financial crisis experienced by the UK since World War Two. It is described in detail in Reid (1982).

The relevance of the financial state of affairs described above to the question of British industry's reaction to North Sea opportunities was largely indirect but real nonetheless. Not only did it attract entrepreneurial talent at what proved to be a critical time for the North Sea but it also absorbed funding on a large scale. The author, then working in a City merchant bank, witnessed the apparent unattractiveness of long-term industrial investments, including those in the offshore sector, in comparison with highly leveraged property ventures, claiming to offer low-risk after-tax returns on equity in excess of 30%; the attractions of exploration company stakes stood up better because of the prospects for rapid capital appreciation following a successful well.

Among the secondary banks that failed was Edward Bates and Sons (Holdings) Limited, which was associated with the Edinburgh investment manager, Ivory and Sime – promoter and investment manger of North Sea Assets (NSA), the largest British vehicle established to invest in the offshore service and supply sector. Bates also held a 51% share in Houston-based Simmons and Company, then newly established but to become the dominant U.S. investment bank for the oilfield service industry and eventually also to establish itself in the UK.

The instability in the British financial markets resulting from the secondary banking crisis was compounded by problems in the shipping industry – particularly tankers – stemming from the October 1973 'price shock' and by the collapse of the Stock Market in late 1974. The combination of these two factors in turn brought about the virtual collapse of the UK's third largest oil company, Burmah, at New Year 1975.

If entrepreneurship and high-risk investment backed by generous funding flourished briefly in the property and secondary banking industries, this was best regarded in a broader context as an aberration. Institutional investment in unquoted companies, particularly small and medium-sized ones was minimal. The UK listing requirements were such that only companies with established profit records could tap the public markets as a source of funds. Listing was seen as expensive, onerous and as according only low ratings to small companies. There was no 'junior' market such as today's Alternative Investment Market (AIM).

Although some investment trusts, pension funds and merchant banks made occasional small equity investments in private companies, the Wilson Committee (1977b) listed only ten specialist private sector venture capital institutions. Only two were prepared to invest as much as £1 million (about £4.7 million in 2008 terms) in a single project. One, Charterhouse Development Capital, would '*not normally*' invest in start-ups, whilst the other – the Industrial and Commercial Finance Corporation (ICFC, now 3i) – was at that time limiting investments in start-ups to one in forty. A sister company, the Finance Corporation for Industry (FCI), existed to make longer-term loans to established companies but was by this time largely inactive due, it was said, to lack of demand. This was an inauspicious background for the financing both of new enterprises established to enter a new and demanding market and established businesses seeking to diversify into it.

ICFC had been formed in 1945 in response to government concerns about the lack of longer-term finance for small- and medium-sized businesses (the so-called 'equity gap'). Its backers had been the clearing banks and the Bank of England itself. It dominated its market. By early 1977, it had investments, net of provisions, of £211 million (slightly less than £1 billion in 2008 terms) in 2,200 companies. Its annual new commitments were running at £25 million – £30 million (or about £117 million – £140 million in 2008 terms), which in only 10% of cases included equity (limited to a 25% holding).

Although the traditional source of equity provided by proprietors and their families and friends was widely regarded as having dried up due to the high levels of post-war taxation and much publicising of the need to substitute term borrowing for overdrafts, ICFC considered its activities were demand rather than supply constrained. This view had been endorsed by an earlier government enquiry, which concluded there was no 'equity gap' (Bolton Committee 1971).

A situation where almost all firms, outside the small minority to which capital market issues were an option, relied on retained earnings (in many cases depressed by falling profitability and rising inflation) and bank overdrafts to finance any growth in activity and where start-up finance was scarce, was unlikely to produce a dynamic response to new opportunities. The scarcity of external start-up finance constrained the initial capital of many ventures to what the proprietor could raise against the security of the family home.

Government-funded investment did little to correct this state of affairs. The National Research and Development Corporation (NRDC) established after the Second World War existed mainly to protect and progress publicly funded innovations, although it did invest in some advanced technology start-ups. By the mid-1970s, it had committed only £27 million (or about £166 million in 2008 terms) to 350 industrial projects. Conservative party direct financial interventions in industry tended to be *ad hoc* reactions to high profile crises, such as RR or Upper Clyde Shipbuilders. Outright nationalisations apart, the Labour Party created state holding company structures, first the Industrial Reorganisation Corporation (IRC) and then the National Enterprise Board (NEB), in order to rationalise/regenerate what were seen as failing elements of the established industrial structure, although the NEB did make some investments that did not match this description. Like the attempts to introduce 'indicative planning agreements' with large industrial companies, neither the IRC nor the NEB survived long enough for their effectiveness to be properly assessed. The development agencies founded in the 1970s for Scotland, Wales and Northern Ireland were longer lasting and empowered to support industry but the amounts at their disposal were small.

The 1960s and 1970s were a period, partly for reasons of punitive taxation and partly as an issue of culture, when the private entrepreneur was generally out of fashion in Britain. Now familiar concepts such as management buy-outs and buy-ins were virtually unknown. Combined with the shortage of start-up funds this was a recipe for a static industrial structure in the unquoted sector.

The situation began to change rapidly in the 1980s. In 1984, the British Venture Capital Association (BVCA) was founded, its members investing £140 million (about £325 million in 2008 terms) in 350 UK companies. By 1995, the figures had risen to £2.140 billion (about £2.960 billion in 2008 terms) in 1030 companies (BVCA 2003).

Venture capital thus became a major feature of the British economy and is increasingly seen as part of the larger sector investing also in established companies known as the private equity business. The venture capital/private equity business of the 1960s and 1970s bore few relationships to what exists today. The number of providers is now vastly larger, many of which manage third-party money rather than simply their own funds. For a time before the 2008 'credit crunch', it seemed as though there was almost no upper size limit to the deals that could be done. With personal tax incentives available to investors in venture capital trusts (VCTs) and enterprise initiative schemes (EISs) plus the development of 'business angel' investor networks, access to early-stage funding for small businesses has also greatly improved. There is a much wider range of financing structures now available and a better choice of 'exits'.

Though this transformation was heavily influenced by the much lower levels of personal and corporate taxation than those of 40–50 years ago, by the general retreat of the state from economic activity and by corporate restructuring driven by the dismantling of now unfashionable conglomerate structures, it also reflects a cultural shift in the prevailing attitude towards entrepreneurs.

This cultural shift in favour of entrepreneurs and the greater availability of the risk capital to support them have added a dynamism and flexibility to the UK economy so notably lacking in the early days of North Sea oil and gas. Although not necessarily apparent at the time, the financial environment then was not conducive to a vibrant response to a major new market requiring investment in start-up losses as well as equipment and training.

The finance sector did not contribute any drivers to the opening-up of the North Sea. It did find good business in providing loans to oil production facilities (where the oil reserves and for long a general perception that prices would rise offered acceptable collateral) and in taking up equity in a long succession of short-lived British independent E & P companies, regularly offered access attractive licence prospects thanks to a supportive DEn but in the main soon sold on to foreign owners. However, for the service and supply sector finance (or rather the lack of it) was, apart from an untypical short period of euphoria in the early 1970s, generally a constraint. This lack of support was less true of the banks, which on occasions were even over-generous in the provision of short-term loan finance, than it was of equity investors. As the true nature of the risks and capital requirements facing firms newly committing themselves to many segments of the offshore market became apparent, equity investors became more notable by their absence than by their presence. Eventually, as the offshore market matured and aided by the 'venture capital revolution', mentioned above, the situation was to change. By then, the major opportunities had largely gone to foreign-owned contenders and many sectors of the market were in decline.

Chapter 4

Before OSO: Offshore Supplies 1963–1972

For there to be an industrial response, there first has to be a market. Whilst there was some earlier spending, particularly on geophysical surveys, serious UKCS expenditure began in 1963, which saw not only further surveys but also a BP well in territorial waters off Dorset (see p. 43). Expenditure grew sharply from 1964 as first exploration drilling and then offshore gas development got underway, before declining at the end of the 1960s due to a loss of interest in the discovery and development of further gas reserves at the prices then on offer (see p. 74). From about 1969, declining expenditure in the southern (gas) basin was increasingly supplemented, first by exploration and then by appraisal and development expenditure, in the northern (oil and gas) basin, where material development expenditure began in 1972.

For these early years, specific expenditure information is very sparse. After the award of First Round licences in 1964, government sources (see p. 71) indicated that licensees had committed some £1.22 billion in 2008 terms to exploration and had already placed orders of about £214 million in 2008 terms with British industry. The larger part of the latter figure represented the five mobile drilling rigs, three semi-submersibles and two jack-ups, delivered to North Sea Operators or their contractors, whereas the former was probably mainly composed of an estimate of the cost of E&P company drilling obligations, for long a feature of licence awards.

There are better sources for UKCS expenditure figures for later in the period. Chart 4.1 shows that in 1972 expenditure reached nearly £170 million (about £1.65 billion in 2008 prices, nearly twice the figure of only 2 years previously). Drilling costs – both development and exploration and appraisal – represented a very large part of this, though there would also have been platforms and pipeline costs in the southern North Sea. Operating expenditure, including maintenance, only applied to the southern gas fields and represented a very small proportion of the total.

Norman J. Smith, The Sea of Lost Opportunity.

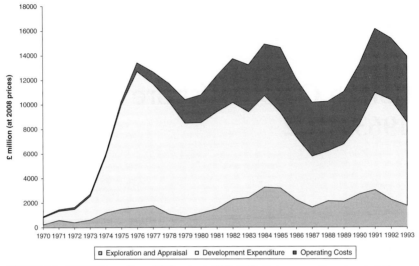

CHART 4.1 UKCS Expenditure (2008 prices) 1970–1993. Data source: Department of Energy and Climate Change.

4.1 OIL COMPANY ATTITUDES TO BRITISH SUPPLIERS

Initially, most oil companies, and particularly the three large British ones – BP, Burmah and Shell – seem to have had a positive attitude towards British industry. British companies undertook North Sea aeromagnetic surveys. British-built supply boats and helicopters owned by local companies were employed and drilling rigs ordered. Foreign Operators too seemed well disposed, as evidenced by Amoco's First Round invitation to tender to potential British rig builders (see p. 76) and the Phillips Group's order of a jack-up rig from a British yard (Arnold p. 38).

When British-controlled companies were not available, use was made of the British subsidiaries of foreign firms. Thus, SSL (see p. 42) took a prominent role in early North Sea seismic surveys. The majority of the mobile rigs ordered were equipped with drilling equipment sourced from OWECo/National Supply of Stockport – IMEG (1972, p. 45).

However, attitudes towards British-owned companies began to change after a series of unfavourable experiences. For instance, a founding director of Shell Expro recalled that contrary to the wishes of Dutch colleagues, the company placed an order for its Staflo semi-submersible rig with a British yard (Furness Shipbuilding) on the Tyne in 1966 whose price and top management it judged satisfactory. After experiencing a change in the managing director, poor labour productivity and an unsuccessful request to have the rig completed in the Netherlands, he had reason to revise his initially favourable opinion of the company; delivery was 2 years late. When Shell Expro was engaged later in developing its initial gas discoveries, the same director also recalled approaching 35 British companies to establish their interest in fabricating shallow water platforms for Shell Expro, of which only three replied. Two said that they had no interest, while the third requested more information and was taken on a 'fact-finding' visit to the Netherlands. It declined to tender.

BP's experience with the construction of its American-designed semi-submersible *Sea Quest*, related in considerable detail in the BP Archive, proved to be an equally depressing experience. When responding to a potential construction contractor, a BP executive confirmed that only British companies were tendering for construction of the rig. The entire project would be wholly sourced in the UK, with BP supplying its own drilling crew and supporting labour. This optimistic approach to British capability was consistent with BP assuring a Minister in May 1964 that shipyards on the northeast coast could '*hardly fail*' to be competitive for the supply of the many drilling platforms needed to explore the North Sea. In April, there had already been a discussion about building a B&R-designed 300 ton crane barge in a UK shipyard using UK materials, perhaps another expression of confidence.

BP's initial intention was to award the contract to Vickers-Armstrong, but when that company withdrew its tender due to an offer of government work, awarded it to Harland and Wolff of Belfast on 21st October 1964 at a price of £2.416 million (nearly £37 million in 2008 terms) for delivery on 21st November 1965. There was a late delivery penalty of £250 per day (over £3800 in 2008 terms); the net cost to BP of each day's delay was about three times greater. BP accepted *Sea Quest* on 6th July 1966 (Figure 4.1). It had paid an additional £100,000 for design modifications (about £1.5 million in 2008 terms) necessitated by the loss of a rig of a similar design late in the construction period, which also led to some delay. Otherwise, delay was attributed to a combination of strikes, imbedded low productivity, lack of relevant experience, the absence of modern planning methods and late deliveries by British manufacturers.

The loss of *Sea Gem* (see p. 56) and the delayed completion of *Sea Quest* impacted on BP's plans for the development of the West Sole gas discovery (Figure 4.2). In response to the unavailability of a mobile rig, BP instructed B&R as its designer and main contractor to procure a fixed platform to support the development drilling. It identified only one British fabricator willing to bid, the Humber Graving Dock, which offered a 4-month delivery. The structure was built in a Dutch yard in 22 days. The production platform itself was also Dutch-sourced and much of the topside equipment was imported. No British firm was able to install either the platforms or the pipeline, though British pipe was used (Pratt et al p. 217). At 17 inches in diameter, it was not considered particularly large; most subsequent pipes were of larger diameter and for long, mainly imported.

Foreign suppliers were, however, not beyond criticism. In November 1967, the BP Archive records that the company expressed its dissatisfaction about

FIGURE 4.1   The semi-submersible drilling rig *Sea Quest*. Reproduced from the BP Archive.

FIGURE 4.2   West Sole gas field. Reproduced from the BP Archive.

the drilling performance on the fixed platform of American contractor South Eastern Drilling Company (SEDCO), though the relationship was obviously repaired. In early 1971, BP was considering a mobile-rig owning joint venture with SEDCO; its Shipping Department advised that it was unlikely that acceptable arrangements to build rigs could be made in the UK. British yards were again busy building conventional ships.

The BP Archive also divulges that by April of the same year, when BP had focussed upon a fixed platform development approach for Forties, two consultants recommended that it should not procure in the UK. Of the 16 UK firms approached, none had expressed interest. As 1971 progressed, British suppliers were expressing interest, but by now BP was adopting a pragmatic position. Responding in January 1972 to Sir Maurice Laing of Laing Construction, Alistair Down, then BP's deputy-chairman, declared that while BP would like to see British firms established in the market and would seek to facilitate this, its actions were constrained by its own need to remain competitive.

When reviewing UK content for the DTI in March 1972, BP was also emphasising the paramount importance of British suppliers meeting delivery promises as well as having '*reasonably competitive pricing*'. In November 1972, the BP chairman (Sir Eric Drake) made it clear to the deputy-chairman (Sir Monty Finniston) of the BSC that his company's preference for British supply sources could not extend to accepting '*unnecessary greater costs*'.

Nevertheless, after spending much of 1971 in considering potential suppliers, BP announced at the beginning of 1972 that it had awarded one platform order to each of two UK yards, one British-American (Highlands Fabricators or Hi-Fab) and the other British (Laing Offshore), with some French support. According to the BP archive record, they emerged as equal first after an extensive evaluation of mostly foreign contenders. Both companies struggled with labour disputes, productivity issues and allegations of poor management. They delivered late, though partly for reasons outside their control, but both emerged with sufficient credibility to win other orders.

4.2 GOVERNMENT ATTITUDES TO BRITISH SUPPLIERS

While from the outset government policy placed great emphasis on the rapid exploitation of the UKCS for the benefit of the UK economy, a specific policy for the support of the British supply industry came very late.

There was certainly some early encouragement of UK sourcing through the use of exhortation at the administrative level (see p. 76). One isolated suggestion that more than simple exhortation may have been involved, was a September 1964 reference in the BP Archive that the Ministry of Power was not concerned about drilling barge late delivery, as BP's 'prime responsibility' was to place equipment orders in the UK. However, nothing else in a similar vein was found during the author's research and given the weight of evidence to the contrary, it is probably best discounted as evidence of any early structured industrial support policy.

In 1972, the DTI advised the PAC that it had not gone beyond encouraging the oil companies to seek UK suppliers (see p. 78). There was also no obvious policy response to early concerns, including those of the FCO (see p. 76), about the deficiencies of British submarine pipe-making capability. Even the realisation after sterling devaluation in 1967 that there were adverse economic implications (on the gas price and the balance of payments) in depending on foreign suppliers, produced no stronger a response than an expectation that devaluation might lead to increased sterling expenditure – TNA: PRO POWE 63/360

Everything changed once the implications of the major northern basin oil discoveries of 1970 and 1971 became apparent and in particular after BP's decision in late 1971 to develop the Forties field – a decision taken well before the upward shift in oil prices induced by the 1973 crisis. It became clear that the British offshore market was about to expand dramatically and that without some policy initiative, foreign suppliers would be the main beneficiaries. The result was the commissioning in May 1972 of a consultant's report, soon leading to the formation of OSO and policies directed at increasing the domestic content of UKCS expenditure. Specific trigger points for these developments appear to include

i. Prime Minister Heath's urging, following a BP presentation on Forties to his Cabinet colleagues at the end of 1971, that UK content in the Forties development should be maximised (see p. 75);
ii. a growing public recognition of the scale of the opportunities likely to be available to British suppliers; this was particularly the case in Scotland where pressure groups on the subject ranged from the political – the Scottish National Party (SNP) – through the labour movement (the Scottish Trade Union Congress, or STUC) to the business establishment as represented by the Scottish Council: Development and Industry;
iii. the investigations of North Sea policies undertaken by the PAC in its 1971–1972 session (see p. 78).

4.3 THE *IMEG REPORT*

Though the *IMEG Report* was submitted in September 1972, it was not published until January 1973. In part, the delay was due to the need to prepare a publication stripped of its commercially confidential content, but it also appears that it required amendment as the DTI found difficulty in accepting all its conclusions and recommendations.

The report is a long and fairly complex document, consisting of a main text in 11 chapters plus a foreword totalling 96 pages and five appendices containing a further 21 pages. Much of the report deals with market research and forecasts and the then current British supply capability. Both were already in a state of significant change, so that in these respects the report's findings became rapidly overtaken by events.

Key conclusions were that the offshore supply market was large, expanding rapidly both at home and abroad and that on the basis of the report's recommendations, a then UK market share of the domestic market of 25–30% could be increased by the late 1970s to some 70%. In value terms, this was to equate to a growth from £75 million to £90 million to an estimated £200 million to £220 million in what are presumed to be 1972 prices, or from roughly £720 million to £870 million to between £1.9 billion and £2.1 billion in 2008 terms. Market growth was, for a variety of reasons, to outstrip IMEG's expectations by a large margin. Significantly, IMEG projected a high level of offshore activity in European waters for at least the next 20 years and globally for much longer.

IMEG's suggested potential UK content of 70% was to become first a target to be met and then a benchmark to be exceeded. It is difficult to believe its choice was not influenced by what was happening with the Forties programme, well underway at the time of the report's preparation. Material from early October 1972 in the BP Archive shows that the company anticipated UK content, development drilling excluded, of 67% in Forties Phase 1 and 70% in Phase 2. Taking both Phases together, the USA contribution was anticipated at about 23%, mainly for offshore construction and pipelaying, with the balance, mostly for pipeline and other special steel products, approximately equally divided between Japanese and European suppliers. It is probable that most of the foreign suppliers benefited from export credit finance and indeed the BP Archive shows that an application for a $90 million credit package (nearly $350 million in 2008 terms) was made in December 1973.

IMEG's recommendations were unashamedly interventionist, though not protectionist, in nature and aimed to avoid the creation of a high-cost protected industrial sector while addressing the fact that foreign enterprise was becoming so entrenched that significant and rapid government support, much of it under the existing Industry Act, was necessary. Hallwood characterised them as '*an infant industry argument*', which was not accepted by any British government (Hallwood 1990, p. 58).

The most important recommendations were

i. the establishment of a Petroleum Supply Industries Board (PSIB) outside the civil service departmental structure but responsible to a Minister and accountable for its results. In addition to overall responsibility for helping to develop an internationally competitive British offshore contracting and supply industry, the PSIB would have responsibility for advising on and/ or implementing specific initiatives arising from other recommendations. It would work closely with the DTI's oil-related departments and with the Industrial Development Executive (IDE) since it was envisaged that government financial assistance was likely to be under the provisions of the Industry Act. IMEG clearly regarded the setting-up of an agency with the same functions within the existing civil service departmental structure as a 'second best' solution (IMEG p. 11).

ii. the provision by offshore operators and contractors of a confidential quarterly return on their purchases, staff employment and use of contractors and subcontractors, together with a less detailed forecast of future requirements and (on request) an explanation of purchasing and tendering practices;

iii. the incorporation into the discretionary licensing system of an assessment as to whether or not an applicant had given British firms 'full and fair opportunity' (FFO) to compete for business;

iv. the setting-up of a wholly British-owned offshore drilling contractor, with government-backed insurance against inadequate returns and, if necessary, direct government investment;

v. the provision of government-backed insurance against inadequate returns for British risk capital invested in contractors' equipment and facilities;

vi. the lease to British contractors of equipment purchased or hired by the British government;

vii. the encouragement of joint ventures between foreign companies with offshore 'know-how' and British partners as a means of technology transfer;

viii. the provision of subsidised credit to British suppliers to counter cheap export credits from overseas suppliers;

ix. the provision of an information and advice service to British suppliers;

x. the establishment of university courses in petroleum engineering and related disciplines;

xi. the provision of government support for relevant R&D projects.

In addition, there were various less far-reaching recommendations dealing with the shipbuilding and steel industries, planning, infrastructure and regional issues, with particular reference to Scotland and Northeast England. The need to address obvious barriers to new British suppliers such as loyalty to established suppliers, standardisation policies and the influence of standards and codes was noted.

As already mentioned, the DTI was unimpressed with the report. Its views were set out on 22nd November 1972 in a memorandum from its deputy-secretary, R H W Bullock, to the Minister for Industrial Development, C. Chattaway – TNA: PRO CAB 184/109. In summary, they were that while the factual information was useful, the recommendations contained '*little of great novelty*' or '*of exceptional significance*', with many already being addressed by existing programmes. The proposal to establish an independent PSIB was rejected out of hand on the basis that it would overlap and come into conflict with the IDE and the Industrial Development Advisory Board (IDAB).

4.4 AN ASSESSMENT OF THE PERIOD

The *IMEG Report* provided little by way of evidence to support their contention that the UK's share of the then current UKCS market was 25–30%. However, it is broadly consistent with an estimate produced nearly 5 years earlier in a brief by the Petroleum Division of the Ministry of Power – TNA: PRO PET 50/469/01.

This estimated that 70% of North Sea supplies were foreign. However, this was on evidence limited to (a) a statement from Esso that this was the proportion applicable to them and (b) the Gas Council's experience that whereas to date roughly 50% of their expenditure had been in foreign currency, this would rise substantially in 1968 as a pipeline would be laid and buried by B&R of the USA using Japanese pipe.

The brief would appear to summarise official thinking on the supply sector at the time. It noted that British industry was as yet unable to compete in many areas such as drilling contracting and the provision of drilling and well services where foreign companies predominated. The installation of platforms and the laying of pipes required specialist vessels where British ownership was absent and British firms had proved unable to offer competitive delivery dates for the fabrication of fixed platforms. Some items of production equipment could only be purchased abroad. Only in the provision of support vessels did UK and foreign firms seem to be competing on roughly equal terms.

There is nothing in this analysis that could be described as incorrect, but – apart from noting the possibility that for '*some items*' the recent sterling devaluation might improve British competitiveness and that KCA Drilling had been trying to enter offshore drilling – nothing was said about what could be done about the situation. Nor was there reference to the long tradition of supplying the overseas operations of the British oil companies, both by British-owned companies and foreign subsidiaries operating in the UK (see pp. 41–47). The fact that much of this capacity, involving companies such as George Wimpey, Power Gas and Costain Process Engineering, was committed long-term to the development of the massive BP-operated Abu Dhabi Marine Areas (ADMA) offshore development (see pp. 37–40) in the Persian Gulf was not mentioned. Nor was it mentioned that the British oil companies themselves had committed substantial internal resources in such fields as drilling and engineering design that might otherwise have been undertaken by contractors, thereby reducing expenditure with third parties whether in sterling or foreign currencies.

Reconciling the Gas Council's estimate of 50% UK content in the pre-development phases with Esso's 30% overall, suggests that the UK content in early developments was very low indeed. No specific estimate of the UK content in southern basin development has been located – save a statement by Arnold (p. 39) to the effect that: '*British industry only picked up about 5 per cent of the work*'. Arnold provided no source.

It is probable that from 1963 to 1965, the UK content of expenditure (then almost entirely for exploration) was high, as British-owned or -based survey contractors were heavily involved and a high proportion of the relatively few wells bored were drilled by British oil company rigs. However, from about 1966 when the number of exploration and appraisal wells rose sharply (see Chart 4.2) UK content would have declined, as foreign Operators brought in U.S. contractors, a trend strengthened by the loss of the *Sea Gem* jack-up in

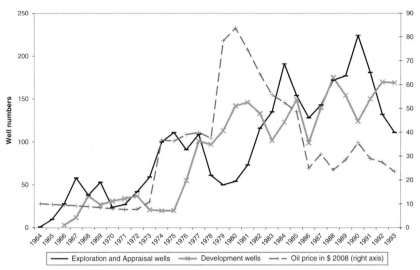

CHART 4.2   UKCS Drilling Activity 1964–1993. Data Sources: Department of Energy and Climate Change, BP (2010), *Statistical Review of World Energy.*

1965 and of the *Ocean Prince* semi-submersible in 1966. Thereafter, the UK content would have further declined as platform design and fabrication, development drilling, most submarine pipe supply and all offshore installation were carried out by foreign firms. The main UK content was in the field of logistics (bases, helicopters and supply boats), although there was also some contribution in the form of oilfield tubular goods, minor fabrications and some topside equipment.

To judge from statements in the *IMEG Report*, some spontaneous increase in UK content in the southern basin may already have been taking place by the time it was written. Thus, it was noted that Farmer Offshore Limited was fabricating a small production platform for Shell/Esso's Indefatigable gas field (IMEG p. 35), while an unidentified British Operator regarded a 75% UK content as typical for a gas field topside, though it was noted that for foreign Operators, a lower figure might apply (IMEG pp. 39–40). Moreover, in the wake of BP's decision to develop the Forties field and a growing recognition that other moves would follow, many decisions that would increase British capacity for the northern basin, particularly in the area of offshore production structures, had already been taken before IMEG completed its study.

The low initial UK content in the mobile and fixed platform drilling areas is not surprising since at the beginning of the 1960s, the offshore drilling contracting industry was still relatively new – having originated only in the mid-1950s – and largely confined to the USA. It was then fast expanding, relied on a limited pool of skilled manpower and was highly capital intensive. Though this did not deter the early rig-owning ventures of the British oil companies, they did not develop the business, moving first to co-ownership positions with selected U.S. contractors and eventually exiting the business. Given the rapid emergence and subsequent decline of southern basin drilling activity, first in exploration and appraisal drilling and then in development drilling (with the two cycles combined totalling about 8 years) and with the largest overseas market (the USA) blocked by protectionist legislation, the chance of new British contractor drilling capacity emerging to meet the initial southern basin demand was small.

By comparison with offshore drilling contracting, the initial North Sea drilling and well services market was small. Indeed this was true of most areas of the world outside North America, where the high proportion of the world's wells drilled provided the best opportunity for specialised service companies to develop. The market, especially in logging, was characterised by considerable barriers to entry based on 'know-how' and established credibility with clients. With a history of a high level of R&D, by the early 1960s Schlumberger was a near global logging monopolist. The limited existing British non-hydrocarbon logging capacity failed to penetrate the market at this time. The UK content in drilling and well services was effectively nil in initial southern basin work.

The British failure to be more successful in the area of production facilities and infrastructure is more surprising. The British capacity problem with large diameter steel pipes was exposed yet again by IMEG (p. 59). It was noted in the DTI's critique of IMEG that BSC did not fully share IMEG's views on '... *what should be considered as desirable and profitable business*' – TNA: PRO CAB 184/109. This remark had potentially large implications beyond pipes since the importance of steel was all pervasive in the offshore oil and gas supply business (see p. 42).

Less attention was paid to the UK's near total early southern basin failures in the areas of design, fabrication and installation and how this arose. During the initial 5-year development phase of the southern basin (1965–1970), all 24 production platforms were designed in the USA and built in the Netherlands. The topsides too were all American-designed and mainly Dutch-built, although at least three British fabricators did participate. Balfour Beatty supplied four of the platforms with their vent stacks. Sea and Land Pipeline (SLP) Engineering built helicopter decks for three platforms and modules for a fourth. William Press supplied at least one module. This early start paid off at least for SLP Engineering (a new company) and William Press, which were to be among the small group of British companies participating in all phases of North Sea development. Another British company, Wilson Walton International Limited, was formed in 1966 on the back of southern basin demand for fabrication and maintenance. It went on to undertake the conversion of the floating production system for the Argyll field and to float on the Stock Exchange in 1976, but failed to survive the remainder of the decade.

There may have been an element of entrepreneurial failure on the part of British business. However, the advantages allowing B&R and its European

subcontractors to establish a virtual monopoly in southern basin gas development in the UK North Sea were powerful and already apparent when BP
awarded it the initial design and development contracts for the West Sole field
in January 1966, the first on the UKCS. They can be summarised as follows

   i. the company was already the world leader in offshore design, construction
     and pipe laying and was pioneering the use of computers in structural engi-
     neering design;
  ii. the company had been established in London since 1959, and was well
     known to BP from work on ADMA; it had designed and installed its first
     11 North Sea fixed platforms (plus a re-installation) in 1964 and 1965,
     one for UK sector use, giving it considerable experience of fabricators
     in the Netherlands and Germany, where all platforms were built;
 iii. it could assemble an unrivalled fleet of installation vessels, including a
     crane-ship with a 200-ton lift and purpose-built pipe lay/derrick barge
     with a 250-ton revolving crane specified for the North Sea, constructed
     in the Netherlands during 1966 and available to lay the West Sole line in
     time for it to deliver gas to the Gas Council the same year (Pratt et al
     pp. 215–216).

Though it probably also saw use for it elsewhere if demand declined in the North
Sea, in the late 1960s B&R invested $20 million (approaching $100 million in
2008 terms) in mainly Dutch-built equipment intended for use in North Sea
waters (Pratt et al p. 212). No other contractor in Europe or elsewhere could
match the offer.

    Since BP's West Sole programme allowed only 2 years from gas discovery
to delivery, B&R had to be the obvious choice as contractor. It was to be so for
most Operators until the early 1970s, by which point the company's lead in
North Sea experience had become difficult to challenge. Indeed by 1970, the
company had engineered and/or constructed some 15 offshore platforms and
had installed around 95% of the 35 platforms by then in place. In pipelines,
its dominance of the North Sea market was even greater; it had laid all five
major lines, about 425 miles in total (Pratt et al p. 220).

    Given these circumstances and market uncertainties, it is easy to see the
difficulties faced by potential British competitors, which, on purely commercial
terms, would probably have to be prepared to face start-up losses as well as their
initial capital commitment. Moreover, perhaps the most obvious potential
British competitor, Wimpey, was already a partner of B&R (see p. 46).

    What seems surprising was the failure of B&R or its oil company clients to
subcontract more elements of the platform and topside structures to British
companies, limiting the scope for British equipment suppliers. This limited
response could reflect a wide range of factors ranging from unwillingness to
restart the 'learning curve' with a new supplier, reservations over labour, productivity and ability to meet delivery dates, price and management competence.
A lack of entrepreneurial zeal among potential suppliers could offer an

alternative explanation. Another factor may have been the deterrent effect of the difficulties that the shipyards building mobile rigs were having. After completing their initial orders none of the yards except for John Brown's at Clydebank persevered with the market (see p. 46).

However, it is easy to overlook what did transpire in these early years. Clydebank's former engine works, which unlike the yard, remained under John Brown's ownership, also became exposed to the oil and gas sector. Known from 1966 as John Brown Engineering, it became a successful manufacturing licensee of General Electric (GE) gas turbines (McKinstry 1997). Although it won some North Sea orders, it was primarily an exporter, with electricity generation as its main market. Already having its Constructors John Brown (CJB) design and project management subsidiary, and with other oil-related engineering design ventures to come, the Group was one of the few traditional engineering companies to re-orient itself successfully to meet oil and gas industry demand.

The existing Decca navigator base stations chain was adopted by the offshore industry for positioning. At least one dedicated telecommunications business, East Anglian Electronics (EAE) was created, as well as at least two new fabricators, Wilson Walton and SLP. There was, however, no general evidence of the commitment of large private funds to the offshore supplies industry. IMEG noted support from the state-owned NRDC to a diving company, Strongwork (IMEG p. 37) and to Vickers Oceanics (VOL) – an underwater vehicle business, founded in 1969, and in which ship owner James Fisher had also invested. This company was considered to have a slender international lead in this area, as well as novel ideas for pipelay, justifying government support (IMEG pp. 80, 81, 91, 96).

Some ambitious plans were abandoned, probably when cost and other difficulties became apparent. Smith's Dock on Teesside, as reported by the shipping correspondent of the *Financial Times* on 21st September 1964, was aiming to specialise in meeting North Sea demand, not only for exploration rigs but also for such other requirements as work boats, well protectors, servicing and repairs. Nor did NORSMEC (see p. 76) persist, although its marine services arm, Offshore Marine, became a Cunard subsidiary and thus eventually part of the Trafalgar House (TH) Group.

Overall, the initial British industrial response to the southern North Sea had been both weak and slow. It took the northern basin oil discoveries of 1970 onwards and the *IMEG Report* of 1972 to win government support and to spotlight the opportunities. As some pointed out at the time, commissioning the *IMEG Report* was late, coinciding with a time when foreign suppliers were not only already entrenched but when demand growth was accelerating and technical complexities increasing. British industry found itself being urged to enter a demanding and still highly uncertain market at a dangerous moment.

Moreover, foreign firms, particularly American, had already been strengthened by early North Sea activity expanding their markets and stretching their technology. Indeed the North Sea was a fundamental factor in the growth of a number of U.S. firms (Smith 1978).

As the BP Archive shows, B&R was the pre-eminent case in point. Its already strong position was further strengthened by the award in August 1972 of a contract for the design, procurement, fabrication, installation and preparation for commissioning of BP's giant Forties field (Figure 4.3). BP provided only a brief description of the works, plus details of national site conditions, but the risk for B&R was limited, as the contract provided for the reimbursement of direct costs plus a fixed percentage fee. It is believed there were two other bidders, both also American.

Design work had already been underway for many months before the contract was signed, as it was BP's original intention for Phase 1 platforms to be installed during 1973. The BP Archives record that this 'fast-track' approach was criticised both by BP employees and external consultants, with one of the former expressing his doubts about the feasibility of the timetable in writing in January 1972. Even fiercer criticism came from U.S. engineering consultant, Earl & Wright in May 1972. It suggested B&R might have foreshortened the design time in order to make a futile attempt at a 1973 installation, leading to a conservative approach, an uneconomic design and a pessimistic outlook on weight and cost.

FIGURE 4.3   Forties oil field Alpha production platform. Reproduced from the BP Archive.

Against this background, plus numerous design modifications, escalating inflation and industrial unrest, it is not surprising that the Phase 1 project suffered greatly from cost-escalation and delay. BP Archive files reveal that relations between BP and B&R were at times strained. Thus, already in May 1972 BP was complaining about excessive referrals to it by B&R. Between September and December 1973, BP made negative comments about B&R's pipelay performance relative to that of Saipem, which shared the contract, about the one-sided nature of the contract and about B&R's 'less flexible and co-operative attitude' ended with a BP complaint coupled with a request for a price reduction.

After the field came into production in 1975, these problems faded into the background, with both BP's and B&R's reputations enhanced from what was seen as the successful execution of the largest and most advanced UKCS development to date.

Chapter 5

OSO's Formative Years 1973–1980

With its high public profile and association with 'headline grabbing' issues, OSO is often seen as the key determinant of British industrial performance in the North Sea. In fact, its importance is easy to exaggerate. Its activities were only some of the forces at work and many key decisions had been made before it came into existence.

It was a Conservative administration that established OSO and most of its life was under Conservative administrations, mainly those of Mrs Thatcher. Conservative hegemony was interrupted by the Wilson/Callaghan Labour government of 1974–1979, a period which combined the most frenzied period of UKCS development with OSO's own formative years. OSO's existence and functions never became a party-political issue. The Ministers of State (MoS) for Energy to whom it was politically responsible pursued a broadly bi-partisan policy of 'constrained support' for the British supply industry, although with varying levels of vigour and commitment. There were occasional variations in how OSO conducted itself, even though the broad themes of activity showed a high level of continuity. The variations reflected economic, political and technological changes as well as the personal inclinations of a particular MoS or Director General (DG).

Constraints were numerous and ever-present, varying in relative importance from time to time. They reflected major policy issues, such as security of oil supply, the balance of payments, employment, industrial policy and the nationalised industries, government revenue, fisheries, the containment of Scottish separatism as well as relations with the USA, Norway, the EEC, and its member states. This list demonstrates that interest in what OSO did extended well beyond the confines of its own host department. The Treasury, Bank of England, Scottish Office and Foreign and Commonwealth Office (FCO), in particular, all saw a need to be involved, especially in these early years. Many other parts of government from the Prime Minister downwards became drawn in on specific issues.

The 1973–1980 period exhibited characteristics that distinguished it markedly from the years that followed. It opened with the tail end of the so-called

Norman J. Smith, The Sea of Lost Opportunity.

'Barber boom', which was soon to give way to deepening economic crisis, rising unemployment and political upheaval. It saw 'real' oil prices reach a peak in 1980 not seen again until 2008, bringing in their wake sharply increasing levels of oil taxation. It was most the intensive phase in North Sea development generating the fastest rise in UK oil production ever experienced. The UK's first offshore oil was produced in 1975, with national self-sufficiency achieved only 5 years later.

It was against this volatile background that establishment of the Offshore Supplies Office (soon almost universally to be referred to as OSO) was announced in January 1973. Crucially, the new organisation was not to have the independence recommended by the IMEG, being located within the Industrial Development Executive (IDE) of the DTI in London. Similarly, IMEG's more radical financial proposals aimed at relieving private sector investors of some of their financial risks (see p. 100) were not pursued. However, modest funding under the Industry Act was envisaged and an Interest Relief Grant (IRG) Scheme, to enable British suppliers to match foreign export credit, was set up.

Inevitably, OSO's early months were occupied with internal issues. Its first Director (soon to be upgraded to Director General or DG) was an engineer, Peter Gibson, previously the Managing Director of Lummus (a UK subsidiary of a U.S. process engineering contractor). After 4 years, Gibson was succeeded as DG by Alan Blackshaw (a career civil servant). In autumn 1978, the writer (with a private sector background in industrial marketing and corporate finance) took on the position, having immediately previously served as OSO's Industrial Director. During this early period, other senior positions were also filled from the private sector on secondment or short-term contract bases. This strong private sector management component was a feature that did not persist. From 1981, OSO was entirely civil service staffed. Nevertheless, by then OSO had developed a structure and methods of operation that were to remain broadly unaltered for most of the remainder of its existence. It had also achieved a degree of maturity and broad acceptance within both the offshore industry and the government machine.

5.1 THE COURSE OF DEMAND

Offshore expenditure grew from around £300 million in 1973 (over £2.7 billion in 2008 prices) to its first cyclical peak of over £2.5 billion in 1976 (about £13.5 billion in 2008 prices), mainly fuelled by an increase in development expenditure, readily apparent from Chart 4.1 (see p. 94). As Chart 4.2 (see p. 102) shows, exploration and appraisal (E&A) drilling activity also expanded substantially.

The pace of expansion reached an unsustainable level; changes in government policy, particularly in respect of taxation and the establishment of the BNOC, as well as uncertainty over the future direction of oil prices provided additional reasons to delay investment decisions. The period ended with a decline in activity clearly entrenched. This was unfortunate for OSO and British suppliers since it meant expectations raised by the initial 'boom' would often not be met. As early as September 1975, UKOOA was complaining that UK suppliers were still basing their plans '*too readily on an extrapolation of the buoyant state of the market a year or more ago*' and would be '*too late to benefit from the initial surge of activity . . .*' It accused OSO of encouraging this.

Reference to Chart 4.1 (see p. 94) shows that expenditure had fallen to about £10.4 billion in 2008-terms by 1979, recovering slightly in the following year. By this time, operating expenditure was expanding rapidly as the UK approached self-sufficiency in oil, reaching over £2.2 billion in 2008-prices by 1980. The overall decline in absolute terms was borne mainly by development expenditure, although E&A activity suffered a greater proportionate decline, with the number of wells drilled more than halving between 1975 and 1979. By contrast, development drilling, still mainly fixed-platform based, continued to expand rapidly, increasing nearly sixfold over the same period (see Chart 4.2, p. 102).

The UK share of orders placed increased substantially, reaching its first peak at 79% in 1979, a figure influenced, according to UKOOA, by ' . . . *some degree of 'bunching' of orders.*' In the development sector (over three-quarters of the total orders), the share was 83% as compared to 70% in operating expenditures and only 40% in E&A – Department of Energy 'Brown Book' (1980). These relative positions were fairly constant for some years from the late 1970s onwards, reflecting the fact that UK firms had been relatively successful where existing labour skills and only relatively modest new capital expenditure were needed to enter the offshore market (as in fabrication) but much less so where specialist 'core' skills in short supply (as in drilling and well services) or large scale new capital expenditure was needed (as in drilling and installation).

Both fluctuating demand and the lack of a serious UK presence in large and important areas of the market were to persist throughout OSO's existence, posing difficult problems for the organisation.

5.2 OSO OPERATIONS IN CONTEXT

Initially, activity was mainly concentrated on the collection of quarterly expenditure returns from Operators, their subsequent 'audit', the definition of a 'British' company and the introduction of venture management, probably the more radical of the novel introductions to the civil service machine. Unlike the audit of quarterly returns, venture managers – largely short-term appointees from the private sector – were not a direct response to the *IMEG Report*.

Their introduction reflected recognition that new entrants to the offshore business would need more from OSO than information on market opportunities, such as contacts with oil companies and lead contractors and, possibly, financial and other assistance. Gaps to be met included 'know-how', which might be filled by foreign licences or joint venture associations, and private capital, particularly for new enterprises where Industry Act funding was likely to be insufficient or impossible to obtain in a realistic time frame. As brokers and advisers, the role of the venture managers was to help British companies fill such gaps, rather as a merchant banker of the time might, though without the direct access to private sector funding. In the event, the limited success of the venture managers in accessing private sector equity was to hamper their effectiveness, which would have been surprising to OSO's new Director, who, in his first month, was reported as saying that there was "... *plenty of money in the City looking for investment in offshore industries*" (*Glasgow Herald*, 26th January 1973).

Gibson may have had in mind the £20 million (about £194 million in 2008-terms) raised by a new Edinburgh-based investment company, NSA a few months previously. This was to be the largest British public issue for portfolio investment in offshore support industries and at the time was considered sufficiently important to be brought to the attention of the Prime Minister by the Chancellor of the Exchequer – TNA: PRO PREM 13/925.

OSO remained part of the DTI for barely a year before being transferred to the newly established DEn. During this period it established working relationships within its host department in respect of the operation of the Industry Act and its regional representation. Thus, its Newcastle venture manager and the Scottish Petroleum Office (in Edinburgh) were within the local DTI organisations. Because Industry Act assistance continued to be administered by the Industrial Development Unit (IDU) and because the DTI remained responsible for foreign trade promotion, the transfer introduced additional management complexity.

At the same time, OSO was continuing to accrue new functions. In this early period, the emphasis in OSO's work shifted to a considerable extent from supporting British industry to helping to expedite oil production. Some longer-term issues continued to be pursued, with an research and development (R&D) and market evaluation branch being added in response to the establishment of the Offshore Energy Technology Board (OETB) in 1975 and with steps to encourage nascent supply industry firms to export.

More functions demanded expansion in OSO's original staffing, something not easy to achieve in its technical grades, given the high demand for such skills at the time. In 1974, the new Labour Government authorised a substantial increase in numbers, combining this with a decision to move the organisation's headquarters to Glasgow and to close the Edinburgh office. A substantial presence was to be retained in London and close links kept up with DTI regional offices, particularly Newcastle. Later, representation was established in Aberdeen.

Headquartering OSO in Glasgow was mainly rationalised on the grounds of the actual and potential importance of oil supply business to the industrial belt of West Central Scotland and the need to be closer to oil industry operations. Such practical reasons were probably less important than for the Labour Party's need to show its Scottish heartland that it would enjoy benefits from North Sea oil, with the intention of easing the electoral threat from the SNP.

Whilst a move to Scotland was probably both unavoidable and justifiable in the circumstances of the time, it imposed additional stresses on an OSO management, already stretched on a number of fronts. The organisation expanded in numbers over the period mid-1973 to mid-1975 from 40 to over 150 (Jenkin, p. 90). The detachment of its senior management from their peers in other DEn divisions, from their departmental superiors and from Ministers – through whom such sanctions as OSO had were largely exercised – could only be partially rectified by continual Glasgow – London – Glasgow travel, a burden unequally borne by Glasgow based personnel. This same small group also had many visits to make to elsewhere in the UK, as well as periodic trips overseas.

The choice of Glasgow as the main Scottish location could be questioned. At the time little more than a forward exploration base, Aberdeen could not then have been considered as an alternative. However, politics apart, there was much to be said in favour of Edinburgh, strategically positioned along the east coast of Britain where the offshore industry was concentrating. The Scottish Petroleum Office had been located there and, importantly, Edinburgh was home to the Scottish Office, with which OSO had much dealing in its early days. In addition, the two most important Scottish banks, the main financiers of Scottish offshore supply firms, were headquartered there. Its transport links were at least as good as those of Glasgow, its industrial base had a greater high technology content than Glasgow's and one of its universities (Heriot Watt) was soon to challenge Imperial College for the role of the UK's premier higher educational establishment for the offshore and petroleum industries. Eventually, OSO settled down in Glasgow but – in the author's opinion – the city was never its ideal home.

OSO's policy continuity was little disturbed when in 1979 the Labour Ministers to which OSO was responsible were replaced by Conservative ones, although changes in other aspects of the new government's energy policies (e.g. the decisions to abolish BNOC and vary oil taxation) had 'knock-on' implications. The continuity stemmed from the fact that the 'ground rules' that defined OSO's operational scope, particularly FFO, definitions of a British company and British content had already been settled within a framework set by the previous Conservative administration. The government, the oil companies and foreign owned, British-based service and supply companies had no desire to re-visit these issues, though this consensus did not extend to all parts of the British-owned industry.

Where important changes did occur, they reflected changes in the market place or external pressures. The most obvious of the latter was the discontinuation in 1979 of the IRG scheme in response to pressure from the European Commission (EC), pressure reinforced by allegations (exaggerated in the author's view) that OSO had administered the scheme badly, resulting in over-payments. It was little noted that the other EEC members agreed not to offer export credit terms in respect of the UKCS and there is nothing to suggest that the scheme cancellation had serious consequences for British industry, a view to which Cook and Surrey (p. 23) subscribed. The concurrent market contraction had a more marked effect. OSO continued to administer repayments under contracts already accepted until about 1987.

The IRG scheme apart, the Commission did not seriously impact on OSO operations in this period, despite periodic threats. According to one OSO official of the time, this was due to the political balance within the EEC. He recalled the Dutch as the main complainants, but without backing from the French with their much larger oilfield support business, they had little influence. The French did not want "*to rock the boat*" because they were at the same time cosseting their own, for the most part 'high-tech', offshore companies and quietly nurturing them with public funds.

OSO's senior management and some of its technical staff remained under great pressure, but it became clear that the move to Glasgow and the simultaneous authorising of a staff complement of 180 had led to an over-inflation of support staff. Given the rigid structure of Civil Service staff relations, reversing this was to prove a long process. The staff in place in 1979 totalled over 120 against a reduced complement of 135. They had been organised into up to seven branches. With a view to increased efficiency and to further staff reductions, one of the writer's early decisions as DG was to concentrate as many activities as possible into three branches, Engineering, R&D and Administration, assigning the venture managers to specific tasks within them. According to one of their number, by this time, the reality was that venture managers could "*neither venture nor manage*", with their effectiveness constrained by a lack of strong direction, an absence of funding and the difficulty in assessing results.

5.3 OSO AND THE MACHINERY OF GOVERNMENT

With probably twenty or so central government departments having an actual or potential interest in offshore matters, OSO's activities attracted the attention of arms of government from the Prime Minister downwards. Thus when the Inland Revenue proposed to take away the self-employed tax status of North Sea divers, with the consequence of sharply cutting their net incomes, OSO's development programme monitoring function required it to report that it anticipated an exodus of experienced saturation divers, with serious programme delays in consequence. When all other efforts to have the decision rescinded had failed, the matter was referred to Prime Minister Callaghan who ensured the proposal was dropped. Whether or not Prime Minister Thatcher intervened as directly in an OSO-related matter, she was undoubtedly well informed about its activities.

Although ultimately responsible, neither the Labour (Tony Benn) nor the Conservative (David Howell) SoSs in position for most of the 1973–1980 period intervened much in OSO activities. This they delegated to their respective MoSs, J. Dickson Mabon and Hamish Gray. Both of these became intimately involved, not only because they were specifically charged with OSO's supervision but also because both were MPs for Scottish constituencies containing considerable numbers of oil-related jobs.

This gave them a close interest in chairing the Offshore Industry Liaison Committee (OILCO), which provided a communication channel between OSO, UKOOA, major contractors and trade unions representing in particular Scottish fabrication yard workers. Separate meetings also took place between two or more of the same parties. OILCO was among the mechanisms that helped to maintain a generally good climate of labour relations in the industry, although it would be wrong to claim that disputes were entirely absent. Completion bonuses continued to be strongly opposed by the oil companies and most were unenthusiastic about the unionisation of the offshore work force, a Labour government aim. In 1977 UKOOA nonetheless agreed guidelines with the Inter-Union Offshore Oil Committee (IUOOC) on how union recognition might be achieved on offshore installations.

Within DEn's internal structure, OSO's DG reported to the Deputy Secretary overseeing the oil and gas divisions, then John Liverman. OSO worked with the other oil and gas divisions and relations were generally good though somewhat distant. Tensions could on occasion arise, for instance, if the Petroleum Engineering Division (PED) suspected OSO wanted to use regulations in the interest of British industry.

That said, sometimes the mere reference by a Minister that the DEn had powers over offshore operations, which, if exercised, might restrict, delay or even stop cash-generating production, could lead to an outcome in line with OSO's view. Where temporary flaring consents were needed to maintain oil production, the companies concerned had often made little recovery of their heavy cash outlays and were thus anxious to be as accommodating as possible towards departmental concerns. Needless to say, not all conversations took place in a formal setting where they could be officially minuted, although civil servants were usually assiduous in their efforts to 'protect' Ministers from making unguarded statements.

OSO participated in the formal departmental discussion of most oil and gas policy issues, but was clearly disadvantaged by the location of its headquarters, which prevented its top management participating in short notice meetings (often the important ones) or in the informal discussions within a 'peer group' that frequently determined the decisions reached formally. As a result, the writer sometimes found himself in the difficult position of raising OSO objections to a decision already taken.

The most intimate inter-action between OSO and the other oil and gas divisions took place during licensing rounds. During the 1973–1980 period, these were confined to the Fifth Round (1977), the Sixth Round (1979) and preparation for the Seventh Round (1981). The succession of (small) licence rounds has been attributed in part to government concern for the work-load of the offshore supplies industry, see Cook and Surrey (p. 25). If true, the writer believes this was misconceived. As long as the DEn was able strictly to enforce licence round drilling obligations (broadly until the 1986 oil price collapse), there was a demand benefit for the (foreign dominated) exploration sector, not for the (British dominated) development sector. Once the first generation giant fields were in production, succeeding development decisions were more likely to be driven by oil price and tax considerations than the results of new exploration, particularly as early exploration campaigns had already identified many marginal accumulations, with few new giants anticipated on post Fourth Round acreage.

Despite the high profile of 'UK content' in government oil industry relations and the government target of 70% plus, its direct influence on new license awards may have been less than the applicants thought. Personal recollection suggests that, certainly at times, OSO representatives on the intra-departmental group that interviewed applicants and made the initial award recommendations commanded only some 10% of the total 'votes'. In order to maximise the influence of OSO, OSO representatives needed to use OSO votes entirely in favour of a single favoured applicant for a particular licence or not at all.

With OSO's role demanding liaison with so many government departments and agencies, it is possible to consider only the more important. With the Scottish Office, after a hectic beginning, few issues demanded much OSO top management attention, but at lower levels co-operation over matters like platform sites continued. There was almost no contact during this period with the newly established SDA. From the viewpoint of relations with English companies, this may have been a good thing. The move to Glasgow had already led some to think that OSO had lost its UK-wide remit and become part of the Scottish administration.

Contact with the FCO tended to focus on two matters. One was the EC, where OSO found the FCO generally supportive. The other was Norway, where relations were less cordial. It was easy to conclude that the FCO was anxious to avoid offending the Norwegians for strategic (military) reasons, even at the expense of UK economic interests.

The other two departments of state most closely concerned with OSO activities were the DTI (and 'outliers') and the Treasury (and 'outliers'). Given the origins of OSO and its dependence on the DTI services in respect of exports, operation of the Industry Act and regional representation, a close relationship was unavoidable, but made the more so by DTI's responsibility for two nationalised industries involved in the offshore business, BSC and BS, where the relationship could reach ministerial level. One DTI 'outlier' whose actions sometimes disturbed OSO was the 'Invest in Britain Bureau', responsible for encouraging inward investment. OSO's concern centred on the Bureau's apparent keenness to encourage (usually with regional aid) foreign new entrants, regardless of the potentially harmful effects on established suppliers.

The Treasury's main interest in OSO was in its use in monitoring the development programme and helping to keep in 'on track'. The latter led to a specific relationship with the Bank of England. This relationship centred on preventing actions by commercial banks impacting on field development 'critical paths' but could be extended to the use of Bank exchange control powers to delay the foreign take-over of 'strategic' companies.

Contact with the Oil Taxation Office (OTO) took two forms. In the first case, OSO was able to advise of changes in oil company procurement practices that seemed to be PRT driven. One example was the shifting of the burden of construction finance on to contractor prices as a means of gaining interest relief against PRT (so-called contractor finance), something OSO feared disadvantaged smaller British-owned suppliers and favoured larger foreign-owned firms, particularly when coupled with a requirement for performance or completion bonds. Another was the manipulation of contract dates to defer or reduce PRT liability.

In the second case, OSO approached the OTO to establish whether or not certain capital expenditure was allowable for relief against PRT by being field specific. An example was the expenditure on multiple-support vessels (MSVs) committed to a sector of the North Sea but 'based' on the dominant field in that sector.

5.4 SOME KEY OSO ISSUES OF THE PERIOD

The main issues that preoccupied OSO's management in its early years were those that were to remain central for most of its existence. One exception was OSO's role as an expeditor and facilitator of offshore developments, which declined in importance after the achievement of self-sufficiency in 1980.

5.4.1 OSO as Monitor and Expeditor

One of the few causes for optimism about the UK economy during OSO's early years was the prospect of substantial North Sea oil production. Both Conservative and Labour administrations were anxious that this should happen as rapidly as possible, an objective broadly shared by the oil companies for commercial reasons. OSO was soon seen by government as a tool to help bring this about, leading at times to a diversion of OSO resources from its primary aim of developing of British offshore capability.

The initial such diversion took place during the state of emergency that followed from the 1973 miners' strike. OSO was given the task of granting exemption certificates to firms engaged in the North Sea development programme from power supply restrictions, allowing it to ease itself a few months later into a more general, 'monitoring' role (Jenkin, p. 79). The appointment of OSO monitoring engineers was intended to keep Ministers aware of sources of potential timetable slippage and in some cases allow government intervention. A Brief for the Secretary of State revealed that 12 of OSO's 40 specialist professional and technical staff were engaged '... *with monitoring the progress of major UK contracts*' – TNA: PRO BT 241/2580. The same document detailed the various ways in which OSO was contributing to the avoidance of programme delays.

During 1974 OSO also worked with UKOOA and BSC to resolve what UKOOA described as the '*UK Tubular Goods Crisis*'. BSC was the major supplier of well casing to the domestic market, as well as having an important export trade. UKOOA members were concerned that BSC could not meet an expected surge in demand. The fear of a crisis proved to be exaggerated due to over-ordering and slippage in platform installation and thus development drilling.

A perceived lack of platform construction sites – particularly for concrete structures, which had been imported initially – led in 1975 to the establishment of OSO's Platform Sites Directorate. During its brief existence, this unit was made responsible for all aspects of the government's platform sites policy, including site acquisitions, preparatory work on site construction and leasing to contractors (Jenkin p. 85). This led to some mainly abortive public expenditure, as within 2 years it had become apparent that platform demand had been overestimated and that concrete platforms had fallen from favour. OSO had to extricate itself from the concrete platform sites initiative, with minimum further cost to the public purse.

The extent to which OSO's efforts were directed to attempts to improve the speed and efficiency of the first stage of North Sea oil development was not confined to the issues discussed above. Confidence in a British firm's ability to meet delivery obligations necessarily became a factor in the extent to which OSO could promote its interest with Operators. An arrangement was set up with the Bank of England whereby no bank operating in the UK could enforce a receivership or liquidation of a supply business on the critical path of an oil project (as confirmed by OSO) without the agreement of the Bank, which would seek a less disruptive solution. Further, according to Jenkin (1981, p. 127), the government put emphasis in choosing projects for Industry Act financial assistance on their relevance to speed of development, impacting on the work of the venture managers.

Often linked to the question of project slippage was costs escalation. Here, OSO played a lesser role, though it did make an input into the report on the subject commissioned by the government in 1975. A key conclusion was that over the period September/October 1973 to March/April 1975 costs escalated at an annual rate of 80%, with unexpected increases in input costs of 20–30% overshadowed by increases in work content of 80–100%. While this experience had parallels in other projects with high development content, their short timescales gave North Sea projects a higher rate of increase – Department of Energy *et al* (1976c, p. 18).

The authors offered an illuminating insight into the environment within which OSO's monitoring function was carried out. It painted a picture of complex interactions between such factors as:

i. a hostile physical environment demanding a major extrapolation of established design and fabrication practices;

ii. exacting timetables requiring fabrication to begin before the design process was complete;
iii. the extremely high rate of spending resulting from upwards of a dozen Operators undertaking developments at much the same time, ensuring an escalation of pressure on technical expertise and other scarce resources;
iv. the need to marshal diverse sources of manpower, materials and services, at times in out-of-the-way locations;
v. the high premium placed on avoiding slippage so that offshore installation could be carried during the target summer 'weather-window';
vi. the general economic background of very high inflation.

Interestingly, the perceived weaknesses of traditional British heavy industry–poor management, labour militancy and poor productivity – did not feature as major causes of costs escalation and the report's authors found '*no evidence*' that the British fabrication industry (its main field of study) was '*significantly uncompetitive*' (Department of Energy *et al* 1976c, p. 81). It also correctly foresaw that the 'learning curve' experienced by all parties in the first wave of offshore oil development, and the probably less frenzied pace of further development, would mean that in future costs escalation would not be as serious a problem. As had been found by NEDO a few years earlier with onshore projects, not having finalised the design before the start of construction exerted a particularly malign influence (see Table 1.4, p. 16).

As the initial over-heated development 'bubble' subsided, the general emergence of excess capacity and the approach of UK oil self-sufficiency lessened the importance of the monitoring function and enabled it to be combined with expenditure auditing. The maturing domestic market and the need to support future export markets increased the focus on British industrial capability, for which a separate unit was established within OSO's Engineering Branch. An early priority of the Industrial Capability Section was to assume responsibility for specific field developments, initially working with the relevant engineers in the Audit Section, though involvement with the field development approval (or Annex B) process would inevitably follow. When it eventually did, the emphasis would be upon 'UK content' rather than upon avoidance of programme slippage, important as that remained.

5.4.2 Quarterly Returns and Full and Fair Opportunity

Adherence to the two key IMEG recommendations of collecting quarterly expenditure returns and establishing Full and Fair Opportunity, or FFO, for British suppliers were central to OSO's operations.

Establishing the Quarterly Return was straightforward and produced few complaints from Operators, beyond those to be expected about 'bureaucracy' and 'time-wasting'. The latter led to 'to-ing and fro-ing' over issues such as the minimum individual order size to be recorded, which in turn could vary according to OSO's interest in a particular market sector. Though introduced in 1973, the returns and their audit did not become fully established until 1975.

The resultant statistics enabled OSO to brief its political masters on overall levels of UK content and to provide useful leads for British industry. During this formative period, the statistics were published separately each year in the so-called OSO 'Blue Books', the formal titles varying slightly between years. Thus the first publication was entitled Department of Energy (1975), *Offshore Oil and Gas: A Summary of orders placed by operators of oil and gas fields on the UK Continental Shelf During 1974*. From the following year (Department of Energy 1976a), it was simplified to *Offshore* (Year): *An analysis of orders placed*. In 1979, in an attempt to reduce media interest in the figures, a decision was taken to incorporate the figures in the DEn's annual 'Brown Books' and to cease to publish them separately.

Their mere collection provoked an angry response from the U.S. government, no doubt anxious to preserve the market leadership its companies held in the oil and gas supply business. Following earlier representations in Washington, on 8th March 1973 the U.S. Embassy wrote to the FCO requesting that the reporting be dropped and not replaced by other measures '...*designed to influence procurement decisions*' and seeking assurances that '...*future drilling rights will not be linked to the question of sources of supplies and equipment*' – TNA: PRO BT 241/2580. Despite U.S. threats to make this a General Agreement on Tariffs and Trade (GATT) issue and pressure on the definition of a British company for FFO purposes, the British government held its ground. However, the episode must necessarily have highlighted early the limitations to what FFO might achieve. The files examined showed no sign that the British side drew attention to the protection afforded to U.S. marine operations by the Jones Act – long-standing legislation reserving U.S. coastal shipping for U.S.-flagged and -owned vessels, which was also applied to U.S. offshore oil and gas operations.

Meanwhile, there is much evidence on the same file (and some elsewhere) that OSO was actively pursuing FFO from the start, often invoking Ministerial support. Perhaps the earliest example, since it took place in February–March 1973, related to the Phillips terminal on Teesside, ironically intended to handle oil from the Norwegian Ekofisk field, the first major North Sea oil field to be developed.

According to Jenkin (p. 181), Phillips's original intention of awarding the contract to a U.S. company when there was demonstrable UK capacity available '...*would have been very damaging to the UK industry*'. In the event, the contract went to a British company, SimChem. Jenkin evidently believed that, although OSO denied putting political pressure on Phillips, it had '...*played a substantial role in encouraging the oil company to*' ... '*give the contract to a UK firm*'.

Records in the National Archives suggest Jenkin's version omits to mention two key issues. First, '*the American firm*' was actually the UK subsidiary of an American firm (Parsons). Secondly, OSO genuinely believed that SimChem had not received FFO – essentially because the competing bids were for different work packages and on 'a like for like' basis the SimChem bid was cheaper – PRO BT 241/2580. The same file also contains accounts of four occasions when OSO intervened in the interests of British firms.

A by-product of the Teesside affair was a request for a Ministerial meeting from U.S. contractor J. Ray McDermott seeking clarification of the position of its own newly established Scottish platform-building subsidiary. At the meeting on 3rd November 1973, the Minister stated that McDermott would be favoured over bidders without UK facilities but that "*more British bidders*" (i.e. UK owned companies or joint ventures involving a UK partner) would be given the opportunity to compete against McDermott.

Interventions were not always successful. OSO's attempt to obtain FFO for British loading buoy supplier, Woodfield Rochester, failed to win an order for the company in Hamilton's Argyll field on the grounds of '*non-operational*' experience. OSO was suspicious of the role played by the U.S. office of Bechtel, Hamilton's engineering contractor. Another unsuccessful case involved an £8 million (about £72 million in 2008 terms) design and project management contract for a compressor station and platform for the Leman Bank gas field. The situation here was that broadly similar bids had been received from Power Gas, a British-owned company, and B&R.

Amoco claimed it preferred B&R because it had recently completed a similar task and because there was more confidence it would complete on time. However, OSO had prevented Amoco from confirming the award to B&R because it suspected that FFO had been compromised by contacts between B&R and Amoco in the USA, which had led to the reversal of an earlier decision in favour of Power Gas. Furthermore, OSO's request to review the bid evaluation was rejected on grounds of commercial confidentiality.

A meeting between the Minister and the President of Amoco Europe was held on 27th June 1973. It was followed on 9th August by a meeting between the Director of OSO and the General Manager of Amoco Exploration UK when, it appears from the written record, the exchanges were somewhat less than cordial Amoco questioned British government policy, OSO's ability to maintain confidentiality, the judgement of OSO's Director and the capability of British industry.

OSO considered it of great importance that procurement decisions relating to the UKCS were taken in Britain. The second meeting exposed that Amoco Exploration UK was not necessarily the arbiter of its own contracts and that contracts placed with Heerema and de Groot of the Netherlands showed Amoco's interest in encouraging '*local*' (*sic*) suppliers. With both those companies having had close associations with B&R, which was well acquainted with Amoco from GoM and German North Sea operations, a picture emerges of the established contact networks with which British new entrants to the offshore supply business would have to compete. For its part, OSO stated that without further information, it could not be satisfied that there had been FFO. Its experience led it to reinforce its view that '. . . *Amoco was the least co-operative licensee*'.

This was a reputation that Amoco was not to live down with the DEn for the remainder of the decade and for which it was to pay a price in terms of poor exploration licence awards. Amoco was by no means the only foreign operator to refer potential UKCS procurement decisions overseas. OSO knew that when TH sought to promote its 'Cleveland Colossus' platform design for Occidental's Piper field, it was referred to Bechtel, the company's engineering consultant, in the USA. The order went to McDermott's Scottish yard, which admittedly offered quicker delivery.

UKOOA records for 1975 reveal that OSO's DG had provided examples of the failures of some of its members to give British firms FFO, going on to state that of the 12 Operators then carrying out developments, one had been '*guilty*' three times and three or four others at least once.

This experience highlighted the difficulty of 'policing' FFO without a frame of reference agreed between the Operators and the government, an issue that came to a head after the formation of a Labour government in 1974 and its decision to expand OSO. At first, it considered legislating to assist British suppliers.

However, this approach was dropped in favour of negotiating a non-statutory MoU and CoP between the DEn and UKOOA, representing the Operating companies. The fact that the government declared its intent to legislate should agreement not be reached, no doubt helped concentrate the mind of the UKOOA Work Group set up to negotiate on the basis of the original government draft. Although the stated position of UKOOA was in favour of FFO for British suppliers, there seems to have been less than wholehearted internal agreement on where the primary responsibility for the support of British industry lay. At least some in UKOOA held the view that rather than ensuring the offshore Operators offered UK firms FFO, it would be better if the government created '. . . *the right financial climate to permit suppliers to compete with those from foreign countries*'.

Little more than a month after the UKOOA Work Group was established, there was a further government threat of legislation. Agreement was reached in November 1975. It dealt with proposed purchases of over £100,000 for goods and over £500,000 for construction and services (respectively, about £600,00 and £3 million in 2008 terms), or such lower figures as might be agreed in cases of special interest to OSO. Although responsibility for implementation lay with individual UKOOA members, the Work Group continued as UKOOA's FFO Committee, maintaining general liaison with OSO and on occasions offering advice in specific situations.

The MoU and CoP, to which Addenda were added in 1977 and 1981, gave OSO not only a greatly enhanced knowledge of procurement processes and an agreed means of intervening in them (though not the power of decision) but also formally committed the Operators to FFO. The Code provided that tender documents were in a form that did not disadvantage UK companies, allowed OSO to suggest additional UK bidders, made bidders provide UK content estimates, established criteria for bid evaluation and required the Operator to inform OSO before announcing a non-British contract or (major sub-contract) award, giving OSO, '*a reasonable time*' for '*representation and clarification*'. Additionally, OSO was to have prior access to information on anticipated procurement programmes, specifications and tender documents, lists of proposed bidders and bid summaries. To reassure the oil companies, the MoU contained an undertaking that it would be employed in a way consistent with the provisions of the EEC Treaty.

The MoU and its code enabled both sides to believe that they had achieved their most essential objectives. For OSO, it saw formal acceptance by the Operators of a key government policy and provided both the right and the mechanisms for it to intercede in procurement decisions. For the Operators, the voluntary code was clearly preferable to legislation and their right to take the final procurement decision and to place the order with a foreign supplier where there were strong reasons to do so was confirmed.

As industry executives and OSO's audit engineers became familiar with the arrangements, their operation soon became embedded in industry practice and involvement by top managements was increasingly confined to a relatively small number of large high-profile contracts. Increased confidence probably contributed to an increase in oil-related product development by British suppliers.

As long as most expenditure was of a capital nature for fields liable to PRT, any complaints about additional costs resulting from FFO were difficult to sustain, since they were (after a delay) largely borne by the British taxpayer. The most common complaint on the part of the oil companies concerned supposed breaches of commercial confidentiality by OSO officials.

The personal recollections of senior OSO officials responsible for the FFO policy throughout its existence were positive, believing it to have become a practical policy tool. When bid lists were prepared, the relevant audit engineer would review them and seek to add additional British companies where necessary. He would also pick up any discriminatory features of the specification that might work against British bidders. The oil company could not place orders above a particular size until OSO was satisfied that there had been FFO. Where a British company had narrowly failed to be preferred, OSO would sometimes make 'behind the scenes' endeavours on its behalf, which might result in a re-bid on the grounds of lack of FFO.

Once the oil companies recognised that UK content was an important factor in licence awards, the FFO system functioned smoothly for many years. It became more difficult to operate when oil companies began to move away from widespread direct procurement in favour of 'bundling' orders through main contractors that were not themselves parties to the MoU or license bidders, as began to happen as the 1980s progressed. Overall, the main problem was oil company resentment of government interference in the 'business process' either for practical or ideological reasons, although problems with the definition of 'Britishness' could sometimes arise. OSO always had to be careful not to imply that the government 'stood behind' a particular British company.

In any event, according to figures published in the Department of Energy's 'Blue and Brown Books', the overall UK content of orders placed rose from 40% in 1974 to 57% in 1976. In 1979, the figure (at 79%) passed IMEG's 70% target for the first time. Thereafter, it was only to drop below it once (1981), eventually rising to a peak of 87% in 1987. However, as will become apparent, OSO's UK content figures were to become contentious.

Although the government never published UK content of orders for individual companies, companies occasionally themselves released them. Shell UK 1982 (p. 4) claimed Shell Expro's UK content had risen from 75% in 1975 to 86% in the first half of 1982. Companies seeking additional licenses were anxious to achieve a UK content that at least equalled, and preferably exceeded, the industry average, thereby exerting upward pressure on the average. At the time of the Sixth Licence Round interviews, for companies already active in the North Sea, the latest UK contents for some individual applicants had reached only the low 40%s, whereas the latest published figure (1978) for the industry as a whole was 66%. Often there were good reasons for a company to have a low figure (e.g. only E&A activity thus far) but where this was not the case, the companies felt themselves to be vulnerable. Nevertheless, it is difficult to pinpoint cases where an award was denied or made solely on OSO considerations.

5.4.3 Defining 'Britishness'

Without a workable definition of what constituted a British company, no measurement of UK content was possible. This issue was to trouble OSO throughout its existence and it surfaced early. Within the first half of 1973, British-owned companies, American subsidiaries established in the UK and the U.S. government all raised the nationality of ownership question.

Perhaps attracted by a desire to emulate the sort of partnership Wimpey had established with B&R, TH had approached the U.S. firm McDermott (with discussions taking still place as late as 1979). However, McDermott remained unaffiliated with any British partner. When TH complained that U.S. firms like McDermott's saw no need for a British partner when they were already in receipt of government grants, the Minister responded that it would be difficult for the government to discriminate against foreign contractors when it was keen to encourage inward investment, particularly as some forms of government support were almost routinely available – TNA: PRO BT 241/2580.

Among other related material on the same file, no record was found of the response received by a much longer-established British subsidiary of an American company – Foster Wheeler. Its UK managing director wrote to the Secretary of State asking whether it would be fair to discriminate on the basis of ownership against his firm, which had only one American (himself) among its 1000–1500 employees in the UK and which was the first UK engineering firm to have received the Queen's Award for exports. In its attempts to help the FCO answer the American government's request for a definition of a British company, OSO made it clear that for the purposes of the quarterly returns, foreign-owned companies registered in the UK and having substantial establishments there would be regarded as British. Nonetheless, in order to develop a potential export capability in the longer-term, it had a particular interest in ensuring that British-controlled companies were not discriminated against in the home market. There was to be little advance on this position. In some market sectors, including such critical ones as drilling equipment and down hole tools, there was little need to maintain the distinction because participation by British-owned companies was either minimal or totally absent.

Cook and Surrey (pp. 68, 69, 87, 88) make it clear that at the end of the day, OSO had no choice but to conform to a general British 'open-door' policy towards inward investment, which precluded discrimination on the basis of ownership. It, therefore, had little or no option in its day-to-day business to do other than to treat all firms with substantial UK employment as domestic businesses, whose supplies to the offshore oil and gas Operators counted towards 'British content'. Needless to say, this policy was not universally popular.

5.4.4 OSO and the E&P Companies

During OSO's early years, state energy companies were important North Sea 'players', BNOC particularly so. By 1977, BNOC was sufficiently developed to begin to influence North Sea procurement, support for UK suppliers being among its aims. Since a new management with purely commercial objectives (culminating in privatisation in 1982) was installed after the return of a Conservative government in 1979, its influence was short-lived.

BNOC itself had arrived too late on the scene (in 1975) to have any influence on major procurement decisions for the first generation of developments. During the period under review, BNOC's own procurement activity was focused initially on the Thistle field and later on the Beatrice field. In both cases, it was the successor to earlier operators, respectively, Burmah and Mesa Petroleum. Thistle came into production in 1978 and most important procurement decisions had been made much earlier. The position at Beatrice, a smaller field in much shallower water, was different. It did not come into production until 1981, so that more procurement decisions were more open to BNOC influence. Assessments of BNOC's UK content made in this period placed it at the lower end of the range at around 40%. Some saw its commitment to FFO compromised by the large number of Americans in senior positions, most formerly employed by Burmah following the latter's acquisition of Signal Oil and Gas of Houston. However, there were clear signs that BNOC's top management were motivated to adopt supportive attitude towards British business.

For example, on arrival in Houston the senior OSO representative to the May 1977 Offshore Technology Conference was asked to meet executives from

U.S. gas turbine manufacturer, GE, which complained that it had been displaced from an order for the supply of U.S.-built turbines for Conoco's Murchison field in favour RR, a matter of which OSO official no prior knowledge. It transpired that OSO had played no part in this but that BNOC exercised its position on the Murchison field operating committee to favour RR, which went on to become a major supplier of offshore turbines.

BNOC also had exploration interests, but did not survive long enough to mount exploration campaigns on the scale of those of BP and Shell. Unlike them, it did not operate its own mobile rigs, though it became involved financially in their construction on two occasions during this period (McKinstry 1998). Thus, when in 1976 Marathon's Clydebank yard faced large-scale redundancies due to a lack of orders, BNOC had placed a speculative order for a jack-up, with the possibility of a second following, although the rig and its possible follow-on were sold to a foreign buyer before delivery. In 1979, the yard received another such speculative order for a rig, again in due course sold on to a foreign contractor. In this case, BNOC had assigned the contract to a company in which the Scottish Office held 50%, BGC having declined to take on this role. These orders fell outside OSO's remit and it was relatively little involved.

Generally of less importance to OSO than BNOC, BGC had limited exploration and no development operations during the 1973–1980 period, although the preliminary stages of the Morecombe Bay development began towards its end. Initially at least, OSO found relations BGC much less cordial and supportive than those with BNOC.

After losing its E&P interests to BNOC, the NCB had retained subsurface activities some of which had developed oil and gas expertise and which included control of Horizon Exploration (established in 1973), the UK's principal domestically owned marine seismic contractor. It therefore was an OSO 'client' as a supplier. At one juncture, it considered establishing a reservoir engineering joint venture with Franlab, an IFP 'spin-off' at the time seeking a UK partner. The idea was dropped when it became clear that many oil companies would not be prepared to disclose reservoir data to what they considered part of the French state oil sector.

Whilst BP had a substantial government shareholding at the time, it always behaved as a private sector company and was treated as such by OSO. Shell was also usually regarded as a private sector British company, but as part of an Anglo–Dutch venture operating on a 50:50 basis with a U.S. partner (Esso), its position was actually significantly different from that of BP. As Burmah all but was eliminated before OSO was fully 'up and running', other purely British private sector participation in North Sea E&P was limited to the British independents, which were generally passive and not involved in operating (except in a limited way in exploration).

Until they came to recognise the importance of UK content to licensing awards, U.S. companies were inclined to be hostile, generally the more so the smaller they were. Total and AGIP were subject to pressure (in favour of their own national suppliers) from their own governments as well as the British but sought to accommodate OSO's requirements. Elf, then the French state oil company, was not an Operator on the UKCS during the period of the FFO policy. It is important to note that whereas the sheer scale and scope of the UKCS operations of BP and Shell Expro demanded an almost continuous engagement with OSO's senior management, with all the other companies there was more of an 'ebb and flow' with major interactions normally associated with individual development decisions.

At the collective level, although complaints from member companies about OSO were not unknown, UKOOA records for the period show relations between OSO and UKOOA after the MoU and CoP had been settled could generally be described as 'settled', with few seriously contentious issues. Thus, when reporting to the UKOOA Council, the Full & Fair Opportunity Committee could state that OSO had agreed to leave the Operator with the initiative to define a supplier as British, whilst Council endorsed the Committee's recommendation that OSO's request for a reduced contract pre-notification level for the special interest areas of 'Diving and Submersibles' and 'Specialised Maintenance' should be accepted. In July 1977, the Council also approved a letter from OSO to Operators drawing attention to its special interest also in 'Feasibility and Design Studies' and 'Consultancy/Appraisal Studies' for long-term structure maintenance and endorsed the continuation of quarterly returns. Some 18 months later there is evidence of continued harmony in the Committee's acknowledgment 'of the excellent co-operation' developed between it and OSO under the direction of Alan Blackshaw, the outgoing OSO DG. It was also noted that Mr Blackshaw's successor as OSO DG (the author) had: '... expressed his intention of continuing the same policy'.

The IRG Committee also reported regularly to the UKOOA Council, covering both the progress of negotiations with the EC and administrative issues. There was usually little difference of opinion with OSO. The Committee subscribed to the view that the PAC's criticisms over OSO's administration of the IRG scheme were 'an overreaction'.

The year 1979 saw an extension of OSO/UKOOA co-operation when there was a joint study of the future demand for inspection, maintenance and repair (IMR) services, broad agreement that UK content (in the context of capacity) was 'approaching the optimum' and a desire to understand OSO's specific areas of interest in creating new capability.

Unlike those with UKOOA, where an essentially 'top management to top management' relationship existed, contacts between OSO and individual Operators were multiple. Different officials handled auditing, the IRG and quarterly returns, with only major issues requiring top management inter-action. Consequently, it is difficult to generalise, particularly as little of the written record seems to have survived. However, the fact that UKOOA was little used as a lever on OSO in pursuit of individual company interests does give credence to personal recollections that, on the whole, relations were best described as 'placid'. Nevertheless, there were considerable differences on a company-by-company basis.

It is perhaps best to consider the major British (or part British) Operators separately. With the effective absorption of Burmah by BNOC, whose special position has already been discussed (see pp. 125–126), these numbered only three, as compared with nine foreign Operators with oil fields in production or under development during the period. If companies undertaking exploration were also included, the preponderance of foreign companies was even greater. At the time UKOOA had over thirty members, only two of which (Cluff Oil and Premier Consolidated) were UK independents. Indeed, even including the two state corporations, there were only six British (or part British) members.

On other measures such as capital expenditure, wells drilled and production, Shell Expro and BP loomed much larger. For instance, they accounted for about a third of oil fields in production or under development, including the two largest. Indeed, in a preface to a 1984 publication commissioned by Shell, the then managing director of Shell Expro stated that his company alone had cumulatively accounted for '*about one third*' of UK North Sea expenditure (The Economist Intelligence Unit Limited). Given the broadly comparable scale of BP's activities, it is clear that these two Operators alone would have accounted for well over half of total UKCS expenditure to that point.

By comparison with this concentration, the expenditure of the foreign Operators was very fragmented. Moreover, by virtue of their massive engineering and other technical resources, their service sector investments and the British nationality of most of their senior managers, Shell and BP stood apart from the other Operators. Success or failure for OSO depended heavily on them and therefore they received disproportionate attention.

Both companies generally sought to maintain an attitude of positive co-operation with the British government and stood high in the 'British content league'. However, neither readily succumbed to pressure to act in a non-commercial manner, particularly if this was in support of what they regarded as uncompetitive British enterprises. BP's resolve to resist was strengthened by a desire to demonstrate to opinion in the USA (where its large Alaskan interests were politically sensitive) that – despite the British state then holding 48% of its equity – it was not a state oil company subject to government direction. In the case of Shell, ultimate control rested in the Netherlands where the government had its own offshore supplies industry to support and was an enthusiastic founder member of the EEC. Moreover, Shell Expro acted as operator of a 50/50 joint venture with Esso, a U.S. company. During the peak of its North Sea development expenditure, which occurred in the 1973–1980 period, the 'Britishness' of its UK based staff became diluted with Dutch engineers drafted in from the Netherlands and Americans seconded from Esso, helping make Shell the more difficult of the two for OSO to influence. Nevertheless, adverse financial consequences from delays and cost overruns in its massive Brent complex made it sensitive to government pressure, particularly prior to the commissioning of the oil and gas export pipelines.

The difference in the approach of the two companies to the construction and operation of MSVs for their respective 'Sector Clubs' in 1978–1979 was clear. The slump in shipbuilding demand and the resultant unemployment gave a strong political incentive to have these expensive vessels – themselves the result of a government/UKOOA initiative – constructed in the UK. Nevertheless, both Shell and BP were rightly fearful of the delays and cost overruns likely to result from placing the orders with yards belonging to the newly nationalised BS. Following lengthy discussions involving the top managements of the companies, the Minister and senior OSO personnel, different ways were found of reaching an acceptable compromise.

In the case of BP, an order was placed with the Scott Lithgow yard on the Lower Clyde for a vessel to be known as the *Iolair*, to be operated in-house. However, as insurance against late delivery of the *Iolair*, BP also asked Salvesen Marine (a British company) to design, convert, and manage a back-up vessel, the *Fasgadair* (Jamieson 2003, p. 140).

Shell remained implacably opposed to ordering its MSV (eventually named *Stadive*) from a UK yard. However, it did respond to an OSO initiative whereby British company Seaforth Maritime, in association with a Norwegian semi-submersible designer, received a contract to assist in the design and construction supervision of the vessel in a Finnish yard and to operate it. At £100 million (about £368 million in 2008 prices), the initial operations contract was worth twice as much as the build contract.

With respect to the other Operators, OSO contacts were focussed on those with current field development programmes, at this time all American. Though never as close as with BP and Shell, on the whole, relations were generally reasonable and in the case of Conoco became especially close during the development of the Hutton field. The smaller companies had limited engineering resources and depended heavily on the major U.S. design and project management companies, benefiting in UK content terms from the progressive transfer by the latter of their North Sea operations to the UK.

An exception was Chevron, which used Anglo–American joint ventures for project management (Taywood-Santa Fé) and jacket design (CJB/Earl & Wright) for the Ninian field, but this choice resulted from the fact that (as at Thistle) it was Burmah that had placed the initial contracts, prior to its loss of the Operatorship, (Jenkin 1981, p. 155).

Amoco was probably the most difficult of the U.S. companies for OSO to deal with. It had a history of difficult relations with the government (see p. 122) and had been the last company to accept BNOC participation. During early 1980, an OSO visit to Amoco's headquarters in Chicago provoked a strong attack on British industrial performance, OSO and British government policies. After lack of success in licence awards, there was a change in attitude and personnel at Amoco UK.

5.4.5 OSO and Two Major Technological Issues

Towards the end of OSO's formative years, the key development decisions were made for two fields based on use of new technology. These were Cormorant Central (onstream 1983), where Shell Expro was the Operator and Hutton (onstream 1984), where Conoco was the Operator. In both cases, the importance OSO attached to ensuring that British industry benefited from projects meant that the procurement issues were mainly handled directly between the company top managements and the DG (then the author), supported by the head of OSO's Engineering Branch. In each case, a critical element of the supply chain was Vickers at Barrow-in-Furness.

Shell Expro's Cormorant Central was developed as a subsea satellite to the South Cormorant platform. Although subsea well completions already had a long history, with both Shell and Esso among thee innovators, Cormorant Central was not only the first field in the northern North Sea to be developed through their use but also for many years was the largest such field in terms of the reserves base, number of wells and peak production. Maintenance was intended to be diver-less, with a sophisticated dedicated ROV and through flow line (TFL) well intervention. Indeed its size and complexity were such that it was more akin to a full subsea field development, for which it could be seen as a prototype, than to a satellite as normally understood.

A key element in the design was an underwater manifold centre (UMC), weighing some 2200 tonnes. The detailed design of the UMC was entrusted to Vickers, then the main UK-controlled source of subsea technology. Senior management in both Shell and Esso independently assured the DG that they fully intended also to have the structure constructed at the Vickers facility at Barrow-on Furness. This intention was determined by the perceived need to ensure total reliability by the highest possible level of manufacturing quality. They believed that Barrow's nuclear submarine experience made it the most appropriate facility in the UK, the more so since Barrow also housed the UMC design team. Shell Expro did not intend to call for tenders from possible alternative providers.

Vickers Barrow prided itself on constructing prototypes, so-called 'first in class', a previous example in the energy field having been the LNG carrier *Methane Princess*. However, on this occasion, Shell Expro's plan was frustrated by the refusal of the Vickers shipyard management to undertake the construction task. The reason given was that the facility did not have the capacity to do so because of defence contract commitments. OSO was in no position to ascertain whether or not this was so, but strongly suspected there were other reasons. In anticipation of shipbuilding nationalisation in 1977, there had been a separation of ownership between Barrow's core shipbuilding and engineering facilities on the one hand, which passed into state ownership, and its offshore activities, which remained in the ownership of Vickers as the Vickers Offshore Engineering Group (or VOEG) on the other. Moreover, the retirement of

Sir Leonard Redshaw, responsible for both the nuclear submarine programme and the decision to enter the offshore business, had deprived Vickers Barrow of strong leadership.

Following this rebuff, Shell Expro advised OSO that it now wished to negotiate a contract with a firm in Rotterdam, with an 18 month construction cycle to be followed by a further 18 months of testing. OSO put forward British alternatives. After much effort on OSO's part, Shell Expro agreed that one other UK company – the Teesside heavy engineering firm of Whessoe (then heavily involved with civil nuclear work as well as offshore fabrication) – was capable of meeting the necessary quality standards. Although Whessoe could meet the client's requirements on price, it was unwilling to commit to the requested 18 months delivery, believing 3 years more realistic.

Consequently, the order was placed in the Netherlands, eventually being delivered about 18 months late. An acceptable UK content had to be achieved by other means, such as UK manufacture of the subsea wellheads by McEvoy and of the subsea controls by TRW Ferranti, both UK/U.S. joint ventures, and by UMC testing being undertaken at Bacton in Norfolk rather than in Rotterdam (Figure 5.1).

The story of the world's first commercial tension-leg platform (TLP) for use in Conoco's Hutton field has many similarities with that of the UMC. TLPs had long been studied as a means of developing fields in water too deep for the use of fixed platforms, without the need for subsea completions. Conoco's real interest in TLPs was for the GoM, but it probably concluded that using Hutton, well within fixed-platform limits, as a test facility had the advantage that the operation of PRT could potentially shield the project from the adverse effects of cost overruns. Several of the field partners continued to favour the use of a fixed platform. Anxious to see the UKCS as the location for another significant technological 'first', OSO helped persuade them to accept their Operator's plan.

Work on the TLP concept had been undertaken in the USA, on the Continent and in the UK. Indeed, according to Conoco's engineering leader on the Hutton project, the first test of a (small) TLP-like structure took place in Scottish waters in 1963 (Mercier 1995). He was almost certainly referring to the test of their 'Tritton' platform concept by International Marine Development Limited of London (McDonald 1974).

The multi-party nature of the Hutton development group demanded a tender process. Conoco is believed to have initially focussed on two preferred bidders for design and construction of the hull and deck – both Continental. When discussing the proposed bid list with Conoco, it became clear to OSO that Conoco was unaware of the fact that VOEG, working with BP and in part EC funded, had carried out extensive development work on TLPs. Conoco quickly became persuaded that VOEG should be on the design bid list, though it rejected BP's offer 'to share and buy into the technology'.

FIGURE 5.1   Central Cormorant UMC at Rotterdam. Courtesy Shell International Limited.

Conoco wanted the design contract to extend to the topside facilities where Vickers was not qualified, leading to OSO initiating a bidding joint venture with B&R. Although the latter was willing to act as sub-contractor, VOEG's management opted for that role. The combination was awarded the contract in early 1980. By this time the ownership of the former VOEG design unit was in the hands of BS. An unconnected Vickers unit, the Design and Projects Division, won a sub-contract to design and supply the tensioned moorings, while yet another Vickers company, Brown Brothers, also provided some of the hardware.

The UK manufacturing subsidiary of Vetco received the contract for the subsea template and the wellheads, with associated development work also

FIGURE 5.2   Hutton TLP in the Moray Firth. Courtesy ConocoPhillips.

carried out in the UK. OSO pressure resulted in a Vetco 100% UK manufactured content, along with an estimated 80% UK technical 'know-how' content. Hi-Fab carried out the construction of the hull structure and McDermott's Scottish yard that of the deck, with the two elements being mated in the Moray Firth (Figure 5.2). Overall, a very high UK content was achieved.

Further TLP developments followed in other parts of the world, particularly the GoM. Unfortunately, the UK never secured other benefits from its pioneering role.

5.4.6 Research and Development

The size and diversity of the offshore supplies industry, the limited public funds available, the speed of development and the relative lack of British companies with the technical and commercial credibility needed to introduce new technology, all militated against creation of effective public R&D policy. The strictures did not apply where the 'drivers' were essentially non-commercial, such as safety and regulation. Commercial spin-offs might (and indeed did) occur but they were by way of a bonus.

An additional complication was Britain's 'open-door' policy towards foreign companies. The fruits of global R&D in the oil and gas sector were readily available for application on the UKCS. Unlike most contractors, large oil companies and oilfield services firms, particularly in the USA, were spending heavily on R&D during this period. Though deficient in large domestic oilfield service firms, Britain was home to three large oil companies. BP alone carried out major upstream/offshore R&D activities within the UK. Burmah as befitted its much smaller size did only a little; work on drilling mud at Birmingham University representing a large part of this. Shell UK's R&D was downstream oriented, with Royal Dutch handling most general upstream R&D in the Netherlands and Shell Oil in the USA concentrating on offshore issues.

Over and above the foreign private sector effort, varying degrees of public research funding existed in France (see p. 51) and several other European countries, in Japan and in the USA. As would be the case with the UK, many of the foreign schemes involved co-financing with industry, often involving repayment in the event of success.

The perceived need for UK offshore R&D aroused considerable interest among politicians and others. Some saw it as a possible means of 'leap-frogging' established competition and others as a means to open up export markets while facilitating the long-term exploitation of UKCS resources. The implication was that these benefits were to be conferred on British-controlled companies, which would exploit them from a UK base. However, the definition of 'Britishness' imposed upon OSO would necessarily lead to public support being also provided to foreign-owned businesses. Fortuitously, this issue rarely arose during the early period because few of the foreign-owned businesses in the sector then carried out R&D in the UK.

The *IMEG Report* (pp. 80–84) devoted only four pages to R&D, playing down its role in achieving market penetration. It concluded that commercially oriented work should be in support of '*large-scale contracting opportunities*'. An emphasis on structures and pipelines was the predictable result, though dynamic positioning (DP), seabed completions and manned submersibles were also mentioned. Although IMEG did not postulate any budget, it recommended selective grants '*of at least 50%*'. It envisaged the proposed PSIB would work co-operatively with the existing government R&D support mechanisms and did not suggest the creation of any new specialised agency.

The Select Committee on Science and Technology for the 1973–1974 and 1974 sessions evidently judged IMEG's limited R&D prospectus as insufficient. Their work resulted in a report on offshore engineering – Select Committee (1974). It took evidence from ministers, civil servants, supply companies, trade associations, research organisations and a journalist. It drew on the *IMEG Report* and more particularly on the Interim Report of a DEn Working Party, which was reproduced in full as Annex 11 (Select Committee 1974, pp. 43–50). The Working Party considered the main objective of research to be the development of a deep-water capability (i.e. up to 2000 m).

Although the Select Committee (1974) dealt with technological issues, it did so in the context of the government's general policy towards the offshore supplies industry. The MPs were broadly supportive of this but not uncritically so. For instance, the Select Committee (1974) saw a risk of too many platform sites, criticised the small size of OSO, considered the value of the IRG Scheme and noted a disconnect between industry and the experience and facilities of the Royal Navy and defence establishments generally. It also believed that government figures underestimated the likely scale of eventual oil production (Select Committee 1974, pp. 7, 8, 13, 26, 29). This last observation is of some interest in the light of the earlier discussion of the resource base (see pp. 53–56).

Committee members were critical of the tardiness of British industry in responding to opportunities both in existing areas like offshore construction and pipelay and new ones such as seabed completions. However, they accepted OSO's view that UK supply companies would have to prove their worth to offshore Operators with current technology before they could expect successfully to introduce the next generation.

Nevertheless, it saw benefits in increasing government support for R&D and in creating a Marine Technology Board, though a plea from the British National Committee on Ocean Engineering (a committee of the Council of Engineering Institutions) for the establishment of a well-funded British Oceanic Authority went unheeded. (Select Committee 1974, p. 209).

Recommendations (Select Committee 1974, pp. 7–8) extended beyond the scope of OSO into areas like education, broad UKCS policy and safety regulation in fields as diverse as diving, platform certification and seabed clearance. OSO-related recommendations ignored by government included further OSO expansion, government industry joint ventures in the fields of subsea completions, subsea reserves enhancement, a more generous replacement of the IRG scheme and government project launch aid. Others, such as greater involvement of defence establishments with the offshore industry, were attempted but proved ineffective. OSO's R&D Branch initiated a dialogue with those branches of the MoD with underwater interests, in line with the wishes of the Select Committee. However, they appeared to have little they could offer in areas of OSO interest, although MoD research establishments did in due course undertake a number of projects receiving OETB funding.

Government reaction to the Select Committee's proposals came in 1975 with the establishment of the OETB located within the DEn. This assumed responsibility for all DEn offshore R&D support, whether existing or new. Among inherited programmes, marine technology was overshadowed by geological and environmental projects.

DEn published its R&D strategy in the following year. It described a Board that was essentially advisory in character with responsibility for ensuring oil industry views were taken into account but decisions consistent with Departmental policies. Of the original 15 members, four were from the British-owned E&P companies, five civil servants, three from the private supply sector, one from UKOOA and another an academic, with DEn's Chief Scientist as Chairman (Department of Energy 1976b).

OETB's responsibilities fell into three areas, only one of which was directed primarily at OSO. They were:

A. *'Acquisition and analysis of geophysical, geological and reservoir engineering data'.*

Within this category were two areas of commercial relevance – enhanced oil recovery (EOR) and reservoir simulation models – both seen as necessary to assist the DEn in exercising proper supervision and control of oil company operations.

B. *'Safety'.*

Work in this area mainly related to acquisition of the metocean data required to improve design codes, of vital interest to designers, certifiers and regulators. Other important areas of interest to the same groups, but also to contractors and manufacturers, were structural monitoring, materials corrosion and fatigue. OETB wished to work with the supply sector to generate commercial spin-off where appropriate.

C. *'Assisting industry in the United Kingdom'.*

This was meant to complement OSO's other activities in developing British capability. It was accepted that the OETB financial contribution would be minor and that the area was a high risk one. The intention was to support commercial initiatives from businesses already established in the North Sea, to place as much work as possible with industry and to encourage cost-sharing and joint programmes. There was to be a major effort '. . . *to ensure that successful work is carried through to the development and manufacturing*' (Department of Energy 1976b, p. 12).

Given that in most cases these later stages were likely to cost a multiple of the research cost, this statement sat uneasily with the recognition that funds were limited and risk high. There was also no explicit recognition that a potentially insurmountable obstacle to commercial application would arise if no offshore Operators were prepared to undertake prototype testing, which was clearly likely to be a problem given the government and commercial pressures to maintain or, better, increase production.

The list of candidate areas for support contained few surprises. It singled out deep-water drilling and production capability, pipelaying, positioning, mooring and anchoring techniques, subsea completions and production, and underwater services including diving, unmanned vehicles, power sources and tools. Nor did it neglect to mention two subjects that would become increasingly important in the years ahead – marginal field production systems and IMR, both above and below the sea's surface. The emphasis on gathering data on the UKCS environment (in the broadest sense) and on the design and maintenance of the current and future infrastructure necessary to sustain oil and gas production is entirely understandable, given the immediacy of the problems in those areas and the scarcity of credible British firms in the traditional 'core' sectors of the industry.

The lack of credible British suppliers did not inhibit support in the areas of reservoir engineering and EOR, both of which formed part of the 'core'. Here, DEn considered that the creation of an independent British capability was vital to the proper execution of its regulatory role. In areas where OSO's research priorities got closest to the 'core' – subsea production and deep-water drilling – there was no comparable imperative for independent national capabilities. There appears to have been little or no consideration by OETB of the implications to British industry of the fact that once the North Sea infrastructure had been put in place, the offshore expenditure pattern would come more closely to resemble the onshore, with a heavy emphasis on traditional 'core' areas, such as well services. This in fact happened in the mid-1980s when exploration and operating costs combined began to exceed development costs on a year-to-year basis.

Establishment of OETB did not end other limited UK government R&D support for the offshore industry. In particular, the Science and Engineering Research Council (SERC) continued to fund university research, mainly through the Marine Technology Directorate (MTD); product and process development and prototype construction occasionally qualified under existing DTI schemes.

The activities of MTD deserve due recognition. Established in 1976, many projects were of a generic nature working on such subjects as fatigue, corrosion, tubular structures and hull design, with published results absorbed into the industry's knowledge base and incorporated into its codes, guides and practices. These university-based programmes also introduced many young engineers and scientists to the offshore industry, thereby helping to increase the industry's stock of technical personnel.

However, the results of a small number of projects were commercialised in a more conventional sense. Of these, the most significant was the 'Vortoil' separator, a space and weight saving device for use on platforms developed at Southampton University. Its considerable commercial significance was better appreciated outside the UK than within it and ownership of the intellectual property rights passed first into Australian and then American hands.

Although formation of the OETB did represent a simplification of government R&D funding, it clearly stopped well short of total unification and the money directed at commercial projects remained small. 'Legacy' programmes, particularly those dealing with the geological and geophysical investigation of the UKCS and with safety, dominated expenditure in the OETB's early years, accounting in 1976–1977 for over 80% of the total of £9.6 million (around £50 million in 2008 terms). This distribution reflected not only the high cost of some of the elements but also the fact that, unlike the industrial support programme where government funding did not exceed 40% of project cost, the UKCS data gathering and safety programmes were in the main totally government-funded. On the whole, they were also undertaken by government-owned research organisations. The existence of such organisations was given as one reason for not establishing a new national offshore technology R&D centre on the lines of France's IFP; other reasons given were the advantages of carrying out applied R&D within industry and the lack of time to establish an organisation capable of making a timely contribution. Many believed this decision was a serious error.

At the European level, there was also a system of support at the 25%–40% level (repayable from a successful outcome) from the EC's Technology Development in the Hydrocarbons Sector (TDHS). According to Cameron (1986, p. 55), from its establishment in 1973 until 1985, some £200 million in 'money of the day' was made available, with a further £84 million budgeted for 1986–1989. In the first period, British firms received 25.5% of the amount available, whereas French firms enjoyed 34%. The high French success rate was widely attributed to the co-ordinating role of the Groupement Européen de Recherches Technologiques pour les Hydrocarbures (GERTH). With OSO encouragement, British participation had improved from a low starting point, although cross-border projects were to become increasingly the norm.

Integrating OSO's R&D and techno-economic appraisal branch in response to the addition of R&D to OSO's responsibilities, into the organisation's mainstream proved far from easy. Unlike those dealing with FFO or venture management, the staff was almost entirely recruited from within the Civil Service, mostly from the scientific grades, and initially lacking either oil industry or business experience. By its nature, their work was unlikely to generate the headlines or overseas trips that made other parts of OSO attractive to politicians.

While the main vehicles for government funding for R&D for the offshore industry remained the OETB and MTD, there was a steady growth of joint industrial projects (JIPs) where a number of parties – oil companies, suppliers, OETB, and sometimes the EC – jointly part co-funded a project.

As a DEn undertaking, the OETB was of much concern to OSO and its DG was a Board member. A large part of its R&D branch's work was devoted to the assessment of projects submitted for OETB financial support. Most projects supported were determined by the original strategy of the OETB. As far as the industrial support programme was concerned, a major change occurred in 1979 when DEn's new Chief Scientist (Professor Sir Herman Bondi) launched an initiative which had as its primary aim the development of robotic technologies that would eliminate the use of man underwater. A secondary aim was to compensate for the lack of British-owned subsea completion suppliers by developing a British competence in subsea systems integration. The formation of an Advisory Group on Underwater Engineering (AGUT) was an early outcome.

Partly as a result of this initiative, the OETB funds available for industrial support rose faster than the total budget, reaching £4.1 million in fiscal 1979–1980 (about £15 million in 2008 terms). Industrial funding nonetheless remained overshadowed by the funding available for the primarily resource assessment and regulatory programmes and it still accounted in 1979–1980 for less than a quarter of total OETB expenditure.

In early 1980, at the point of his leaving OSO, the author made an evaluation of the OETB industrial support programme. It was critical of the complex organisational structure underpinning OETB. The OSO and/or PED sponsored projects, but the Programme Committee (PC), or OETB itself for large or special projects, denied or granted financial support, subject to DEn approval. The Offshore Technology Unit (OTU), also responsible for financial control, provided administrative support to the OETB and the PC. Actual project supervision was sub-contracted to the Marine Technology Support Unit (MaTSU) at Harwell. The cost per man-day of MaTSU staff was twice that of DEn staff. As well being the DEn's agent, MaTSU was also represented on both the PC and the OETB itself. With more than a dozen internal interfaces, this structure was cumbersome and not without conflicts of interest. The swift decisions so valued by industrial applicants could rarely be achieved. Moreover, much of the system's administrative controls involved inter-action with the 'client' companies, absorbing much management time and acting as a disincentive to small companies in particular from applying.

As well as being inefficient, the arrangements were expensive. The combined annualised expenditure of OSO and MaTSU on OETB industrial support in early 1980 amounted to £525,000 (about £1.9 million in 2008 terms); total expenditure would have been higher, given costs elsewhere in DEn, and not less than £550,000. This figure represented some 11% of new commitments for the year in question or about 6.5% of expenditure on projects in hand. Since the same management methodology was applied to small as well as to large projects, the cost burden applied disproportionately to the latter. For the 72% of projects receiving support of £100,000 or less, the cost of administration could have added an average 30% to the direct cost of support.

Another criticism related to whether it made sense to have the PC and OETB handle both projects driven by regulatory and data gathering needs sponsored solely by PED along with those having industrial/commercial drivers sponsored solely by OSO. It was also pointed out that by virtue of their memberships, advisory groups suffered from internal conflicts of interest and that their independence was questionable.

The report found that up to the point it was written, out of 160 offers of OETB support for industrial projects, 140 had been taken up involving a total DEn commitment of about £14 million (probably around £59 million in 2008 terms). It was noted that among the cases where an offer had been made and declined, OETB's endorsement had sometimes raised the confidence of commercial backers to the point where the development could go ahead with all private sector finance, the Gall Thomson marine beak-away coupling being a notable example.

Although the average scale of support was £100,000 (probably about £360,000 in 2008 terms), the median was only £43,000 (probably about £155,000 in 2008 terms). The smallest project received support of £3400 (around £12,000 in 2008 terms). The 28% of projects of above average size accounted for 75%

of total expenditure, with the average support of the four largest being nearly £820,000 (probably around £3 million in 2008 terms).

Of the 99 organisations that had to that point received support, just 11 had received about 73% of the total. Three companies (Vickers, BP and Taylor Woodrow) received between them more 37% of the total. Vickers alone (with 14 projects) accounted for about 14%. Its withdrawal from the industry shortly before the report was written was a serious set-back for OSO as it appeared to be the one UK company with the potential to become a contractor capable of competing with the U.S. majors in innovatory activity.

About 80% of projects – representing about 20% of funds committed – were with SMEs or operating units, including roughly 25% with companies employing less than 100 people. As OSO came to recognise, it was important to establish that such small companies had the financial and managerial resources to undertake a project successfully and without compromising routine commercial activities. The risk of small companies becoming externally funded R&D 'junkies' was seen as a real one.

Nevertheless, while projects of less than average size had tended to have a higher probability of commercial success, no relationship was found between likelihood of success and size of organisation. However, since at the time most work was still ongoing or had only very recently been completed, such judgements were then difficult to make. Of the minority where the outcome was already clear, success and failure were fairly equally balanced.

None of the larger projects completed had been judged a success and problems were apparent with the ongoing ones. Most such large projects were production system design studies, some 'stranded' by other technological and market developments, others lacking any credible means of market entry. The case for continuing with them depended on other factors, such as the benefit in keeping engineering teams in being during a period of slack demand and/or the potential value in individual component parts of the overall project.

The report suggested there should be a simplified form of administration for small contracts, including the abolition of the recovery levy, and a closer integration of OETB activities with OSO industrial capability objectives. An important part of achieving this aim depended on OSO adopting a 'holistic' approach to a company and its OETB funded project(s), which required bringing OSO's Engineering and R&D branches more closely together, which proved very hard to achieve, in part due to the personalities involved and in part to different perspectives.

Paradoxically, one of the earliest and most successful commercial pay-offs from OETB funded R&D came from its regulatory side rather than the industrial support programme. The government needed to be able to monitor reservoir performance and in the mid-1970s was reviewing the then commercially available numerical simulation models (Department of Energy 1976b, p. 10). At that time no such models were of British origin and the commercial sensitivity of reservoir simulations meant that the most advanced models were internal to the major oil companies. Simulation models were indispensable to the formulation of field development plans as well as performance monitoring and equity determinations in situations where reservoirs crossed licence or international boundaries, making it essential for DEn to have unrestricted access to independent models of at least comparable quality to those of the major Operators.

OETB, with support from BNOC and BGC decided to use the resources available in computing, mathematics and physics at the Atomic Energy Authority (AEA)'s Harwell site to develop an entirely new simulation model known as PORES. It was not made available commercially until the mid-1980s when PORES was marketed through a British consultancy, Robertson ERC (Smith Rea Energy 1990a). The same source refers to the ECLIPSE simulation programme emerging as a '*world leader*' and PORES having become '*obsolete*'. ECLIPSE was developed at ECL Petroleum Technologies – a British company – by part of the former PORES team.

The AEA's Winfrith site in Dorset went on to develop other reservoir engineering software and to build up laboratory facilities capable of undertaking pressure volume temperature (PVT) analysis, geochemistry, logging tool calibration and experiments on high pressure core flooding, well stimulation and EOR. It thus became an important centre within one of the most technically sophisticated elements at the 'core' oilfield service industry, perhaps its very heart. From a commercial viewpoint, the strengths of AEA's Winfrith Petroleum Services remained largely unexploited since they were for long generally available only to government.

Not all R&D benefited from the OETB, MTD, or other public programmes. Some projects went ahead at a firm's sole risk. For instance, BGC's development of its intelligent pipeline inspection pig, ultimately to form the basis of a significant independent service company (now American owned), was internally funded.

5.5 THE SUPPLY INDUSTRY

OSO interacted both with oil and supply companies. The former was the far easier. The oil companies were few in number and licensed by the DEn. The even smaller number of active Operating companies, around a dozen at this time, were in almost continuous dialogue with the DEn. Although the companies differed in many ways, the nature of what they did and the requirements of the licensing process imposed some degree of homogeneity and all belonged to a single trade association – UKOOA.

With suppliers, the situation was entirely different. Given fluctuations in activity and the prevalence of subcontracting, it is impossible to know exactly how many firms were involved at any one time. Some firms became permanently and totally dedicated to the industry, whereas others were involved only to a marginal extent and/or intermittently. At the bottom of the supply chain, it was even possible that firms could be involved without their knowledge.

Nevertheless, it soon became clear that the number of British suppliers was very large. Jenkin (p. 17) quoted a 1974 government estimate of nearly 3000, of which 55 were major contractors, 800 major sub-contractors and 2000 supporting suppliers. There was no homogeneity of function among the suppliers. At one end stood the designers, fabricators, and installers of offshore structures and at the other the providers of catering and cleaning services. In between were a myriad of suppliers of standard industrial goods and services, logistics operators and a relatively small number of businesses engaged in the subsea and subsurface areas, including the manufacture of equipment and provision of subsea, drilling and well services. The latter could be decidedly 'high-tech' in character and the spectrum of technological sophistication ran across the supply sector from the highest to the most basic. All in all, the offshore supplies market proved to be a 'market' only in the narrow sense of having a common group of ultimate end-users, the oil and gas Operators.

Although a close relationship eventually developed between OSO and many major suppliers, particularly in politically high-profile activities where credible British contenders existed, suppliers generally were under no compulsion to contact OSO and many never did. However, after the introduction of monitoring and the signing of the MoU, OSO sometimes had incentives to initiate contact itself.

Monitoring compelled OSO to become acquainted with difficulties oil companies faced with their suppliers, though OSO's ability to intervene directly was very limited. At various times, price, delivery, industrial relations and steel supplies were all issues, with delivery dates being the issue to which the oil companies and governments attached the greatest importance. By developing direct relations with trade unions as well as management, OSO helped both sides focus on the importance of meeting delivery dates as a means of winning further contracts. As a result, in the author's opinion, industrial relations within firms in the supply sector, although not perfect, were generally better than those in comparable, longer-established industries.

A traditional channel of communication between government and industry is the trade association. Though it was gradually to change through greater familiarity by all parties and the coming into being of new more focussed bodies – of which the Module Constructors Association (or MCA) was among the first – this did not function efficiently for OSO in its early days. The heterogeneity of the offshore supplies industry was the cause, as various established trade associations took on board the offshore interests of their existing memberships.

As a result, Jenkin (p. 17) identified some 16 trade associations with a significant number of members involved with the offshore industry. Ten attended a meeting on FFO with UKOOA, which noted that the most protectionist trade association representative (from the Process Plant Association) did not receive 'much support' from his peers.

OSO tried to focus its attention on the more responsive and relevant associations, such as the long-established (see p. 41) CBMPE/EIC, the Association of British Oceanic Industries (ABOI) and the British Marine Equipment Council (BMEC). It was nonetheless not much relieved of the need for direct contact with individual firms. An additional complication was that some associations, CBMPE being a prime example, had many U.S.-owned members.

In an attempt to assess the attitude of supply firms to OSO and their level of contact with it, in 1976, Jenkin carried out a postal questionnaire survey of the membership of an anonymous trade association. A hundred responses were received, mainly from small and medium sized enterprises or SMEs (Jenkin pp. 136–145). Jenkin, while admitting that his survey could not be regarded as statistically representative, concluded that a large number of firms had had contact with OSO. They mainly sought market intelligence, introductions to oil companies and FFO. The level of 'customer satisfaction' was not high, which Jenkin – unlike the respondents who blamed a lack of experience and expertise on the part of OSO staff – attributed to its limited powers.

Jenkin's conclusions on why companies contacted OSO can be compared with what OSO's management itself thought. The originator of an internal OSO memorandum of January 1975 wrote of the 'steady work-load' of enquiries from firms interested in supplying the offshore market. He went on to list seven headings under which most enquiries could be classified. Two related to marketing, one to material supplies and one to technical standards. The remaining three lay in the areas where Jenkin had found little demand – licenses/joint ventures, government financial assistance schemes and R&D support – TNA: PRO EG10/64.

As to the criticisms made by Jenkin and his survey respondents of OSO's effectiveness, the author would make three observations. First, 1976 was still very early in OSO's history and staff inexperience was unavoidable. Second, few beyond those directly involved would know of OSO's successful interventions with Operators or of the ability of its Ministers occasionally to threaten sanctions to bring a recalcitrant Operator 'into line' with government policy. Third, it was too early to judge the effectiveness of applying FFO and UK content criteria to licence awards.

During its early formative years, OSO developed close relations with a small number of large British-based supply and contracting businesses. Contact was infrequent, if not absent, with the majority of the medium sized and small companies, which vastly predominated numerically. Overall the number of companies contacting OSO probably declined with time, reflecting not only 'market saturation', but also the fact that the North Sea was widely seen as having lost its 'growth status', with doubts about the longevity of the domestic offshore industry re-surfacing.

As the market turned down from its first cyclical peak in the mid-1970s (see p. 111), excess capacity emerged and margins as well as volumes declined. A number of companies, particularly in the design, fabrication and mobile drilling sectors, were faced with the possibility of being forced out of the offshore business altogether unless they could obtain further remunerative contracts. Inevitably, there were accusations that foreign competitors (particularly Continental) were in receipt of subsidies, leading in 1979 to OSO commissioning an external report, which failed to find conclusive evidence of subsidies.

In some cases, companies decided voluntarily to withdraw from parts of the offshore business. Prominent examples included McAlpine Seatank (concrete platforms), Laing Offshore (steel platforms) (Figure 5.3), Foster Wheeler–John Brown (modules), and Salvesen (drill-ships). In other cases, such as Weldit (specialist fabrications) the companies failed. Often, companies in difficulties turned to OSO for increased support. This was especially true of the fabrication business, one of particular political resonance, since the yards concerned tended to employ large numbers of unionised workers in areas with little alternative employment. OSO – with considerable Ministerial input – was remarkably successful in ensuring that the 'lion's share' of such fabrication work as was available went to UK yards, which increased their share of orders in this sector from 63% in 1977 (Department of Energy 1978 'Blue Book') to 84% in 1979 (Department of Energy 1980 'Brown Book').

In drilling, there was much less OSO could do. The Salvesen vessels were unsuitable for the North Sea. The companies operating new-build semi-submersibles were inexperienced and highly geared new entrants to a shrinking market, offering equipment little in demand outside the North Sea. The incumbents

FIGURE 5.3   End of an Era: Thistle jacket at Laing Offshore Yard. Reproduced with permission of Energy Institute and BP Archive.

were on the whole more experienced, financially stronger and frequently held term contracts let in an earlier period of stronger demand. Even with FFO, the British companies were seriously disadvantaged when bidding for those contracts that did become available. A policy much stronger than FFO, which would focus not only on increasing the chance of a British contract award but which could also ensure a cash break-even rate, was needed to help the drillers; it was not available.

The result was consolidation. Since it took place among British companies, OSO had no reason to become involved. In 1978, Ben Line became sole owner of Atlantic Drilling's two rigs and the following year BP acquired the rig it was already chartering from Celtic Drilling (Jamieson p. 140). At 40% Norwegian owned Kingsnorth Marine Drilling (also with two rigs), there was a struggle for control between KCA and Houlder, culminating in 1980 in the latter assuming both management and voting control.

In late 1978 or early 1979, OSO correctly foresaw the prospect of an up-turn in mobile drilling. The number of E&A well starts nearly quadrupled between 1979 and 1984, (see Chart 4.2, p. 102). In an attempt to stimulate British mobile drilling capacity, the Minister sought to enlist the co-operation of the state-owned BNOC and BGC. BNOC's chairman (Frank Kearton) was supportive. His opposite number at BGC (Denis Rooke) was not, claiming the British industry lacked entrepreneurial leadership. Though not formally abandoned, the initiative languished.

The depressed home market focused attention on export opportunities, although in the mid to late 1970s both the nature of the British service and supply industry and the then relatively limited overseas demand for the 'harsh environment' products it offered, restricted what might be achieved. Nevertheless, OSO – again with considerable Ministerial support – worked closely with the trade associations, prominent suppliers, British diplomatic posts and the British Overseas Trade Board (BOTB) to generate more export business. A major problem with many foreign buyers was to convince them that it made sense to deal with British suppliers rather than go direct to the U.S. industry, not helped by the fact companies participating in the missions were sometimes U.S.-owned.

The greatest export success of the period was probably for the Brazilian offshore development programme being undertaken by its state oil company, Petrobras. CJB Offshore won a structural design contract. Worley Engineering (a William Press subsidiary) was awarded a major topside design contract and McDermott (Scotland) a construction contract for a large steel jacket. Whilst the design contracts were successfully executed, the jacket sank 'en route', showing the hazards of long-distance international trade in such products.

Many British companies had routinely sold into the Norwegian market but increasingly faced protectionist barriers in the oil sector, not solely due to the growing role of Statoil, although this became ever more important in 'strategic' areas. During the late 1970s, at a succession of routine meetings between the British minister and his Norwegian equivalent, the same item always figured on the agenda – the use of a Norwegian language test to exclude Scottish electricians working on the contract which the Glasgow company, James Scott (another Press subsidiary), held from Phillips Petroleum on its Ekofisk complex. Another 'bone of contention' was the Norwegian requirement for British supply boats to carry a Norwegian pilot when on the Norwegian Shelf; the UK did not require a British pilot on foreign ships on its Shelf. British rig owners also complained that the Norwegians used national regulations to exclude them. There was little beyond raising such issues that OSO could do for British suppliers in the light of the FCO's unwillingness to support a more robust response.

Despite the closure of the Laing and McAlpine yards, it still seemed as though the British civil engineering industry would establish itself successfully in the offshore scene. In the contracting sector, three companies in particular stood out. One was George Wimpey with early exposure to the sector, a platform-building joint venture with B&R (Hi-Fab) as well as supply boat and offshore survey activities. The second was Taylor Woodrow, a potential builder of concrete platforms, with a successful project management joint venture, Taywood-Santa Fé, and for a time, management control of logistics base and offshore support vessel (OSV) operator, Seaforth Maritime. The third was TH, strongly established in module fabrication, but with a volatile record in other areas ranging from geophysics and drilling to supply boats and design engineering. Of the consulting engineers, the most committed were probably Halcrow and Atkins. Though Lloyds Register remained important, its early prominence faded, perhaps due to the loss of key personnel and increased competition.

None of these companies had refocused their interests on the North Sea as wholeheartedly as did other long-established businesses such as Mathew Hall (topside design), William Press (fabrication, craft labour supply and topside design) and John Brown. Although the last withdrew from module fabrication and exported most of the GE gas turbines it manufactured, it became a force in North Sea design. In addition to topside work, it was in joint venture with U.S. company Earl & Wright, becoming the first British company to be recognised for its offshore structural design capability. It also established a short-lived underwater design consultancy, CXJB, in association with French diving firm Comex.

5.5.1 Creation of New British Enterprises

Of the 3000 or so British-based firms supplying the UKCS, the great majority were existing businesses, for which offshore sales represented an incremental market. The proportion specifically set up to serve the offshore market is unknown, but the author's assessment is that it was probably less than a third, including new subsidiaries of established businesses, joint ventures, inward investments and true start-ups. It is probable that the peak rate of new entrant formation occurred in 1971–1976, between the initial northern basin development decisions and the first cyclical demand downturn.

The first 2 years of this period pre-dated OSO. It is clear from the *IMEG Report* and other sources that there was a high rate of enterprise formation during these 2 years. Since both OSO itself and the surge in company formation were in response to the early northern basin oil discoveries, it is fitting to include new enterprise formation in 1971 and 1972 in this section.

The *IMEG Report* (1972, p. 35) noted that there were already in 1972 five groups with sites or facilities for the construction of large steel platforms and that only one (J. Ray McDermott) did not have British participation. In the related area of module fabrication, it had little to say beyond mentioning that William Press had added U.S. 'know-how', probably meaning its association with Worley Engineering in the topside design field. Several more module construction facilities were created or expanded, among them Whessoe, RDL and Cleveland Bridge and Engineering on the Tees, John Brown on the Clyde, Press's new yard and Charlton Leslie on the Tyne and Burntisland Engineers and Fabricators (BEFL) on the Firth of Forth. BEFL was a partially owned subsidiary of Dundee shipbuilders, Robb Caledon. The minority shareholders were a London merchant bank (Baring Brothers), a London shipping company (Ellerman Lines) and an Aberdeen motor vehicle distributor (J. G. Barrack).

Another area where existing British businesses were active was logistics. In the case of Scottish supply bases, port owners either moved into supply base operation themselves or more commonly leased facilities to oil companies or specialist providers. In Aberdeen (where the local Harbour Board invested heavily), a fishing company – the John Wood Group – set up a supply base and utilised its existing facilities to offer a repair and maintenance service for oil equipment, initially in a joint venture with the Weir Group. Shell, Seaforth Maritime and the Aberdeen Service Company (ASco) also established bases in Aberdeen. ASco set up additionally in Peterhead, eventually becoming the largest supply base operator in the entire North Sea. Its parent, Sidlaw Industries, an old Dundee jute company, also invested in the Skean Dhu hotels, a business targeting oil industry personnel.

Shipping companies became involved in new enterprises. In 1971, Ocean Transport and Trading of Liverpool formed a 50/50 joint venture (Ocean Inchcape or OIL) with the Inchcape Group to operate North Sea supply boats and a base in the Shetlands. A year later, two Glasgow shipping companies, Lyle and Hogarth, established Seaforth Maritime as a supply boat and base operator. They were joined, as minority shareholders by other Scottish investors such as James Finlay and the Bank of Scotland, as well as the specialist investment trust, NSA. From 1973 another Glasgow ship owner, Harrison (Clyde), operated North Sea supply vessels through its Stirling Shipping subsidiary.

A more ambitious entrant was Star Offshore Services. After discussions involving stockbrokers Cazenove, the Blue Star shipping line, United Towing (the UK's only deep sea tug operator) and an experienced offshore manager, it was launched in 1974 with an initial equity capital of £10 million (over £78 million in 2008 terms), mainly subscribed by some 200 institutional investors. It was a rare example of such a funding approach. Blue Star and United Towing had small stakes, and initially provided management services. Star quickly developed a fleet of supply boats, anchor-handling tugs and diving support vessels. After entering the diving services business through acquisition, it became the first British company to undertake saturation diving (Wilson Committee 1978, pp. 55–56).

The first British shipping company to become involved in offshore drilling, Edinburgh-based Christian Salvesen, established Salvesen Offshore Drilling in 1971. Strangely, it selected as equipment moored drill-ships unsuited to the North Sea. However, with its Salvesen casing crews, the company also entered another 'core' oil service field, an activity at the 'low-tech' end of drilling services but one capable of deployment in the North Sea. Later, it added coiled tubing services.

During 1972, London-based Furness Withy, through its Houlder Brothers subsidiary, entered the mobile rig business by taking a 20% interest in Kingsnorth Marine Drilling, which ordered two new semi-submersibles; among the other shareholders was Berry Wiggins, an entrepreneurial Kent-based company, whose subsidiaries came to include both BW Muds (drilling mud) and KCA Drilling (mainly involved in platform drilling).

In 1974, Edinburgh shipping company Ben Line entered the offshore drilling business through the creation of Ben Ocean Drilling and Exploration Company (Ben ODECO). ODECO – a major component of the U.S. offshore drilling industry – sold a 50% interest in its UK subsidiary, which owned a Clydebank built jack-up, *Ocean Tide*, to Ben Line Offshore Contractors (BLOC). Ben Line owned 60% of BLOC, the remainder being held by NSA and the Royal Bank of Scotland. Ben ODECO soon went on to acquire from ODECO the conventional drill-ship *Ben Ocean Typhoon* and to order from Scott Lithgow on the lower Clyde, the sophisticated dynamically DP drill-ship *Ben Ocean Lancer*. This represented a very major investment £24 million (around £148 million in 2008 terms), of which 87.5% was borrowed from the Royal Bank (Wilson Committee 1978 p. 41). In 1976, Ben Line further extended its interest in the drilling business by acquiring Sheaf Steam Shipping of Newcastle, which had an interest in two semi-submersible new builds via its investment in Atlantic Drilling. The co-investor was Cardiff ship owner, Reardon Smith, which had also ordered a new-build semi-submersible for its Celtic Drilling subsidiary.

The UK's most famous ship owner, P&O, abandoned its own plan to enter drilling in early 1975, pulling out of an international consortium planning to acquire leading U.S. drill-ship operator Global Marine, as a direct result of deteriorating economic conditions in the UK (Simmons, 1979).

Although new enterprise formation and the ensuing increase in UK content was clearly not dependent on OSO activity, both specific OSO initiatives and general venture management activity added additional impetus. The years 1973–1976 marked the highpoint of joint venture formation, OSO's main

TABLE 5.1 UK–U.S. Joint Ventures with Probable OSO Involvement

British Partner	U.S. Partner	Activity	UK Location
Ferranti	TRW	Subsea controls	Edinburgh
John Brown	Earl & Wright	Platform design	London
Hattersley, Newman Hender	McEvoy	Subsea well heads	Gloucestershire
Marconi Avionics (GEC)	Koomey	Subsea controls	Rochester
Taylor Woodrow	Santa Fé	Project management	London

means of encouraging inward technology transfer. Some examples where the author believes there was an OSO involvement in an Anglo–U.S. joint venture are shown in Table 5.1.

OSO did not originate all Anglo–American joint ventures; one prominent one that it did not was the diving company 2W Diving, where experienced British managers provided personnel to complement saturation diving equipment owned by Taylor Diving and Salvage, part of the B&R/Halliburton Group. While it was for long widely believed in the UK that the ownership interests were roughly 50/50, in fact the American partner initially owned 90% and the British (Wharton Williams Limited) only 10% (Swann, p. 560). The joint venture became very successful and absorbed its U.S. partner in 1985, becoming Wharton Williams Taylor. By then British ownership interest had '*significantly increased*' (Swann p. 561).

Not all joint UK-North American ventures were with U.S. firms, though the majority were. English China Clays began its drilling mud subsidiary, International Drilling Fluids, as a joint venture with a Canadian partner. Energy Resource Consultants (ERC) – the first British-controlled reservoir engineering consultancy – began as a co-operation between Imperial College academics and a Canadian firm. British Oxygen Company (BOC) had a joint venture with a Canadian firm, Nowsco Well Services.

However, it is probable that the second largest group of joint venture partnerships in this period involved the French, including both of the firms that built concrete platforms in the UK (McAlpine Seatank and Howard Doris). One of the four enterprises that built large steel jackets in the period (Laing) had French support. Houlder Brothers provided marine support to the UK subsidiary of the French diving firm Comex. In 1979, Houlder acquired a 50% interest in the Comex 's UK subsidiary, which latter became known as Comex Houlder Diving Limited, as well as a 16% share in the Comex Group holding company. According to Swann (p. 555), the total cost was some $24 million (approaching $60 million in 2008 terms).

Whether or not any of the earlier Anglo–French ventures resulted from the determined pre-OSO attempt by IFP to partner French oilfield technology with British companies is not known. However, it can be safely said that the London investment company Flextech was certainly a consequence. IFP had enlisted the help of 'blue-chip' City institutions, such as Baring Brothers and Cazenove. Having failed to find a British industrial company prepared to take a stake in Coflexip, IFP's flexible subsea pipeline 'spin-off', Cazenove raised funds from City institutions to establish Flextech as an investor in oilfield service and supply companies, with a 17% equity stake in Coflexip as its first holding.

OSO was not overly concerned with the activities of foreign entrepreneurs setting up British businesses. Ramco (oilfield tubular inspection and maintenance) was one such with an American promoter, while Rig Design Services (specialist design) and Arunta (a supply base in Peterhead) both had Australian founders.

At this stage, OSO was mainly concerned with ensuring that the main high value and high-profile capital items, such as platform jackets and modules were sourced from UK facilities. Government funding was often involved through regional aid. OSO's focus on fabrication reflected the political imperative as much as issues of industrial strategy; here it could make the most immediate and largest impact on UK content and employment. Nevertheless, there could also be a reasonable expectation that the UK was capable of developing a broadly competitive position in this sector and indeed for many years this proved to be the case. This was less true of concrete platform facilities, where direct government investment led to over-provision.

Expansion of the fabrication industry led to an increase in demand from specialist providers and sub-contractors in fields such as welding supplies, stress relieving heat treatment, non-destructive testing (NDT) and inspection. Mostly, this demand was met by existing businesses. However, new firms also came into being, serving new demand niches. One such was Weldit of Scunthorpe – a classic owner-managed business – which became an indispensable element in the steel platform-building supply chain, providing a back-up source of critical structural elements. Another specialist small entrepreneurial company was Bruce Anchor, based on its deep-water anchor design; it was a rare early example of a British company with a defensible patent position.

Meanwhile, parts of Scotland were benefiting from inward investment by U.S. oilfield equipment suppliers. This is made clear in a 1974 interdepartmental briefing. It refers to Smith International and Vetco Offshore setting up in Aberdeen, where Baker Oil Tools was to expand its labour force from 85 to 400–500. FMC Corporation had announced its intention to set up in Dunfermline – TNA: PRO INF 12/1298.

The initial years of OSO saw more activity in the supply sector by specialist investment institutions than was the case in the rest of the period covered by this book, though they did not confine their investments to British companies. Flextech and NSA have already been mentioned. Another was New Court

Natural Resources (a Rothschild vehicle), notable for backing Expro, a British drilling and well services start-up in 1973.

Of these, NSA had the 'deepest pockets' but initially chose to invest more heavily in foreign than in British service and supply companies. Shortly after its foundation in late 1972, it announced its intention to take a 20% interest (costing U.S. $25 million, or around $100 million in 2008 terms) in the *Viking Piper* semi-submersible pipelay barge (Figure 5.4). The Bank of Scotland and Norwegian and French groups held the balance. The vessel proved highly effective when laying the Ninian to Sullom Voe pipeline, but thereafter suffered long periods of unprofitable operation as a construction support vessel, or of idleness. Other significant 'non-British' investments were made in Oceaneering International (a diving company) and Northern Offshore (a support vessel operator). Among unsuccessful UK investments was Marine Oil Industry Repair and Maintenance (MOIRA), a rig repair venture.

NSA had as its investment manger the Edinburgh financial house Ivory and Sime described by Hall (1973) as being at the centre of '*an intricate financial web*', involving the Bank of Scotland, Noble Grossart, Edward Bates, and various investment trusts. A lower profile financier of offshore supply and service sector in 1973–1976 was the mining finance house Selection Trust. It acquired the crane ship *Thor*, which it leased to Heerema for 10 years.

Regional aid apart, central government was not an important source of finance. Jenkin (pp. 125–128) showed that over the 1973–1976 period, OSO made only four grants of selective financial aid under the 1972 Industry

FIGURE 5.4   The *Viking Piper* Pipelay Barge. Courtesy Dr. J. Bevan.

Act out of twenty-four applications processed. He attributed this low proportion to the difficulties surrounding the application process, such as its bureaucratic nature (with 14 decision points) and consequent slowness, the small sums on offer and differences in approach between OSO and the IDU, which administered the grants.

The total aid approved totalled £3.369 million (say £23 million in 2008 terms). Over half was a loan (never fully utilised) to Seaforth Maritime, enabling it to complete two ships under construction when the (British) builder went into receivership. IRGs were made to MOIRA (see above), which failed, and for *Ben Ocean Lancer* (see p. 117), which was successfully completed, although late. The fourth payment was a loan for the launch of a new flow meter by a firm that subsequently went into receivership. No further grants of selective financial assistance were made between 1977 and 1980 period, mainly because suitable cases did not come forward. After the change from a Labour to a Conservative government in 1979, use of selective financial assistance was in any case out of favour.

Industry Act funding had been of no help to OSO in addressing one of its major strategic objectives – the creation of a British capability in offshore construction. In the second half of 1975, OSO was still considering, as recommended in the previous year by the Select Committee (1974, p. 22), the creation of a major UK offshore capability, possibly to be named as the British Offshore Construction Corporation – TNA: PRO EG10/64. It was almost certainly too late by then to found a venture with much chance of success. Firms from the USA, the Netherlands, France, and Italy were already in the field. A BP witness to the Select Committee (1974, p. 17) pointed out that the Continental entrants to pipelay and heavy lift were no larger than some British contractors, but cumulatively they had made large capital commitments to what was likely to be a cyclical market with its first construction peak already in sight.

Indeed, a downturn in rates in 1976 was forecast by Department of Energy *et al* (1976c, p. 115), which had also noted that some vessel operators would be moving equipment from the North Sea, which they saw no longer attractive for investment. Their reasons included uncertainty over future North Sea demand, the very high capital cost of capital equipment for North Sea use, with limited market opportunities for its use in other areas, and a background of rapid technological change, which increased the risk that investment might not be recovered (Department of Energy et al 1976c, p. 80). The opportunity had passed. The initiative was allowed to lapse.

In the other strategically key area, offshore drilling, the outlook appeared brighter. KCA Drilling had been awarded a large development drilling contract for BP's Forties field, almost certainly with OSO backing. Several private British ventures in mobile rigs had been established, adding to the rigs owned by Shell and BP and in August 1975 an OSO Press Release could claim that '... *there should shortly be eight large semi-submersible drilling rigs and four drill-ships substantially under UK ownership*'.

No mention was made of the fragmented control of this 'fleet', exposing the individual enterprises to dependence on one or two contracts in contrast to their larger and more broadly based U.S. competitors with a portfolio of rigs and a spread of contract maturity dates and rates, of growing European competition, of the lack of operational experience or of the heavy reliance on debt. When the E&A market turned down, as it soon did for the rest of the decade (see Chart 4.2, p. 102), these weaknesses were to be cruelly exposed.

During this period many large established companies set up divisions or subsidiaries whose title included words such as 'energy', 'offshore', or 'oil', intended to imply commitment. In some cases, this was genuine, as in the case of P&O Energy, with interests in supply boats, a supply base and subsea services. In others, it was merely an attempt to co-ordinate fragmented marketing efforts within a diverse group – an example being ICI Offshore. Sometimes offshore activities were under the charge of a knowledgeable executive with a clear vision, but who was unable to have it 'championed' at main board level where capital allocation decisions were made. This was recognised by the Select Committee, which had noted that the main boards of some British companies did not always back the offshore enterprises within their groups with capital investment (Select Committee 1974, p. 22).

Only a strong-willed and committed leader, some of the more notable of whom are shown in Table 5.2 which follows overleaf, could take a business successfully into the offshore field.

The above list is by no means exhaustive, but the names shown were all involved with businesses that came to have more than a purely local or short-term significance. Interestingly, over a third of those listed were involved in the underwater sector.

Given the risks involved, the difficult financial circumstances of the time and the perceived novelty of the offshore business, the evident scarcity of main board 'champions' in public companies is not surprising. In more recent times, with its greater availability of venture and development capital (see p. 91), this situation would probably have led to management buy-outs or the foundation of more dedicated start-ups along the lines of Star. Such new enterprises were disadvantaged in that unlike profitable established businesses they could not offset start-up losses against other profits or demonstrate an existing cash flow and asset base. These factors gave a bias towards building the new domestically owned offshore supplies industry on the pre-existing industrial structure, including sectors suffering severely from the industrial problems discussed on pages 10–29.

No doubt influenced by the relatively favourable financing conditions of 1972–1976, the Wilson Committee (1978, p. 44) found '... *no evidence to suggest*' that a shortage of funds had generally inhibited British involvement in the offshore supplies market, although it did admit some '... *potential opportunities involving large initial risk*' had not found financial backing. It failed to recognise that these '*opportunities*' included OSO's initial priorities of establishing a large British presence in offshore drilling and installation.

TABLE 5.2 Some Early British Offshore Champions and Entrepreneurs

Company	Status	Individual	Activities
Aberglen/ Balmoral	Private company	Jimmy Milne	Buoyancy products
ASco (Aberdeen Service Co)	Private company	Jimmy Simpson	Supply base
Berry Wiggins	Public company	Paul Bristol	Drilling, drilling mud, inspection and testing
Bristow Helicopters	Private company	Alan Bristow	Helicopter Services
Bruce Anchor	Private company	Peter Bruce	Anchors
Expro	Private company	John Trewhella	Drilling and well services
Furness Withy	Public company	'Uncle' John Houlder	Drilling, diving support, maintenance and repair
Global Diving	Private company	Martin Deaner	Diving, diving equipment
Oceonics	Public company	Bob Aird	Underwater equipment hire
OSEL	Private company	Doug Hampson	Underwater vehicles
Petrocon	Public company	Peter Hodgson	Pipes and valves
SLP Engineering	Private company	Dennis Abbott	Manpower agency, fabrication
Star Offshore	Private company	Brigadier Parker	Supply boats, diving support, diving
Trafalgar House	Public company	Nigel Broackes	Fabrication, supply boats
Sonardyne	Private company	Dave Partridge	Underwater acoustic equipment
SB Offshore	Private company	Steve Buxton	Fabrication, supply base
UMEL	Private	Mike Borrow	Underwater vehicles and equipment
Subsea Surveys	Private company	Roger Chapman	Underwater services
Wharton Williams	Private company	Ric Wharton	Diving
Weldit	Private company	Gordon Cryne	Fabrication
John Wood	Private company	Ian Wood	Supply base, maintenance and repair
Vickers	Public company	Leonard Redshaw	Underwater services and equipment; design

The Committee recognised that the commercial banks were the main source of short and medium-term finance to the industry. It was also noted that direct government financial assistance had so far been small. While generally commenting favourably upon investment activity by financial institutions and on Stock Exchange listings, the Wilson Committee (1978) did not attempt to quantify the funds resulting for the offshore supplies industry. In the author's view, they would have been modest compared to identifiable bank lending, put at £277 million for August 1977 (over £1.3 billion in 2008 terms).

For ship and rig owning companies, the banks could also facilitate access to longer-term and cheaper finance via the Ship Mortgage Finance Corporation (SMFC) or its foreign equivalents. The relatively easy availability of bank finance (including 'off-balance sheet' finance leases) compared to equity led to many cases of new enterprises becoming more highly geared than was prudent in a volatile market. This was particularly true of businesses to which ship mortgage finance was available. The small equity element in the funding of the *Ben Ocean Lancer* has already been mentioned (see p. 148). In the case of Kingsnorth Marine Drilling, the entire enterprise rather than a single unit seems to have been funded in a broadly similar manner. Unfortunately for such British new entrants, their competitors were mostly established foreign firms with balance sheets already 'fattened' by retained profits built up during earlier periods of successful trading in strong markets.

During this period of generally tight supply, both Shell and BP continued to invest in service provision, maintaining sole and/or joint interest in drilling rigs. Shell also held for a time a stake in an Italian diving company, Subsea Oilfield Services (SSOS), an affiliate of the Italian offshore construction group, Micoperi (IMEG 1972, p. 36). BP owned supply vessels and had an interest in Strongwork, a UK diving company.

In early 1974, BP established its own in-house venture capital unit, BP Ventures. Though its remit was not confined to offshore activities, the fragmentary information contained in the BP Archive shows that it was active in the offshore field. Investment proposals rejected included a crane barge joint venture with Heerema, a rig repair facility and the subsea production system in part developed by BP in its ADMA operation (see p. 40). It did, however, invest in oil spill control (Vikoma) and in underwater services, where a major move came late in 1975. At the suggestion of ODECO, the controlling shareholder, which felt in need of a UK partner, BP Ventures acquired a 47.5% interest in Subsea International (SSI), a U.S. diving company active in the North Sea and valued by its owners at $20 million (nearly $65 million in 2008 terms).

In mid-1977, it was proposed that BP acquire Sonamarine, an operator of remotely operated vehicles (ROVs), and merge it with VOL, thereby creating a 50:50 joint venture with the Vickers Group, capitalised at £14.362 million (nearly £68 million in 2008 values), of which debt would have represented nearly 80%. In the event, Sonamarine was bought but the Vickers deal never materialised. Subsequently, Sonamarine would be merged with SSI, which in turn ultimately reverted to full U.S. ownership.

Differences between the1973–1976 and 1977–1980 sub-periods were very marked, with British companies now leaving sectors perceived as over-supplied or otherwise un-remunerative. To some extent, their places were taken by Continental firms, which concluded that their chances of winning work in a declining market would be improved if they operated from within the UK.

This happened particularly in the fabrication and drilling sectors. In 1979, the well-established French fabricator, Union Industrielle et d' Entreprise (UIE) acquired the old John Brown shipyard at Clydebank from the U.S. rig builder, Marathon. It successfully achieved its aim of adding more general offshore fabrication to the yard's activities. This was not its owner's first interest in UK acquisitions; in 1974 the author had considered KCA Drilling as a possible target.

Also in 1979, the Dutch fabricator, de Groot, the original pioneer of North Sea offshore fabrication (see p. 53), became a shareholder in BSC's RDL (North Sea) platform construction yard at Methil, renamed Redpath de Groot Caledonian (RGC). Having Continental partners was not a novelty for RDL. In 1972, three Italian companies had held a combined 45% interest, with the (unfulfilled) intention of creating a turnkey contractor with design, construction and installation capability – see Select Committee (1974, Minutes of Evidence pp. 301, 311).

Two Norwegian drilling companies established British entities in the late 1970s. In mobile drilling, ship owner Jebsen brought two of its existing fleet of semi-submersibles under the British flag, to trade as Jebsen Drilling (UK) Limited. Another Norwegian driller, Smedvig, chose a different route, forming a platform drilling joint venture company, Dan Smedvig, with Davies and Newman, a British shipping and aviation firm.

Norwegian entrepreneur, Fred Olsen, was responsible for several investments in the UK offshore sector. These included British-flagged mobile drilling rigs, probably part of the basis for joint venturing with the Vickers Offshore Engineering Group (VOEG) to establish Vickers Aker. Although this never obtained an operational contract, it is believed to have been the world's first business specifically dedicated to floating production and undertook some of the development work for the world's first purpose-built mono-hull floating production vessel (FPV), BP's single well offshore production system (SWOPS). Other, wholly owned businesses, also set up were a fabrication yard in the Hebrides (Lewis Offshore, later sold to Dutch firm Heerema) and Aker Offshore Contracting (AOC), which provided platform hook-up and maintenance services, being ultimately acquired by B&R.

Despite the upsurge from Norway, the formation of new joint ventures and the establishment of new businesses, whether British or foreign controlled, was less during the second half of the 1970s than earlier in the period. Although some activity continued in the equipment field (e.g. the establishment of a joint venture between Australian crane maker Favco and Northern Engineering Industries of Newcastle), it tended to focus on services, particularly underwater services. A new British-owned saturation diving company (Global Diving Services) emerged, as well as small British-owned ROV operators. Prominent among these was R. R. Chapman Subsea Surveys Limited (usually known as Subsea Surveys or SSS) specialising in pipeline surveys and in which the NEB held 47.2% of the equity (its first offshore sector investment; Wilson Committee 1977b, Volume 1 p. 77). At much the same time, the SDA (later Scottish Enterprise or SEn) made its own first offshore supply investment, in a flange bolt manufacturer. It contributed £60,000 (about £240,000 in 2002 terms) for a 33% interest (Wilson Committee, 1977b, Volume 6, p. 174).

The arrival of limited public sector equity coincided with a decline in activity on the part of private venture capitalists. For the 'generalists' the combination of a cyclical demand downturn, high entry thresholds and continued rapid technological change, presented an unattractive picture. Among the few specialists, NSA, the largest such investor, was pre-occupied with its heavy exposure to the *Viking Piper* lay barge. Another quoted investment trust, East of Scotland Onshore, could develop a portfolio of only small investments. Flextech, however, added a major British interest to its existing French one by acquiring control of Expro.

OSO itself continued in the late 1970s to seek ways to involve British industry in the elusive offshore installation and pipelay fields. It made an unsuccessful approach to interest Dutch heavy-lift entrepreneur, Pieter Heerema in a British joint venture, following the announcement of his firm's order for two semi-submersible crane barges, so large and stable that they would revolutionise module and deck design while also greatly enlarging the installation 'weather-window'. Heerema correctly concluded that having a British partner would be of any real value to him. A tentative discussion also took place with the Norwegian government about the possibility a joint attempt to ensure that the *Viking Piper* stayed out of U.S.-control but this too quickly died against a weak demand background.

5.5.2 Some Supply Sector Corporate Issues

The depressed market conditions of late 1970s and the scarcity of new equity funding meant that OSO was in these years more often concerned with the problems of existing British-owned companies, sometimes extending to their very survival, than about the creation of new enterprises. Sometimes it could help but often it could not, or occasionally would not. There was also the ever-present issue of overseas take-overs.

Against this unpromising backdrop of the late 1970s, the 'traditional' activities of OSO's venture managers declined and their numbers were allowed to fall. They did, however, become involved in trying to find a 'British solution' when foreign take-over bids arose for what were seen as strategically significant

British service and supply companies or promising innovations. Sometimes such approaches were easily deterred. On one occasion a U.S. corporation approached the DEn to enquire whether a bid for Mathew Hall – by then a major force in topside design engineering – would be unwelcome. On being told that it would, the bidder withdrew.

On other occasions, the prospective foreign purchaser (at this time almost invariably American) would not give prior notification of its intention but nevertheless required exchange control approval before the transaction could be completed. OSO's understanding with the Bank of England (see p. 117) allowed approval to be withheld until the venture managers had exhausted the prospect of finding an alternative UK buyer. Jenkin (p. 24) mentions such a case where the U.S. takeover '... *of a promising UK diving company which had a series of new products under development*' was prevented.

Although it was an underwater equipment manufacturer rather than a 'diving' company, the British firm concerned was probably Underwater and Marine Equipment Limited (UMEL). Its 'Jim' suit (a one-man atmospheric diving suit, or ADS- see Figure 5.5) was of interest to Oceaneering, a major U.S. provider of diving services in the North Sea and elsewhere. If so, Jenkin's account is perhaps too flattering towards OSO's efforts to thwart Oceaneering. OSO certainly attempted to prevent control of 'Jim' (development of which had been partly financed by a loan from the NRDC) passing abroad and succeeded in delaying that outcome while it

FIGURE 5.5   'Jim' atmospheric diving suit. Courtesy Mr. M. Borrow.

sought a 'British solution'. It was unsuccessful and in 1977 Oceaneering acquired a 50% shareholding in UMEL and the exclusive global rights to market 'Jim' to the offshore oil and gas industry. It chose not to sell 'Jim' to other underwater service providers. An element of 'British face-saving' was achieved by the acquisition of the remaining 50% of UMEL shares by NSA, the Edinburgh investment trust, itself also a shareholder in Oceaneering.

Two other British companies went on to develop ADS equipment – OSEL and Slingsby Engineering, which built its prototype with the support of a DTI product development grant. In 1985, Slingsby acquired UMEL, restoring British-control for a few years before Slingsby in turn succumbed to foreign ownership. Although the ADS approach to deep-water work initially looked attractive, it did not succeed in replacing saturation diving. As time went, both technologies were increasingly displaced by ROVs.

Generally, OSO was not successful in attempts to find a 'British solution' and for good reasons. Too few British offshore suppliers had the market credibility successfully to launch innovatory products and services and those that did were often constrained by capital scarcity and/or risk aversion.

There remained, nonetheless, a certain amount of corporate activity not initiated by OSO among large British companies, particularly those focusing upon the emerging IMR market. Thus Taylor Woodrow and James Findlay jointly took control of Seaforth Maritime with this in view, freeing Lyle Shipping (a founder shareholder of Seaforth) to develop an independent offshore strategy through the Lyle Offshore Group (LOG). There also continued to be smaller and longer-lasting specialist initiatives, such as the establishment in 1978 by Plenty (a pumps and filters subsidiary of the Booker Group) of Oil Plus, a water injection and produced water treatment consultancy. Oil Plus was originally a joint venture with BP, which sold its shares in 1982.

An example of a case where OSO was asked to intervene but was unable to was when John Trewhella, founder of Expro, sought OSO assistance in locating funding so that he could exercise his personal pre-emption rights in order to prevent control of the company passing to KCA. As long as this remained a strictly British issue, OSO's position had to be one of neutrality and it declined to become involved.

A case where OSO would have liked to help but could not, was that of Kingsnorth Marine Drilling. The company was a heavily indebted newcomer without a track record in a market experiencing a sharp cyclical downturn. The best that OSO could do was to insist that the company's name was on the bid lists; it was not enough. Fortunately, a 'British solution' emerged when a stronger British company, Houlder Offshore, took control (see also p. 145). OSO was even less able to assist another British drilling venture, Salvesen, whose equipment was unsuitable for use in North Sea. Salvesen resolved the problem by disposing of its drill-ships.

It should not be thought that OSO was never able to intervene constructively. Interactions with Seaforth Maritime, Weldit, and Vickers, showed otherwise.

As a Scottish company created specifically to serve the North Sea, Seaforth was a business OSO tried hard to support. OSO financial support has already been mentioned (see p. 152). The final vessel delivered under the financial rescue package, the *Seaforth Clansman*, was completed as a diving support vessel with a fire fighting capability, with the intention of avoiding a crowded supply boat sector (Jamieson p. 149). Unfortunately, the diving support market in turn developed over-capacity as construction activity declined whilst other (mainly foreign) owners also entered the market with higher specification vessels, either semi-submersibles (of which Houlder's *Uncle John* was the first) or DP monohulls. Faced with the damaging consequences of a lack of employment for the *Seaforth Clansman*, the company turned to again OSO for help. OSO persuaded the MoD to charter the ship to support Royal Naval diving operations. Freed from this potential liability, the company was able to go on developing, further diversifying into manufacture of saturation diving equipment and other projects. Recognising the company's potential in the emerging IMR market, Taylor Woodrow and James Findlay bought out the existing shareholders and – underpinned for a time by the Shell *Stadive* MSV contract (see p. 129 and Figure 5.6) – it remained under UK control until acquired by B&R in the late 1980s.

Weldit was an entirely different case. The main business was highly efficient and became for a time a critical component of the heavy fabrication supply chain. The owner (Gordon Cryne) became a 'tax exile' and developed a 'mini-conglomerate', openly owning air charter and electronic security businesses and clandestinely, it was alleged, other small 'high-tech' companies. The company had a highly developed 'market intelligence' system, a by-product of which was (unfounded) accusations from oil companies that OSO had 'leaked' commercially sensitive information.

The company was personally managed and internally financed, supplemented by bank overdrafts. In order to further improve its high productivity, the company invested heavily in the most up-to-date equipment. On one occasion, the company sought an increase of £2 million (about £8 million in 2008 terms) in its overdraft to finance the import of new machinery, committing to the purchase on the basis of what the owner believed was a verbal 'green light' from its bank, the Midland. However, the increase was not forthcoming, the company breached its borrowing limits and the bank sought to appoint a receiver. On confirmation from OSO that Weldit stood on critical development paths and that several oil companies valued its services highly, the Bank of England (see pp. 116–117) intervened to prevent the receivership and found an alternative bank in Barclays. Under pressure from the bank and OSO, Cryne agreed to the employment of management consultants, professional managers and external equity. The extent to which this programme was ever fully implemented is unclear, but within a year or so of the 'rescue' Barclays had itself appointed a receiver.

At the time of its failure (1980), Weldit was heavily engaged on the construction of the platform jackets for BNOC's Beatrice field. The work-in-progress was removed by Edinburgh entrepreneur, Steve Buxton from

FIGURE 5.6   *Stadive* MSV. Courtesy Shell International Limited.

Weldit's Scunthorpe works to a site using an itinerant labour force at Barrow-in-Furness, where under a highly 'incentivised' contract, it was completed early. With the profits generated from the contract, Buxton acquired the Arunta supply base at Peterhead, subsequently selling it on to ASco.

To the extent that OSO ever had a 'chosen vehicle', it was Vickers (see p. 140), whose activities for a time opened the possibility of British leadership in the fields of underwater engineering and floating/tethered production systems, seen from the late 1960s onwards as representing the long-term future of the offshore industry. It was not the only naval shipbuilder to diversify into the offshore industry. Yarrow, through its Y-ARD subsidiary and Vosper-Thorneycroft with David Brown Vosper Offshore (DBVO) also did so, but had narrower technology bases and never matched Vickers in importance.

Vickers had suffered severely at the hands of government policies in the 1960s and 1970s, not least from the contraction of the demand for traditional armaments. It had also lost its steel making interests as a result of nationalisation. Government-induced rationalisations had eliminated its interests in computing while leaving it with a 50% stake in the British Aircraft Corporation (BAC). Nevertheless, it still entered the 1970s as one of the major elements of the British engineering industry, with extensive overseas interests. Its activities mostly fell within either its Shipbuilding or its Engineering Groups. It held, along with Foster Wheeler, an interest in a nuclear submarine propulsion system joint venture led by RR. There was also a corporate R&D unit, to which a contract R&D business (International Research and Development or IRD) was added.

Engineering Group had facilities across the UK whereas Shipbuilding had become very much Barrow-in-Furness focussed. At the start of the 1970s, the vast Barrow complex was all brought within the control of Shipbuilding's management, which at much the same time acquired Slingsby, a glider manufacturer in North Yorkshire, with the intention of introducing its composite materials technology into submarines. Engineering meanwhile was nurturing a new business in medical equipment, with the intention of developing a major Vickers Medical Group.

While parts of Vickers took small incremental steps into the offshore industry, such as fabrication of heavy structural components at various locations and the supply of riser tensioning and heave compensating equipment for drill-ships from Brown Brothers, the main entry point was centrally initiated in the late 1960s, an event in which the author was personally involved. The period coincided with the initial development of the southern North Sea gas fields and U.S. navy's 'Sealab' underwater habitat programme.

The writer brought to the attention of top management the possible opportunities arising from North Sea gas production for Vickers to enter what promised to be a growing new market, first in natural gas transmission equipment, an opening not pursued. The writer's second attempt – offshore and underwater engineering where there was a strong existing technology base, particularly submarine building and life support systems – was more favourably received. It resulted in the formation of an intra-Group working party. This never proceeded very far because Sir Leonard Redshaw, Shipbuilding's managing director at the time, evidently decided that any new venture in this field would be entirely within his Group. This was a decision difficult to dispute, as Shipbuilding then operated with a very high degree of autonomy. It was the principal profit contributor to Vickers, largely as a result of its prime contractor role in the 'Polaris' nuclear submarine programme. Dependent on U.S. technology, the vessels were extremely sophisticated for the time and gave access to advanced underwater navigation and positioning systems and 'state of the art' project management techniques.

Having left Vickers in 1969, the writer had no further direct involvement with it for nearly a decade, but 1972 saw the launch of VOL, one of the first British ventures in oilfield subsea services. The *IMEG Report* (1972, pp. 80, 81, 91, 96) written in the same year made several generally favourable mentions of VOL and recommended increased financial support. Two years later, Vickers executives gave a much fuller account of the venture as it then stood to the Select Committee (1974, Minutes of Evidence pp. 187–206). Like IMEG, the Select Committee was impressed by what it learnt. First, VOL had been founded with an initial capital of £1.9 million (nearly £15 million in 2008 terms) and a further £2.5 million (nearly £20 million in 2008 terms) had been approved; the initial ownership was Vickers 63%, NRDC 26.5% and James Fisher & Sons 10.5%.). In these early years, VOL had been able to generate a regular income and obtain valuable operational experience by recovering torpedoes on a naval trials range. Though VOL's progress might depend on the extent of government funding, it did not see NRDC as a further source of capital and indeed Vickers soon became sole owner.

VOL was then operating from bases in Barrow and Leith, three submersibles, built in Canada (Figure 5.7) by a company initially supported by Vickers, together with two 'mother' ships. Potential markets had been investigated by IMEG. As part of heavy on-going R&D expenditure, VOL was also proposing – in co-operation with Oceaneering International of the USA – to develop a 'diver lock-out' service based on a fourth submersible under construction in the USA. 'Lock-out' involved transporting a diver in saturation state to and from an underwater work-site by submersible, which also provided life support and

FIGURE 5.7 *Pisces 2* VOL-operated Canadian-built submersible. Courtesy Mr. M. Byham.

diver observation. It was perceived as safer than surface support and somewhat less vulnerable to poor sea states.

In November 1972, the Minister for Aerospace and Shipping had invited Vickers to put forward suggestions for projects aimed at establishing a substantial role for British industry in support of offshore, particularly underwater, operations. The response had been three proposals.

Subsequently, one proposal (*'use of a nuclear power plant on the seabed'*) had been abandoned. With respect to another, *'laying and burial of pipes on the seabed,'* as a result of a positive report undertaken by CJB and BP, a proposal had been submitted to the Shipbuilding and Marine Technology Requirements Board (SMTRB) but evidently was not accepted. As far as the remaining proposal for *'an underwater operations command ship'* was concerned, whereas Vickers was looking for a £2 million (over £19 million in 2008 terms) loan, an offer of only £350,000 to £500,000 was made, with unacceptable conditions attached. Vickers had decided to proceed on its own. It is probable that the vessel concerned was the conversion of the *Vickers Voyager*, for diver 'lock-out' service. The Vickers representatives noted that whereas DTI support could work effectively with small proposals, it was unable to respond to larger ones.

Finally, the Vickers representatives made it clear that Group activity in the offshore market was not confined to VOL. Shipbuilding Group itself was involved in design and development work on underwater equipment as an extension to its military submarine work. Engineering Group, Brown Brothers, IRD and Slingsby were also mentioned. The role of Slingsby in particular could well have been explained more fully. It was being developed as a specialist developer and builder of underwater vehicles based on composite material hulls, reliability-critical underwater components and systems and underwater communications. The representatives were probably unaware that BAC, an associate company of Vickers, was also developing an ROV.

Vickers did well in the first phase of northern basin construction, which peaked in 1976–1977, with VOL gaining credibility, particularly in pipeline route survey and inspection. In engineering design and R&D; an impressive client list was acquired, including BP, Shell, Esso and Statoil, as well as DEn and the EC. However, as the market contracted, the narrowness of the commercial business and the early stage of the venture, left Vickers very exposed. To a shrinking market were added problems of severely increased competition, with another British company, P&O Subsea, as well two French ones, Comex and Intersub, entering the market over a period of about 2 years.

The Vickers management also made some serious errors. It failed to recognise that new technological developments meant that the manned submersible market, with its high cost base, was unlikely to survive long. The cost-saving and safety advantages of the ROV and its great potential for technical advances were not accepted by senior management as a serious threat to manned submersibles, with grave consequences. The arrival of DP diving support vessels deploying their saturation divers from 'state of the art' bells showed them to be more cost-effective than diver 'lock-out'. Vickers found itself committed to the expensive conversion of the *Vickers Voyager*, which over-ran both in terms of time and budget, with little prospect of being able to trade out of a misjudged investment. Meanwhile, there was too little investment in other areas, such as survey equipment.

In 1977, to avoid their nationalisation, Vickers had placed its dedicated offshore interests into a separate VOEG, comprising:

- VOL – manned submersibles and support ships, Leith.
- Vickers Offshore Projects and Developments (VOPD), including Vickers Aker – designers and consultants, Barrow and London.
- Vickers Slingsby – manufacturer of aerospace and underwater equipment based on composite materials (Kirbymoorside).
- Vickers Underwater Pipeline Engineering (VUPE) – development company for deep-water pipeline repair (Barrow).
- Vickers Intertech – development company involved in encapsulation of subsea production equipment (Liverpool).
- Brown Brothers and Hasties– manufacturers of marine steering gear and drilling rig ancillaries (Leith and Greenock).

By 1977–1978, VOL was losing money heavily. There was no offsetting profit stream from the engineering activities, then mostly still in an early development phase. Overall, losses were running at about £7 million per annum (about £33 million in 2008 terms), see Smith (1982). The initial driving force of the business, Sir Leonard Redshaw had retired.

At the Group level, there was an even bigger problem to face – the loss of its two main profit earners, naval shipbuilding and heavy engineering at Barrow and the 50% interest in BAC due to their nationalisation in 1977, damage compounded by a delay to the compensation settlement. The weakened financial state of the parent provided some explanation of its subsequent increased risk-aversion.

Although Vickers had successfully prevented the nationalisation of its offshore activities, shipbuilding nationalisation had serious consequences. VOL was able to concentrate its activities at Leith and Slingsby remained in North Yorkshire. The balance of the business was mainly 'marooned' within the BS complex at Barrow, with its prospective heavy manufacturing capacity now in the ownership of a separate organisation. Inevitably, dealing with these problems diverted much management time from the market place and complicated issues like dealing with the UMC and TLP (see pp. 130–133).

There were also many changes in management. Indeed, poor management was an important component in the downfall of VOEG, with misjudgement of the market a particularly serious failing. By late 1977, it was clear to OSO that VOEG was in difficulty, leading to a meeting between the DEn's PUS, the writer and the managing director and the deputy-chairman of the Vickers Group. While admitting that there were problems, Vickers expressed confidence that they were manageable, declining an offer of help from the DEn. It emerged separately that an offer of investment by the NEB had also been rejected.

In April 1978, Vickers requested another meeting to advise OSO that the Vickers was quitting the offshore business, the manufacturing elements apart. This news posed an immediate and serious problem for OSO, since it meant the prospective loss of the most credible British contender in both underwater engineering and services and floating production. Moreover, the business had been the largest recipient of public sector R&D money in these fields.

The immediate OSO response was to provide a list of potential purchasers for the business or parts of it (Smith 1982). The list included, along with the state-owned NEB and BS, foreign as well as British companies, though all parties professed to favour a 'British solution'. It did not include BP Ventures, of whose former interest in VOL (see p. 155) OSO's management was unaware.

The eventual outcome was complex and time-consuming. Brown Brothers and Hasties were quickly transferred from VOEG to another part of the Vickers Group. VOPD was soon acquired by BS, where it did not in the long-term prosper. The fate of the other units depended on the outcome of negotiations between Vickers and the NEB (the only serious potential buyer). OSO was not party to these negotiations, although it did second a senior member of its R&D branch to the NEB as technical advisor, see (Smith 1982). One of his roles was to make NEB aware of the importance of including Slingsby, which Vickers wished to retain, in any deal since it was the ultimate repository of most of VOEG's exploitable proprietary technology as well as having a sound ongoing business. Since Vickers was anxious to shed the heavily loss-making VOL (by now non-operational) and the R&D companies, it eventually agreed to relinquish Slingsby (Figure 5.8).

A new company, British Underwater Engineering Limited (BUE), was formed to acquire most of the remaining activities, although VUPE was not initially included due to the interest of Weldit. BUE's initial owners were the NEB (89%), B&R (10%) and Wharton Williams (1%). The initial capital of BUE was £6 million (about £22 million in 2008 terms), with an undertaking that this would be doubled as expansion opportunities became available. The capital structure was very complicated and provided for Wharton Williams to 'earn' a 24.5% stake; B&R had an option to acquire a further 14.5% stake from the NEB (Smith 1982).

The presence of B&R reflected NEB's perception that it needed a partner already well established in the offshore industry. OSO was given no opportunity to put forward UK alternatives. At the time, OSO viewed the involvement of Wharton Williams positively, as means of providing a link with the diving sector (then about 80% of the underwater services business) through the 2W joint venture, although 2W had close commercial links with Intersub, a French manned submersible operator. The NEB appointed Ric Wharton of Wharton Williams as BUE's first chairman and gave his company a management contract, with B&R also seconding management personnel (Smith 1982).

FIGURE 5.8   Slingsby-built LR class submersible. Courtesy Dr. J. Bevan.

By the end of 1979, it was clear that the initial arrangements were unworkable, principally due to conflicts of interest. The NEB consequently bought out Wharton Williams.

In mid-1980, having left OSO earlier that year, the writer became Chairman of BUE, in a general management restructuring. The future development and eventual demise of BUE is dealt with in Chapter 7.

These events did not entirely eliminate Vickers from offshore industry activity, which continued in both manufacturing and in engineering design. More important was the dispersal of the former Barrow labour force, at the time one of the most specialised and highly qualified groups of subsea engineering specialists in the world. Some left the industry but many took advantage of high

TABLE 5.3 Companies Founded by Former Vickers Personnel During the 1970s and 1980s

Company	Activity	Key Individual
Douglas-Westwood	Market research consultancy	John Westwood
Furness Underwater Eng.	Subsea engineering	Peter Redshaw
HMB Subwork	Underwater services	Peter Messervy
Hoad Design	Subsea and specialist design	Kevin Hunt
Mentor Engineering Conslts	Subsea production system design	David Pridden
Orcina	Specialist subsea consultancy	Mike Isherwood
Orcina Cable Protection	Cable and flexible pipe protection	Orcina spin-out
Remote Marine Systems	Subsea connectors and robotics	Gordon Robertson
RUMIC	Underwater services	Roger Chapman
Subsea Surveys	Underwater services	Roger Chapman
System Technologies	Underwater sensors and tools	Marcus Cardew
Tronic	Underwater electrical connectors	John Alcock
Ulvertech	Underwater acoustic equipment	George Colquhoun

demand for their skills from oil companies, engineering consultants and designers (including those in Norway), often rising to senior positions. Others became entrepreneurs, the Vickers restructuring *'eventually produced no less than 20 specialised companies serving the offshore oil and gas industry'* based in the Barrow area (Cross 1986, p. 2).

A dozen or so such companies, some very small and not all located in or near Barrow, are identified above in Table 5.3 Some did not survive long, but the Barrow area has remained an important centre of specialist subsea oil and gas industry expertise.

Chapter 6

OSO's Long March into History 1981–1993

OSO as originally conceived existed for just short of 20 years. Its role in UKCS procurement and licensing policy was brought to an end in anticipation of the European Single Market Act of 1992 and the introduction of European Union (EU) Directives for Utilities and Services. Like the rest of DEn, OSO was reabsorbed in 1992 into the DTI, where for a time a distinct organisation continued some OSO functions – particularly in the field of export promotion and to a lesser extent R&D support. In 1994, OSO's remit was expanded to include the downstream oil and gas and petrochemicals industries and it was renamed the Oil and Gas Projects and Supplies Office, though it continued to use OSO acronym. In 1997, OSO ceased to exist as a separate organisation and was absorbed within a broader DTI export promotion unit known as the Infrastructure and Energy Projects Directorate (IEP). Use of OSO acronym was discontinued within a year or two. Thus, 1993 is no more than a convenient termination point of the main narrative about an organisation that had already lost its 'raison d'être' but would still linger on the scene for a little longer. Symbolically, it also happens to be 30 years, conventionally a generation, after UKCS exploration began.

During the decade or so from 1981 until the early 1990s OSO underwent further modest organisational changes and staff numbers fell by more than 30%. The political and economic environment within which it operated and its relations with other arms of government changed substantially as did its international stance.

Ronnie Custis (a career civil servant) was in post as DG when the period began but was soon succeeded by John D'Ancona (another career civil servant), who remained in the position until 1994. The position of Deputy DG (formerly Industrial Director) was abolished, along with the surviving venture manager post. The effective 'number two' became the Director of Engineering, directly overseeing both Audit and Industrial Capability. Until 1991, Bill Allison, originally appointed in 1976, held the position of Director of Engineering until 1991. There was, therefore, a marked change in OSO management style.

Norman J. Smith, The Sea of Lost Opportunity.

There were to be no more private sector recruits and a stable – some might say static – top management team superseded a regularly changing one.

The importance of the Industrial Capability Section grew. It became responsible for Quarterly Returns, the residual Monitoring function, financial assistance under the Industry Acts and some residual aspects of the IRG scheme, as well as continuing to provide assistance for the Audit, Export and R&D units. With support from the accountant employed by the Administration Branch, it took on the former functions of the venture managers. It was much involved in pre-Annex B approval activities; as early as 1981, its analyses were enabling Ministers to ask at the time of approval for indications of prospective UK content. Later, the linkage of Annex B approvals to UK content became formalised, with oil companies giving UK content undertakings to Ministers. An analysis of the potential for British suppliers in a particular Annex B became the basis for direct discussion between oil companies and OSO top management. It also brought OSO closer to PED and thus integrated more tightly into DEn.

During 1983–1987, OSO operated more aggressively in the domestic market than at any time since its foundation. This mainly reflected industry conditions but partly the character and objectives of the then MoS – Alick Buchanan Smith, who represented a constituency close to Aberdeen. Subsequently, changes in the way oil companies did business and in the relations between Britain and the EC, to say nothing of the growing maturity of the British supply sector, began to undermine OSO's rationale. Activities became increasingly dominated by export promotion work serviced by a reinvigorated Export Branch.

The balance of payments constraint became a thing of the past, thanks to oil revenues, though national economic problems did not. The economic background necessarily changed over so long a period, opening with a fall in inflation accompanied by steeply rising unemployment resulting from so-called 'de-industrialisation' as large parts of British manufacturing were rendered uncompetitive by an appreciation of sterling resulting from the rise of North Sea oil production. An economic recovery began in the mid-1980s, leading at the end of the decade to the 'Lawson boom', ended by an increase in interest rates necessitated by resurgent inflation and currency market turbulence. A sharp recession followed, succeeded at the period's end by the beginning of a sustained growth in the now service-based economy and a steady fall in unemployment.

Privatisations of state assets changed the environment within which OSO operated. BNOC was unwound, culminating in its upstream equity assets being privatised as Britoil, with its residual trading functions being undertaken, until they were wound-up, by the Oil and Pipeline Agency (OPA). Acquired by BP, after the government had disposed of its own large holdings in BP, Britoil's independence was short-lived. BGC was first stripped of its oil assets, with the offshore elements floating on the Stock Market as Enterprise Oil, which became an Operator. The balance of BGC was floated intact but subsequently dismembered under regulatory pressures. Most of the nationalised shipbuilding industry was simply closed but one or two profitable units, notably Vickers at Barrow (subsequently floated on the London Stock Exchange as Vickers Shipbuilding and Engineering – VSEL) were privatised. BAC (subsequently British Aerospace and now BAESytems) was also privatised, in due course taking over the military shipbuilders, VSEL and Yarrow. Prior to privatisation both BSC and the NCB experienced long strikes as their capacities were severely reduced.

The 'real' oil price fell sharply to low points in 1986, from which it recovered only weakly until 1990 when it rose sharply, following Iraq's invasion of Kuwait and the subsequent international intervention, thereafter declining again (see Chart 4.2, p. 102). There was an increase in interest in gas exploration and development, first in the dry gas fields of the southern North Sea and then in the gas condensate fields of the central North Sea as producer prices were raised to more realistic levels and additional Norwegian supplies were denied.

Apart from a 2-year delay in the Clyde field development and in the period immediately following the 1988 Piper Alpha disaster, when safety issues dominated, the government allowed market forces to determine the rate of depletion, with the interests of the domestic supply industry sometimes cited as a reason for favouring rapid exploitation. An account of British depletion policy can be found in Kemp and Stephen (2005).

Oil taxation fell for a decade from 1983 as the government realised that the taxable capacity of the North Sea had been exceeded, risking loss of investment. From 1983 to 1993, companies paying PRT were able to reduce their tax liability by carrying out UKCS exploration, generating until the end of the period a generally upward trend in exploration and appraisal drilling activity. The all-time peak in exploration and appraisal drilling was seen in 1990, largely as a result of BP completing Britoil's drilling obligations (see Chart 4.2, p. 102). After the 1986 price decline, DEn's ability to enforce drilling obligations was weakened. Development drilling, like development as a whole, fluctuated in line with such factors as oil prices and taxation changes.

As predicted, most new discoveries were small and there was no repetition of the run of giant discoveries of the early 1970s. As a result of improved technology, better knowledge of the metocean environment, the emergence of spare capacity in the infrastructure as output from the first generation fields began to decline and a decreasing tax burden, it proved possible to develop much of the large inventory of small discoveries.

Although 'light weight' platforms built of higher quality steel and designed (in an increasingly computerised environment) to more informed criteria than their predecessors and FPVs featured, there was a predominance of small subsea developments 'tied back' to existing infrastructure. When combined with the development of long-reach, horizontal and other advanced drilling techniques as well as improved seismic, to say nothing of the effects of long periods of excess supply capacity, the effect was to reduce the unit capital cost of oil developed, which was further enabled by new types of construction and support vessels.

The demand for large semi-submersible lay barges for trunk pipelines and for heavy-lift vessels declined. For fixed installations, the availability of the ultra heavy-lift vessels conceived in the previous periods allowed more of the labour intensive 'hook-up' work to be carried out onshore, with a marked reduction in cost. Smaller diameter steel pipelines for tie-backs of satellite fields to infrastructure and infield lines were laid by reel barge or substituted by flexible lines laid by specialised monohull vessels. Diving support vessels evolved into light construction or well work-over mode. Less steel was employed and the existence of surplus steel fabrication capacity became more and more apparent.

Developments were often perceived as 'marginal' in economic terms, leading to a quest by the oil companies to seek additional cost reductions. This began in 1984 with a call from Shell Expro for a 15% capital cost reduction, together with indications of how suppliers might achieve this – The Economist Intelligence Unit Limited (1984). By early 1987, the Shell Expro was already claiming to have achieved 30% development cost reductions in the northern North Sea and 15% in the southern North Sea; see Select Committee on Energy (1987, p. 238).

Any immediate results from industry initiatives were masked by cost reductions from market forces following a sharp decline in activity induced by the 1986 oil price slump and the 'knock-on' effects of Piper Alpha disaster 2 years later. This period saw widespread labour force reductions, both in the oil companies and their suppliers.

Partly reflecting reduced technical and management capacity within the Operators and partly to reduce costs, there was a move towards giving a main contractor an EPC (engineering, procurement, and commissioning) or an EPIC (engineering, procurement, installation and commissioning) contract. Such 'turnkey' contracts removed the need for oil companies to transact directly with great numbers of largely small suppliers, an objective in its own right. An approach pioneered by BP and aimed at sharing both risk and reward between the oil companies and their suppliers were development 'alliances' or 'partnering' contracts. The theme of what might be described as 'institutionalised' cost reduction was taken up by the UKOOA membership in aggregate and (probably with less enthusiasm) by the main contractors with the establishment in 1993 of a formal Cost Reduction in the New Era (CRINE) Network, which extended the search for more cost-effective supply into areas such as contract documentation and technical standardisation.

The cost reduction ethos spread from capital expenditure into operations, where the increasing maturity of the province put upward pressure on unit costs, leading to further substitution of oil company staff by cheaper contractor personnel. The effects on the exploration segment were more limited, perhaps because this was plagued by excess capacity and low prices for much of the period.

Whether or not the gains from the 'institutionalised' cost reduction fully reflected the effort put into it is hard to judge. However, a study undertaken for the EC on the sources of reserves gains on the North-West European Continental Shelf (NWECS) in the period 1990–1997 concluded that its contribution was only 7.4%, compared to 37.5% for improved drilling technologies and 22.5% for improved seismic technologies – AEA Technology et al (1999, p. 9). During the course of that study, the CRINE secretariat had – when asked – been unable to substantiate its claim to have reduced costs by 30%, despite the fact this supposed achievement had been reported in several government publications from 1993 onwards, see, for example, Department of Trade and Industry 'Brown Book' (1996, p. 3).

Oil company R&D budgets began declining in 1993 (AEA Technology et al 1999, p. 12). The philosophy appears to have been that the supply chain would provide the new technology, as well as simultaneously providing cost reduction, which many suppliers regarded as being at the expense of their margins.

'Institutionalised' cost reduction had profound effects on the relations between the parties in the industry. It strengthened the position of the large contractors and service companies strong enough to accept more risk and which benefited from larger work scopes relative to smaller companies. The effects on small suppliers, particularly those without strongly differentiated products, were generally negative. Direct contact with Operators was lost and with it a source of vital feedback, while there was pressure on margins from the procuring contractors, in some cases potential competitors, anxious to demonstrate cost-cutting credentials. Such negative effects were far-reaching in their impact because of the predominance of small firms in the supply chain. In 1982, 68% of suppliers employed 50 people or less, according to The Economist Intelligence Unit Limited (1984, p. 27). Consolidation, of which the period saw a great deal, was encouraged by the cost reduction initiatives and new contracting forms as well as by excess capacity.

Increased delegation of procurement to contractors also made life more difficult for OSO. While MoU/CoP required extension of FFO to sub-contracts, contractors were not themselves party to the agreement and it was not always easy for OSO to carry out its role, particularly where a foreign contractor was concerned. Leverage on the oil companies through exploration licensing also declined as the perceived prospectivity of the UKCS became less.

6.1 THE COURSE OF DEMAND

Peaks in 1984 and 1991 (when total expenditure, respectively, reached nearly £15 billion and over £16 billion in 2008 terms) were separated by a trough in the wake of the 1986 oil price collapse, centring on 1987 (when total expenditure was only a little over £10 billion in 2008 terms). Development expenditure was the main source of the cyclicality, although exploration activity also fluctuated. Operating expenditure grew sharply at first, declined following the 1986 price slump to around £4 billion a year in 1987, recovering only slowly prior to declining again on renewed oil price weakness (see Chart 4.1, p. 94).

While offshore oil and gas activity increased around the world, the North Sea remained a key theatre of operations. As late as 1983, the NWECS, though well down on its peak share of around 50%, was estimated still to be employing around 35% of the global resources of offshore equipment and manpower, with the UKCS alone representing about 20%, see Smith Rea Energy/Hoare Govett (1983a, p. 1). Three years later Cameron (p. 27) pointed out that from early in the 1970s, the UKCS had '. . . *been the world's largest single theatre of operations for the offshore industry'* with '. . . *over 30% of the total world offshore market for goods and services'*.

The North Sea also played a prominent role in the development of new 'core' offshore activities, particularly those related to underwater operations. It was estimated that in 1988 the UKCS would account for about 30% of global capital expenditure on subsea production, with the adjacent Norwegian market only slightly smaller, see Smith Rea Energy (1987, p. 17).

The annual 'Brown Books' showed a fluctuating UK content. A low point (67%) was experienced in 1981, whereas the period 1985–1989 inclusive consistently saw figures in excess of 80%. The all-time peak (87%) was achieved in 1987, a year also characterised by the all-time peak, on OSO definitions, in the British content of exploration and appraisal drilling (92%), perhaps reflecting what UKOOA saw as '. . . *increasing pressure'* to use UK rigs.

This pressure from OSO no doubt reflected the parlous state of British drilling contractors as revealed the by the British Rig Owners' Association in its Memorandum of Evidence to a Committee of MPs and the particular interest the latter expressed in British-owned companies – Select Committee (1987, p. 102).

6.2 OSO OPERATIONS IN CONTEXT

Thanks primarily to the oil industry's achievement of removing the UK's balance of payments constraint, but also to the growing competence of the supply industry and its generally stable industrial relations, the Monitoring function had been absorbed first within Audit and then Industrial Capability.

The growth of the service industry in Aberdeen as it began to support production as well as exploration and to generate entrepreneurial 'high-tech' companies, together with the progressive transfer to the city of oil industry jobs previously located in London and elsewhere, amplified the city's importance.

Aberdeen was home also to the SDA/SEn's increasingly ambitious Energy Group. By the early 1980s, this body had come fully to recognise the importance of the Scottish offshore supplies industry and developed its own strategy. It saw helping innovatory Scottish companies, encouraging inward investment and promoting exports as key functions, as they were also to OSO and successfully campaigned for the transfer of more government and oil company managerial functions to Aberdeen. Although OSO expanded in Aberdeen, the expansion probably failed to match the growing importance of the city as a decision centre. Frictions between OSO and the SDA were sometimes unavoidable.

Unlike OSO, the SDA/SEn was able to undertake capital projects, such as the National Hyperbaric Centre and the Drilling and Downhole Technology Centre, both in Aberdeen, and intended to offer local companies shared-use access to experimental facilities they could not themselves afford. Neither was successful in commercial terms. The SDA/SEn also encouraged others such as AEA Petroleum Services (APS) and the British Hydromechanics Research Association (BHRA) to locate test facilities in business parks it had sponsored in Aberdeen. In some cases, it helped small companies by funding product development or very occasionally investing in the company.

OSO's increasing concentration on export promotion demanded close relations with the DTI's export support functions and the overseas posts of the FCO. Complete alignment of objectives between the parties was not always self-evident.

As a result of the impending introduction of the Utilities Directive as part the Single European Market legislation, 1992 saw OSO withdraw the MoU and abandon the collection of procurement statistics. The UK order content figures for 1991, the last published, appeared in 1992's Department of Energy 'Brown Book' (p. 99), showing an overall figure of 78%. Later the same year, DEn (and hence OSO) was reabsorbed into the DTI.

The Utilities and Works Regulations implementing the relevant European Community Directives came into force in 1993, with services added the following year. The new regime introduced transparency through the publication of oil industry calls for tenders and contract awards, banning all forms of differentiation by country of origin. References to benefits to the UK economy and industry had already been removed from licensing criteria in 1986. These events brought to an end a 20-year long bipartisan policy of support for the UK offshore supplies industry, by then itself some 30 years old and thus scarcely any longer an infant industry.

As already mentioned (see p. 169) OSO survived for a few more years, subject to reorganisations and name changes. It continued its roles as a facilitator of exports and R&D, as well as acquiring new ones devoted to UKCS competitiveness and commercially fair opportunity in international markets.

6.3 SOME KEY OSO ISSUES OF THE PERIOD

Except in respect of R&D, OSO gained no new tools in the period but it had to pursue the issues of FFO and UK content in a rapidly changing environment, The rise in importance of the SDA, further difficulties with Norway, increased pressure from the EU, institutionalised cost reduction and radically changing contracting practices all raised important issues for OSO that often had to be addressed against a difficult economic background.

6.3.1 Research and Development

Far reaching changes occurred in this area. Initially, OETB remained the focus. Its expenditure continued to be dominated by the collection of subsurface data and studies into matters such as metal fatigue and the behaviour of concrete in the ocean, related to the DEn's safety and regulatory responsibilities. Only limited support for commercially directed R&D continued to be reported upon in the annual 'Brown Books'. For 1982, it was noted that projects included downhole pumps, fire detection systems, non-destructive testing, ROVs and seabed production equipment. In 1983 the report went further by mentioning the commercial application of a specific product, the 'Hydra-Lok' steel tubular connecting system.

It also referred to seeking to involve Eighth Round licence applicants with UK organisations for research, design and demonstration of new concepts, equipment and techniques ('Brown Book' 1983, pp. 22–25). Since late 1982, the government had sought to impose an R&D criterion for the Eighth and subsequent licensing rounds, with the intention of having new technology developed in the UK and tested on the UKCS. UKOOA papers show that this proposed extension of OSO powers was unwelcome, particularly to Operators with global R&D programmes, as was OSO's intention to introduce an additional MoU. UKOOA agreed to accept the additional criterion, but only in a 'watered down' form, without an additional MoU.

UKOOA resisted OSO's attempt to compel Ninth Licensing Round applicants to engage in discussions about 'R & D aspects' in advance of their applications. UKOOA's concern over the Ninth Round's new technology support criterion continued into 1985. In particular, there was objection to the requirement that companies should undertake R&D '... *anchored in the UK with indigenous British companies'*. The UKOOA view was that it was preferable to make use of capable U.S. subsidiaries established in the UK than to try 'to squeeze them out', while OSO saw further joint ventures with the British partner control as an acceptable compromise. Despite such difficulties, it was noted that OSO discussions with member companies about the new technology criterion had gone well, with operators 'inundated' with approaches from companies, academics and others.

All this new activity imposed a significant additional workload on OSO's R&D Branch. Only BP and BGC maintained substantial British R&D facilities as well as participating in JIPs, commonly undertaken by a supply company or a university. Shell, with its main upstream R&D facility in the Netherlands, also participated in many UK JIPs. Many other companies found themselves in a difficult position. If they had R&D facilities at all, these would be located abroad, needing the appointment of UK 'R&D co-ordinators' whose main tasks were to liaise with OSO and manage participation in UK JIPs.

Whilst the emphasis shifted towards oil-company sponsored R&D, the OETB did not disappear. In 1984, the 'Brown Book' (p. 37–39) noted that the Weir hydraulically-driven pump had been employed in production, a precedent in mentioning a company with an OETB project by name, and that two (underwater) neutrabaric encapsulated well completions had been installed. It also pointed out the long-term nature of development projects, as well as their importance for future exports. There was renewed commitment to underwater technology.

The 1985 and 1986 'Brown Books' did not mention OETB and references to commercially directed R&D were brief. Funding for safety R&D was estimated in the 1985 edition (pp. 34–38) at £3.7 million (about £8.2 million in 2008 terms) whereas that on the subsurface resource base totalled some £8.8 million (over £19 million in 2008 terms), with '*reservoir engineering and enhanced oil recovery*' accounting for about £3.3 million – about £7.3 million in 2008 terms. The 1986 edition (pp. 20, 51, 62) showed that a focus on underwater work continued, with mention of AGUT's role in recommending projects for the Underwater Initiative. Expenditure on safety R&D had risen to £5 million (about £10.7 million in 2008 terms), reckoned to be roughly equivalent to what the offshore industry was spending on the same topics. No overall figure was given for expenditure on the subsurface resource base, probably because the geological data gathering and mapping programmes were by then largely complete. However, EOR expenditure was given as £3.3 million (about £7 million in 2008 terms), with concentration on field specific issues '*resulting in increasing collaboration with the oil industry*'.

As part of the government's response to the 1986 oil price slump, in the 1987 budget, tax relief against PRT liability incurred against R&D expenditure for a specific oil field was extended to PRT relief related to the UKCS. It was easier to demonstrate the required specificity in respect of reservoir related topics than many others. PRT relief against such R&D was to last until 1993, with tax relief potentially in excess of 80%. The same year saw also a House of Commons Report (Select Committee 1987) examining the effects of the oil price collapse on North Sea activity. It praised the efforts of the Minister to give the OETB a higher profile (including assuming the role of Chairman himself, which proved only temporary with the position being subsequently filled by industry figures) and the significance the department attached to R&D both in respect of the future of the North Sea and exports of offshore technology. However, the Select Committee also recognised that fostering supply sector firms demanded more than OETB could provide, specifically mentioning the requirement for training and financial services tailored to the needs of innovative firms in highly competitive markets.

The term OETB returned to the 'Brown Book' in its 1987 edition (p. 66), without explanation for its disappearance or what its budget was. However, it did offer as its revised priorities of subsea systems and equipment, weight and cost reduction through improved design and lighter materials, greater precision in locating and evaluating oil reservoirs, the improvement of cost-effectiveness in drilling and production technologies and increased reserves recovery.

While DEn's reservoir engineering and EOR research programme continued (always outside OSO's control) throughout the period with EOR being re-designated as improved oil recovery (IOR), responsibility for safety, and thus related R&D, was transferred from DEn to the HSE in 1991 in response to the Cullen Report (1990).

The resultant 'stripped-down' OETB was now in effect OSO's R&D vehi-cle. Its priorities now focused on a UKCS where new developments would mostly be small and economically marginally and overall production heavily influenced by the extent more could be 'squeezed' from the large early discov-eries. Many of the early problems associated with the behaviour of materials and the appropriate inspection and repair techniques had by now been solved or were at least well understood.

Of particular interest in the 1987 priorities is the inclusion of subsurface activities at the 'core' of the established international service industry, previously seen as largely 'off limits' to British companies. Their inclusion may in part reflect the arrival of a new generation of British innovators, producing new ideas in old 'core' areas on the basis of experience largely gained on the North Sea. It also recognised that the incumbent foreign firms had probably failed to identify and/ or pursue all of the many technical niches available, sometimes no doubt because they were considered to be too small or speculative. The way was now open to support a larger number of small product or software focused projects.

The 1988 'Brown Book' (pp. 78–80) recorded progress in all priority areas, while making it clear that other high-risk areas were also being supported. Again, no expenditure figures were quoted. The report in the 1989 'Brown Book' (pp. 91–92) had many similarities to that of the previous year.

However, it differed in two important respects. First, it referred to the success of British applicants in winning 24% of the funding available in the 14 round of awards of the EC's programme for Technology Development in the Hydrocarbon Sector (TDHS). Secondly, it reported on the establishment in Edinburgh of the Petroleum Science and Technology Institute (PSTI). Jointly funded by the DEn and a number of oil companies, this was a joint initiative by the Universities of Edinburgh and Heriot Watt. Its very title was an additional confirmation of growing British confidence in areas, mainly subsurface, of gen-eral, rather than specifically offshore, oil industry relevance.

It also distinguished PSTI from the existing university R&D vehicle, MTD, which faced an uncertain future as many of its areas of activity had declined in importance and whose London HQ seemed outmoded as Operators concen-trated operational management in Aberdeen. Both found it necessary to open Aberdeen offices and when the organisations merged (in 1997) as the Centre for Marine and Petroleum Technology (CMPT), it located in Aberdeen.

In addition to listing on-going R&D projects, the 1989 'Brown Book' referred to certain 'infrastructural' developments (where OETB's role was small, if any) such as the creation of The Drilling and Downhole Technology Centre in Aberdeen, and of a national multiphase flow database.

OETB's financial commitments were revealed as very small in the 1990 'Brown Book', which disclosed 1989 expenditure as about £3 million (or nearly £5 million in 2008 terms) – about a third of the budget for a decade earlier (see p. 138). With so little OETB money available, DEn was at pains to point out that most R&D funding came from oil companies, with some also from supply companies. Other public sector funders mentioned were the SERC, the Natural Environment Research Council (NERC) and the EC. With respect to the latter, it was noted that in the 15 TDHS round (the last), British companies had '*again*' received 24% of the available funding. It failed to note that co-operation with an organisation in one or more other Community country was becoming a condition of European funding. In addition to listing on-going projects in the priority areas, the 1990 'Brown Book' (pp. 88–89) again drew attention to a major 'infrastructural' development in Aberdeen. In this case, it was the construction of the EUROPA nuclear logging facility – a commercial initiative by APS.

The 1991 'Brown Book' (pp. 90–91) noted 37 new OETB projects; expenditure had remained at about £3 million (somewhat under £5 million in 2009 terms), in other words slightly falling in real terms. Nevertheless, it claimed evidence of long-term UK success, specifically in subsea technology, by citing UKCS and overseas contracts. Reference was made to a prototype sale of a TIGRESS reservoir and geological analysis software system, then OETB's largest project. Its developer Robertson-ERC, like many beneficiaries of public R&D funding, was eventually destined to fall into foreign ownership. There was also some modification to the priority areas. Weight reduction was dropped and the subsea area divided between production systems and advanced technology for underwater operations. A new subject, '*technology for reservoir appraisal and management*', was added. This included a LINK programme involving UK universities via the PSTI with a service company partner and part-funded by NERC.

The 1992 'Brown Book' (pp. 101–102) reported in similar vein. In 1991, 47 new projects were started, while initial commercialisation was mentioned in respect of geological workstations, crack detection equipment and the detection of shallow gas deposits potentially hazardous to drilling operations. It was noted that a prototype automated drilling rig was being erected at the International Drilling and Downhole Technology Centre in Aberdeen and that favourable review of the OETB programme had been received from external consultants. Expenditure increased to about £3.4 million (over £5 million in 2008 terms). However, it was also noted that OSO had estimated that in 1991 total expenditure on '*offshore R & D*' had been between £145 million and £160 million (between £224 million and £248 million in 2008 terms), with over 85% provided by the private sector.

The oil companies would have been the main providers and probably the largest supply company spenders would have been subsidiaries of foreign-owned firms, particularly Schlumberger, which had opened a major R&D facility at Cambridge in 1985. Of the public sector funding, the OETB was clearly only a small element, overshadowed by EC and other UK government programmes.

A French-led study concluded that in 1984 the UK had invested about $60 million (about $109 million in 2008 terms) in oil-related research, see Association Scientifique et Technique pour Exploitation des Océans, or ASTEO (1986, p. 75). Taking both this estimate for 1984 and OSO's for 1991 at 'face value' suggests that the introduction of R&D into licensing (see p. 176) had boosted spending significantly.

The 1993 'Brown Book' was the last to mention OETB, showing that although 33 new projects were supported in the previous year, expenditure actually fell. The extent to which public sector R&D was now dominated by EC programmes was revealed in that commitments to UK companies in Round Three of the Thermie programme were given as £22 million (nearly £32 million in 2008 terms), with modest additional funding from the Joule 2 programme. Project funding would generally have been spread over 3–4 years and part may have flowed to European collaborators. Nevertheless, it was additional to on-going projects under earlier awards. A working group was established to consider the scope for UK/Norwegian collaboration in the R&D field and renewed support announced for the PSTI. There was an unusually long listing of on-going projects (including nine in subsea production, three in underwater engineering and three in drilling), as well as of new commercialisations. These included an expert system to monitor and control gas turbines, a real time photogrammetry system, a solid seismic streamer for use over ice, an underwater excavator, memory-based logging equipment, and computer equipment to introduce parallel processors to seismic processing. An automated downhole drilling guidance tool was in its demonstration phase and seven LINK projects underway (pp. 110–111).

6.3.2 UKOOA and the Oil Companies

Initially, the smooth working arrangements established in earlier periods continued undisturbed. An addendum (dated 2nd February 1981) was made to the MoU after UKOOA had agreed with OSO that 'Feasibility, Design and Maintenance Studies' were areas of OSO 'special interest', the rationale being that they often influenced the direction of subsequent contracts. The returns themselves were simplified. The cut-off for orders to be included in the quarterly returns remained unaltered at £100,000, but new pre-notification levels were agreed, £250,000 (about £700,000 in 2008 terms) or orders and service contracts and £1 million (about £2.8 million in 2008 terms) for construction contracts. A number of special case exceptions were eliminated.

In late 1982, UKOOA papers show that not only were the oil companies resisting the proposed extension of OSO powers to R&D, but also were becoming aware that OSO was getting involved in Annex B submissions. This last issue was an important element in the difficult relationship that developed between UKOOA, its members and OSO. UKOOA had initially noted that an *'understanding'* on a UK content figure (usually 70%) in new developments had been sought at time of Annex B approval since 1982. By mid-1984, UKOOA members believed that such *'understandings'* were coming to be seen as *'... minimum commitments by operators'*.

Thus, there was scepticism of OSO claiming to regard the understandings *'as "gentlemen's agreements" or "best endeavours" levels'* rather than as commitments. Such scepticism may not have been unrelated to the vigour with which OSO pursued a 70% UK content understanding in respect of Sun Oil's Balmoral development, the most publicly contentious procurement decision in OSO's history and dealt with separately in below. Whether or not it referred to Sun Oil or was more broadly based, there was another new UKOOA complaint voiced – the difficulties experienced by an Operator asked to *'... "reconsider" or "modify" procurement strategy once the competitive process is underway'*.

By the end of the year differences had obviously been patched up and UKOOA could report *'assurances'* from the government that it remained committed to the MoU and CoP. In reaction to the UKOOA complaint about frequent late intervention by Audit Engineers in the bidding process, OSO subsequently accepted it should not interfere with a bid list once agreed.

A more fundamental issue was to arise in 1985. For a second time (the ending of the IRG scheme being the first, see p. 113), the EC forced a major retreat on British policy towards the offshore supplies sector. In response to EC pressure following an initial complaint from the West German mechanical engineering association about the discriminatory nature of the Ninth Licensing Round FFO and R&D criteria, the British government agreed that for subsequent licence rounds no detailed criteria for assessing applications would be published, though Guidance Notes for Applicants were introduced. It also gave assurances that firms from all member states would be treated impartially as required by European law (Cameron p. 53).

As it did not directly relate to the arrangements between the government and UKOOA, the EC intervention resulted in little immediate overall change in OSO domestic policy. OSO make it clear to UKOOA that the FFO policy would not change for 'normal' purchases and nor would the Annex B stage UK content *'gentlemen's agreement/best endeavours understandings.'* Moreover, OSO now required more detailed information from Operators in order to be able to substantiate its UK content statistics as valid in the face of *'considerable criticism'*. Finally, UKOOA was reminded of the political sensitivity of large fabrication contracts being placed entirely abroad.

Nevertheless, despite this 'hard-line' stance, an OSO policy re-assessment was triggered. Resources were increasingly channelled towards exports and

from the Tenth Licensing Round it was intended to enquire what applicants had
done to assist British firms in securing business abroad. The powers of the audit
engineers were severely curbed, but Ministerial intervention at the Annex B
stage compensated. R&D was less of an issue with the EC than procurement
and was still mentioned in the application Guidance Notes for the Fourteenth
Round (1992–1993).

The following year (1986) saw the collapse of the oil price, triggering general reductions in oil company operating expenditure, delays in drilling programmes and development decisions, with consequential widespread distress
in a UK supply industry, still overwhelmingly dependent on its home market.
This is well documented in evidence to a Parliamentary enquiry – Select
Committee (1987).

An OSO response was unavoidable. UKOOA reported that OSO was
'...requesting extremely detailed bid analyses', applying pressure in favour
of UK rigs and introducing new procedures in respect of contracts for supply
and anchor handling vessels. The last mentioned was to become known as
'the supply boat initiative', a major OSO operation which was recorded by
UKOOA as an OSO 'Area of Special Interest' continuing until 31st March
1989. It is more fully considered below.

In 1987, OSO was expressing renewed anxieties '...about the dilution
of the UK content of lump sum contracts' and '...particular interest in
sub-contracts to be let outside the UK by the primary contractor'.

Such concerns were soon to be overshadowed by the first mention (in 1988)
of the issue that was to end FFO and eventually OSO itself, EC activity in
the broad area of 'public procurement'. Both the industry and the British
government (no doubt for different reasons) sought to have the offshore oil
and gas industry excluded from planned directives aimed primarily at utility
and government procurement. Their efforts were to prove unsuccessful.

6.3.3 OSO and Two High Profile Issues

The case of the FPV for Sun Oil's Balmoral field placed OSO operations into an
uncomfortably high public profile. It affected UKOOA member attitudes
towards OSO and possibly those of the EC. Sun was inexperienced as a UKCS
operator and its decision to use an FPV was an unusual one for the time, factors
probably contributing to the acrimony that arose.

The furore over whether or not UK industry had received FFO prior to the
placing of the FPV design and build contracts in Sweden had the unprecedented
result of being subject to enquiry by a Parliamentary committee – Select
Committee on Energy (1984).

The resultant report was brief, consisting mainly of the memoranda submitted, respectively, by Sun and DEn, the former in full and the latter – 'with
reluctance' – minus a commercially confidential annex dealing with '...assur-
ances given by North Sea Sun Oil Ltd' (Select Committee 1984, p. v).

The three key issues were whether or not Sun Oil had or had given British yards FFO to tender, the strength of assurances on UK content given as part of field development approval and the sanctions available to DEn. The Committee diplomatically decided not to adjudicate, since the MoS had already made his own position clear in the Commons on 13th March 1984 by that stating Sun had '... *not given British yards a full and fair opportunity to tender for this work*' (Buchanan Smith 1984a).

As presented to the Select Committee by Sun, the position was entirely different. With respect to FFO, the original list of potential bidders (developed by Bechtel of Great Britain as Sun's management services contractor) contained six British and six Continental yards. None of the Continental yards were prepared to bid and only three British expressions of interest were received – Hi-Fab, Cammell Laird (BS) and Howard Doris/UIE.

When the three British fabricators were asked to bid, Hi-Fab declined. Sun found both the bids it did receive unacceptable on the grounds of price or contractual issues, but continued to negotiate. A design and licence contract meanwhile had been placed with Gotaverken Arendal (GVA) in Sweden, OSO agreeing that there was no British alternative. Sun also requested GVA to provide a price, delivery date and maximum UK content for a Swedish build.

Negotiations with Cammell Laird and Howard Doris continued on the basis of joint bids with GVA and Hi-Fab re-entered the list on the same basis, although an approach from TH was rejected as too late. By 10th February 1984 the field partners, no doubt concerned by the effect these protracted negotiations on the date of first oil from what was perceived as a marginal field, had agreed to place the contract with GVA. The Minister requested on 16th February that Sun did not place the order with GVA until Hi-Fab and Howard Doris (Cammell Laird by now 'out of the running') had had more time to re-bid with GVA as partner. Both re-bids were judged commercially unacceptable by Sun and the contract with GVA was signed on 23rd March 1984.

Between 6th July 1983 and 26th March 1984, there had been 13 meetings between Sun and OSO, five of which involved the Minister. Sun regarded a meeting on 2nd February 1984 (at which the Minister was not present) as of particular importance, since although it was stated that the Minister expected a UK content from Balmoral of at least 70% and that any joint venture must have '... *erection taking place in the UK*' (Select Committee 1984, p. xiv), it was also said that Sun could not be forced to use any contractor or joint venture in which it had no confidence.

On the question of the validity of UK content assurances given by Operators during the field development approval process (70% in this case), the Balmoral affair did not remove ambiguity. According to Sun's memorandum to the Committee, it and its partners did not accept that '... *the assurances constitute a binding agreement*', with the UK content achieved arising from '... *the effort of the operator as directed by the Group, acting in good faith under the relationship between the Group and HMG'* (Select Committee 1984, p. vii). In other words, everything ultimately depended on the competitiveness of British industry.

In its memorandum, DEn put thing differently. Assurances were '... *given under the Code of Practice*' and '... *expected to be honoured'*, although approving a development did not depend upon achieving a particular UK content percentage (Select Committee 1984, p. xvi).

When presenting the industry's view at an OILCO meeting, UKOOA suggested that the Minister seemed to have seen Sun's '... *70% UK content understanding*' as a '*minimum commitment'*.

On the question of sanctions, DEn made it clear that oil companies were well aware of the connection between UK content and both the approval of development plans and the granting of exploration licenses. The Minister had put it more directly, with *The Scotsman* on 22nd March 1984, quoting him as saying future exploration licence and development applications by Sun would "... *obviously be considered in the light of the way that they have performed"* (Buchanan Smith 1984b).

Sun Oil received no exploration licences in the subsequent licensing round, was never again to seek a development approval and eventually withdrew entirely from the UKCS.

Sun expressed confidence to the Committee that the Balmoral development would achieve a UK content of 70%+, partly as a result of a commitment obtained from GVA to achieve '*approximately £40 million in UK content'* (Select Committee 1984, p. xiii). Though it is believed that these objectives were broadly met, this was not without considerable difficulty, as OSO had anticipated in the event that the FPV order was placed outside the UK (Cameron 1986, p. 50). GVA was not party to the MoU and has been accused of using specifications biased against British suppliers.

However, the root cause of the whole problem was probably the unwillingness or inability of UK yards to offer a keen price. This may have been because at least in part, the size of the proposed FPV was such that any UK yard would have needed additional capital investment in the form of civil engineering and new cranes (Select Committee 1984, p. xiii). Clearly, cost apart, there was also a risk that this investment need would also have impacted on delivery times.

On the grounds of technology transfer, OSO persuaded Bechtel to enter into what proved to be a short-term joint venture with the British construction group, John Laing. Alleged discrimination against U.S. companies like Bechtel and the relative lack of success of U.S. applicants in the Ninth Licensing Round led to revival of the U.S. government's opposition to DEn policies, with an official complaint being made in 1985 direct to the Prime Minister by the U.S. Secretary of State (Cameron p. 52). Allegations, however, circulated that Bechtel had itself obtained the Balmoral management contract via a principal-to-principal deal in the USA, overturning the choice of a UK company by Sun's British management.

For the remainder of the period, there were no further new-build FPVs for the UKCS. However, there were cases of vessels being converted to FPVs, starting with the *Rob Roy/Ivanhoe* field vessel for Amerada Hess, successfully carried out on the Tyne, which went on to establish an outstanding reputation for work of this nature.

Another high profile issue for OSO was what OILCO described as '*. . . concern about Norwegian competitiveness for supply ships*'. Such concern was to lead to a second major OSO intervention – its Supply Boat Initiative.

This took the form of a UKOOA/OSO agreement that the offshore support vessel (OSV) sector was to be an 'area of special interest', which lasted from early 1986 until early 1989. All OSV charter awards were made subject to a new weekly reporting system, with OSO having the right to clear all contracts; for spot charters (of 15 days or less) verbal clearance could be granted.

By requiring clearance for all charters, OSO sought to prevent short-term contracts below the £250,000 cut-off for services escaping scrutiny. OSO's aim was to ensure fair competition, rather than 'dumping' by Norwegian owners, to safeguard the survival of what had previously been regarded as a relatively robust sector of the British offshore supplies industry.

The roots of the problem went back a long way, but were much exacerbated by the collapse in demand associated with the oil price trough of 1986. They were presented in the detailed verbal and written evidence submitted by the British Offshore Support Vessels Association (BOSVA) to the Commons committee investigation of the effects of the price fall on North Sea activity – Select Committee (1987, pp. 94–101) and broadly supported by a report jointly commissioned by BOSVA and OSO in 1986 – Coopers & Lybrand (1986).

The main factors involved were the uniquely open nature of the UKCS market, virtually all other countries protecting their own seafarers and vessel owners and the highly supportive fiscal policies of some foreign governments such that building decisions were driven by tax, rather than market, considerations.

OSO initiative was specifically directed at Norway where over-building dated back to the 1970s and had little or no effect on other non-British flags. The number of vessels working on the UKCS (then about two-thirds of the total North Sea market) had fallen from 157 at the end of 1985 to 104 a year later, with Norwegian owners enjoying up to a 30% share while all foreign owners held at best 3% of the Norwegian market (the second largest North Sea market at about a quarter of the total).

Although brought to an end because of UKOOA opposition on grounds of commercial principle, that organisation recorded that the initiative had '*. . . not created any significant operational problems for Operators*'. It was widely attributed with preventing Norwegian domination of the market and marked a change in the previous policy of dealing with the Norwegians by means of dialogue and persuasion rather than anything more vigorous. Needless to say, it was initially greeted with hostility from the FCO, though on this occasion it proved to be short-lived.

6.4 THE SUPPLY INDUSTRY

OSO had to cope with considerable industrial reorganisation and changes in procurement practices before having to abandon its core policies in response to the European Single Market.

In the period 1982–1988, OSO attempted to use FFO more forcibly. First came the initially informal process of extracting a minimum UK content commitment from Operators at the time of field development approval. With the Eighth Licensing Round in 1982 came an attempt to boost development of new technology tied to the UK, perhaps the closest OSO had come since its inception to differentially favouring UK-owned organisations, since foreign-owned ones were less likely to have a UK R&D capability (though it was far from unknown). The Tenth Round introduced the idea that UKCS licence holders should help the export efforts of UK supply and service firms. The Sun Oil affair demonstrated in 1984 that Annex B commitments on UK content would be treated as binding. The notion that an Operator placing a large construction contract entirely abroad during a period of low demand did so at its peril was voiced in 1985. In the depression following the 1986 price slump, OSO worked particularly hard at maintaining utilisation of British supply boats, very publicly, and drilling rigs, less so.

Such activities were often justified by charges of unfair competition in both the UKCS and other markets by foreign suppliers made by companies, trade associations and trade unions. Norway was nearly always prominent in any list of offending countries.

Accusations were rarely supported by hard evidence. An exception was the evidence submitted by the BOSVA to the 1986–1987 Session of the Select Committee on Energy (Select Committee 1987, p. xxv), provoking it to recommend that OSO investigate such claims and take their findings into account in implementing FFO policy. This no doubt made it easier for OSO to run its Supply Boat Initiative.

A few years later in 1990, the government committed itself to ensuring 'Fair Commercial Opportunity' in the domestic and international oilfield supplies markets ('Brown Book' 1991, p. 67). According to the respective 'Brown Books' for 1992 (p. 99) and 1993 (p. 111) OSO pursued 41 specific cases of suspected unfair competition in 1991 and a further 36 in 1992. Thus late in its life, OSO incongruously found itself working in concert with its old 'sparing partner' the EC, in developing more open and competitive markets.

6.4.1 Trade Associations and the SDA/SEn

OSO's task in liaising with its client base continued to be dogged by the fragmentation of representative bodies. Among the 39 organisations whose Memoranda were published by the Select Committee (1987), the largest group (nine) were trade associations, although not all bodies with offshore interests made submissions.

One that did was the British Indigenous Technology Group (BRIT), which potentially represented a challenge to existing government policy by confining its membership to British-owned/controlled companies, with the aims of increasing their participation in both the domestic and international offshore supplies market and strengthening their technology. It was established in April 1984 with about 30 members. Its Chairman was J. Dickson Mabon, the last Labour MoS. By 1986, BRIT's membership had increased to over 90 (Cameron p. 68).

In its evidence to the Select Committee (1987), BRIT implied that it had influenced government policy and had widespread oil company support. Other evidence casts doubt on this. UKOOA clearly decided not to support it and had no doubt been encouraged when in 1985 an OSO representative had stated that BRIT's aim of having its definition of 'Britishness' introduced into the regular procurement process had "... *no support at all from OSO and Government*". Insiders have confirmed that OSO did not look upon BRIT with favour, feeling uncomfortable with a pressure group to which it could not respond because of wider policy constraints.

Although BRIT survived for a number of years, its impact was limited. However, though seemingly more influenced by criticism of OSO statistics on UK content, the Select Committee (1987, p. xxvi) did recommend that future 'Brown Books' should show '... *the performance of UK-owned companies*'. Between 1988 and 1992, 'Brown Books' observed this recommendation, although only in a highly generalised manner, without statistical support or any reference to the frequent loss of British businesses with proprietary technology to foreign control.

Another organisation with which OSO had to come to terms was the SDA/SEn. Having popularised the view that in the mid-1980s, the UK had only about 3% of the export market potentially open to it and contrasting this unfavourably with a supposed 10% market share held by the French service and supply sector, the SDA/SEn devoted much effort in trying to increase this, see its evidence to the Select Committee (1987, pp. 6, 139–140, 143, 263). It was able to employ methods not available to OSO, such as the setting up sales offices in regionally important oil centres from which representatives of Scottish companies could operate with local support. This was particularly helpful for small companies with niche products and services.

6.4.2 The Export Question

OSO did not need the SDA/SEn to emphasise the importance of exports to the future of the British offshore supplies industry as its home market matured. The scale of the opportunity was clearly large. It was, however, difficult to quantify for reasons ranging from defining what sections of expenditure were open to overseas suppliers (which varied from country to country), coping with exchange rate fluctuations and project timetable changes, absence of, or lack of clarity in, accessible information and the definition of market sector boundaries. With respect to the latter, it gradually came to be recognised that in some market segments it made more sense to think in terms of an oilfield service or supply market rather than an offshore service and supply market.

ASTEO (1986, p. 3) concluded that in 1984 global expenditure on offshore E&P had been about $64 billion (about $116,000 billion in 2008 terms), of which about $43 billion (over $78 billion in 2008 terms) would benefit the service and supply sector, a decline of 23% from the turnover that the sector enjoyed in the previous year. No attempt was made to assess the proportion open to full international competition, which the SDA assessed at about a third.

The study further estimated that the 1984 turnover (virtually all exports) of French oil-related industries represented nearly 5% of the global offshore market, a decline from an estimated 8% held at the end of the 1970s. By comparison, it estimated that the 1984 turnover of the British service and supply sector represented approaching 7% of the world market and was entirely absorbed on the domestic market. Since some export sales certainly did exist, the study may have underestimated the British industry's size, but the fact that such a view could be reached tells its own story. As far as Norway was concerned, ASTEO reckoned that the turnover of the local offshore industry for 1984 accounted for over 4% of the total world market, of which exports represented around a sixth (ASTEO, pp. 78, 84, 107).

OSO put great efforts into helping UK suppliers to export, as can be judged from the references in the annual 'Brown Books' from 1985 onward, often showing considerable Ministerial commitment. OSO worked closely with the BOTB, British posts in oil producing countries and trade associations, particularly the EIC. It employed trade missions and participated in exhibitions, as well as maintaining contact with national oil companies and their governments. It used its contacts with oil companies operating in the UK and with the World Bank to organise seminars on potential opportunities for British suppliers. Promotional literature was produced, market surveys undertaken and databases on opportunities setup, with advice readily available to potential exporters.

All this could not overcome the many difficulties faced in making the British offshore supplies industry more export focused. Two of the largest segments of world markets, offshore installation and mobile drilling, were largely closed to the UK due to lack of capacity. There were many small and financially weak companies to which the resource implications of export marketing were daunting, even when an exportable niche product existed. Only Scottish firms of this nature could receive help from the SDA/SEn.

Many larger firms were engaged in fabrication, supply bases or the supply of labour for hook-up, maintenance and repair – activities that did not travel well beyond neighbouring North Sea markets where local competition was fierce. The heavy penetration of foreign-owned companies was a further complication. When there was a corporate decision to service Eastern Hemisphere markets from a UK subsidiary, a relationship with OSO could be mutually beneficial as was the case with U.S. subsea equipment manufacturers such as Cameron, McEvoy, National Supply, and Vetco. Otherwise, although case-by-case co-operation was still possible, conflicts of interest had to be avoided.

The combination of OSO and SDA promotion and support, industrial restructuring and, above al, sheer economic necessity did slowly increase the export orientation of those sectors with internationally tradable products and services. Such sectors were increasing in relative importance with the passage of time as growing experience and maturing R&D programmes fostered product development.

6.4.3 Restructuring at Home

For much of the 1980s, the UK market was more buoyant and active than its U.S. equivalent. A weak oil price-driven contraction of U.S. oil field activity, which began earlier and was more intense than in the UK, led to a consolidation of the U.S. oilfield service sector, with repercussions felt in the UK. Indeed, a substantial UK presence would often be seen as a positive attraction in a merger. Among the consequences was the reorganisation of some U.S. controlled businesses in the UK. The main overall result was reinforcement through acquisitions of the positions of Schlumberger and Halliburton (parent of B&R) as the number one and number two oilfield service companies globally and the emergence of a clear number three – Baker Hughes, born of a merger between Baker Oil Tools and Hughes Tools. All had substantial UK operations.

Early in the period, OSO continued to be criticised for 'unfinished business', particularly the lack of British representation in the offshore platform and pipeline installation fields. Its position was an uneasy one since on the one hand further increases in UK content were held back by the gaps in these areas, whilst on the other it knew the investment 'window' had been closed. An OSO representative emphasised to OILCO in 1982 that the "...*UKCS requirement alone would not warrant the heavy investment*".

This realistic judgement did not prevent further proposals coming forward. In 1984 UKOOA noted that the Minister had welcomed a B&R heavy-lift initiative, which it is understood involved British interests. Shell Expro's then head inferred that the British share was to be 51% and that the Dutch would also be involved (Jennings 1984). It did not materialise and B&R was subsequently to place its semi-submersible pipelay barge into a joint venture with Saipem of Italy (European Marine Contractors, or EMC) and otherwise withdraw from heavy construction in the North Sea. It did, however, become a major force in subsea construction through its Rockwater affiliate, formed in 1990 by the merger of Wharton Williams Taylor (see p. 149) and the marine contracting interests of Smit International.

A little later came a fully domestic initiative, in the form of the *Challenger*, a DP crane ship with a lift capacity of over 4, 000 tonnes completed in 1988 and aimed at a niche market in southern and central North Sea. It was built at Sunderland for ITM of Middlesbrough, previously mainly involved in moving modules, with an OSO brokered finance package. Though the concept may have been sound, it was a 'step too far' for ITM, which had limited resources and no operational credibility in offshore heavy lift. *Challenger* never worked on the UKCS. Ownership of the vessel reverted to BS, which leased it to McDermott for use in the GoM and similar benign environments where it proved very successful.

McDermott also acquired the *Viking Piper* pipelay barge following the failure of an attempt to build and finance a proposed central North Sea gas gathering pipeline system by an all-British group comprising BSC, BUE, Taylor Woodrow and the Bank of Scotland. Had this proceeded, it was intended that Taylor Woodrow should acquire the barge.

After these disappointments, British aspirations in heavy offshore construction were effectively at an end. In view of heavy investment by American, Dutch, French, and Italian companies and changes in market structure diminishing the need for new trunk pipelines and, even more, heavy lift, this was recognition of reality. For a time, the prospects looked more promising in the field of smaller and less expensive vessels required for the maintenance of the existing infrastructure and, increasingly, installation of subsea production equipment. In addition to saturation diving facilities, such vessels had light lifting and, in some cases, well service capabilities. They included both semi-submersibles and monohulls, commonly DP.

A pioneer of such vessels was Houlder Offshore, with the *Uncle John* semi-submersible and the *Orelia* monohull, both dating from the 1970s. In the mid-1980s, BUE entered the field with a multi-function field support vessel, the *British Argyll*. Lyle Offshore Group (LOG) acquired two expensive vessels under construction in the Netherlands when it took over Global Diving. There was also new building by Dutch (Smit), Norwegian (Stolt Nielsen and others) and Swedish (Stena) owners, with inevitable accusations of subsidisation given the poor state of European shipyard order books in the mid-1980s. Oversupply ensued, with the already difficult market conditions deteriorating further after the 1986 price collapse. Before the end of the decade both the BUE and LOG vessels had passed into foreign ownership. Houlder Offshore, Hong Kong controlled since 1980, was sold to the Swedish company Stena Offshore in 1989. At much the same time, Stena also absorbed BUE's diving arm, by then the UK's only remaining significant oilfield diving company.

No UK company attempted to enter the fast developing market for laying flexible pipelines, where the market leader in steel reel lay was Sante Fé, a

U.S. company subsequently acquired first by Kuwaiti interests and then by Stena Offshore. In flexible lines, for many years the Norwegian company Ugland, worked with the French supplier, Coflexip and its Flexservice subsidiary. Ultimately, Stena Offshore was to merge with Coflexip, becoming the general market leader in both the manufacture and installation of flexibles.

Coflexip had been the monopoly supplier of flexible pipelines. However, from the mid 1980s onwards, competitors began to emerge. One of these was the UK's Dunlop, which established a manufacturing facility on the River Tyne. The product failed to achieve market acceptance. The company entered into a joint venture with Coflexip known as DUCO, the facility being converted to manufacture subsea control umbilicals.

Though UK ownership (ROVs apart) was effectively eliminated from subsea support, where demand growth became strong, both Stena Offshore and its main rival, Rockwater, established operational headquarters in Aberdeen. They became the core of a vigorous underwater community in the city, many of the smaller members being locally owned. In 1991, SEn set up a Scottish Subsea Group to provide advice and support to this cluster.

In mobile drilling, the situation also deteriorated, although it took until 1996 before British ownership was totally eliminated. It was easy to see the unattractiveness of this highly capital-intensive business during a period when a cyclical demand pattern and over-supply resulted in long periods of excess capacity.

The position of the fabrication yards was at times as bad as at that of the rig owners and represented a more urgent problem for OSO. The yards and their orders (or lack of them) had a high political profile, given the lack of alternative employment opportunities in the areas where they were located. Since 'Brown Book' figures showed British yards commonly took 80–90% of available orders there was relatively little extra help OSO could offer. The yards' problems stemmed as much from a shift in the nature of demand – away from large northern basin deepwater steel and concrete platforms and their associated decks and modules towards lighter northern basin platforms, small southern basin platforms, underwater structures and eventually FPVs – as from reduced development activity.

In Scotland, the surviving concrete platform builder, Howard Doris collapsed after unsuccessfully attempting to enter the steel fabrication business. On the Tyne, BTR closed both its Charlton Leslie module yard and the former shipyard where it had successfully converted the *Rob Roy/Ivanhoe* FPV.

Conditions in the supply boat market were particularly depressed after the 1986 price crash. OSO was able to intervene with its 'supply boat initiative' (see pp. 146–147). Although OSO intervention may have stabilised the situation, most of the original UK OSV operators sold out to mainly foreign interests in the years that followed. Notable exceptions were Stirling and in particular OIL, which became 100% owned by Ocean and even purchased a major German competitor (Jamieson p. 154). This helped it end the period reviewed as a major global player, though its parent too ultimately sold it out of British control. The George Craig Group of Aberdeen, through its North Star Shipping subsidiary, retained its dominant position in the stand-by/safety boat business.

There is little readily available information on the aggregate financial performance of firms in the sub-sectors of the North Sea market. However, rates of return on 'capital investment' for 11 sub-sectors the years 1985–1989 inclusive were published by Mackay Consultants (1990). The methodology underlying the calculations was not disclosed, beyond indicating that the data were inflation-adjusted and based upon information supplied directly by clients and from company accounts. Although Norwegian and Dutch companies were included, it is probably safe to assume that the main weighting was from companies based in the UK.

Chart 6.1 is based on Mackay's data. It shows that the more successful sectors were those where 'know-how' or proprietary technology rather than capital or labour intensity underpinned the activity, as was the case with the worst performers – rig owners and fabricators. Examples were design engineering, engineering contracting and drilling tools and services. Helicopters are probably an exceptional case, since at the time contracts often included a 'retainer' element, with many fixed assets probably leased.

All sub-sectors enjoyed their peak profitability for the period in 1985, followed by a sharp decline to a trough in 1987, when both Rig Owners/Operators and Fabricators made significant losses, with only a partial recovery by 1989. This trend is much in line with market the conditions experienced at that time.

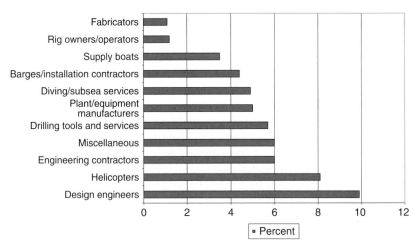

CHART 6.1   Median Rates of Return on Capital Investment (per cent) 1985–1989. Data source: Mackay Consultants (1990) *North Sea Oil & Gas Commentary.*

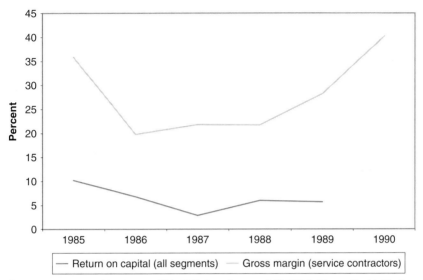

CHART 6.2   Profitability Trends 1985–1990. Data Sources: DTI (1996) 'Brown Book', Mackay Consultants (1990) *North Sea Oil & Gas Commentary.*

Margin statistics relating mainly to drilling and drilling and well services contractors published in some of the later 'Brown Books' show a more immediate response to the 1986 oil price collapse than do Mackay's rates of return, as can be seen in Chart 6.2 above.

Poor financial returns and the difficulties in keeping abreast of changing technology and contractual forms that had the effect of shifting greater responsibility, risk and uncertainty on to contractors and suppliers, encouraged both rationalisation and the withdrawal of many early entrants to the North Sea market.

There was a continued exodus of civil engineering contractors, extending to Taylor Woodrow and eventually Wimpey. Costain remained in low profile engineering role, while TH absorbed the John Brown Group. Fairclough acquired both the William Press and Mathew Hall businesses, thereby creating AMEC, which became the UK leader in the fields of topside engineering design and project management. BP and Shell completed their withdrawal from oilfield service activities, as did such industrial stalwarts as BOC and ICI.

Even in the prosaic field of supply bases, there were to be mergers and foreign take-overs in response to declines in activity, changing contracting practices and margin pressure applied by the Operators. Having already given up its marine operations, Seaforth Maritime was acquired by B&R, possibly in response to the changes in contracting practices. This continued a well-established transfer of British businesses to foreign control, with, at the end of the period, Scandinavians joining the Americans as buyers.

Normally, the motive was to acquire or strengthen a position in a technologically advanced market sector, often by buying out the UK partner. In Smith (1984), six acquisitions are mentioned; four buyers are American, one Canadian and one Norwegian. Three of the deals involved North American principals buying out British partners, one in underwater services, one in subsea completions and one in well services. Of the three 100% sales, two were in well services and the third in seismic processing. Many other sales followed, including those of Salvesen's casing crews and coiled tubing services, completing the exit of another seemingly well-established British early entrants to the offshore supplies industry.

Having reviewed the recent transactions, the author wrote, '...*U.S. buyers place a higher value on oil-related companies of technological or strategic market significance than do their British counterparts*' (Smith 1984, p. 1). By 1984, the government had largely relinquished such powers that it had to restrict take-over activity, while the UK venture capital industry remained largely uninterested in the sector.

There were few, if any, matching outward capital investments in the period, though British companies did use the joint venture route where necessary – in Norway, Canada, or further a field. A few Anglo–American joint ventures were quietly wound up, in some cases leaving the British partner in a position to continue independently. In the subsea control fields, this was the case with GEC Marconi Avionics and Ferranti Offshore Systems, though both later passed under Norwegian control.

Some mergers took place between foreign partners driven in part by difficult conditions in the North Sea as their key market area. An example was the merger in 1992 of French diving contractor Comex Services with the Norwegian underwater support vessel and services contractor Stolt Nielsen Seaway to form Stolt Comex Seaway, which was to join Rockwater (later Subsea 7 following a merger with a Norwegian company) and Stena Offshore as the dominant firms in underwater contracting in the North-West Europe.

OSO had to cope with substantial change resulting from the privatisation and/or rationalisation of the nationalised industries. With BNOC and BGC, the main change was the removal of at least theoretical partners in the development of the British offshore supplies industry. However, greater challenges were undoubtedly faced with BS and BSC, the former wound up in 1989 and the latter privatised in 1987.

BS inherited an antipathy towards the offshore market because of heavy losses on early rig contracts and because subsidies, on which sales increasingly depended, went with ships. Few offshore opportunities were apparent to yard managements, who were generally uninterested in building unconventional vessels with the constraint of fixed prices. The oil company approach of 'on time', 'to specification' and 'to budget' was unappealing to an industry whose main problem was unreliable delivery.

Sites and facilities were usually inappropriate. Tubular rolling capacity and cranage were limited. Semi-submersibles had to be assembled and not launched, Cammel Laird using two dry docks and Scott Lithgow multiple berths. Investment for offshore was piecemeal and mainly at Scott Lithgow and Cammel Laird. Although BS closely observed foreign competitors, the scale of their investments in new facilities was beyond it. BS could only envy Harland & Wolf – state-owned but not part of BS – because of its high level of politically driven funding. Its new yard was the best in the UK and it completed several large offshore contracts for BP. Subject to privatisation in 1989 through a management buy-out supported by Fred Olsen of Norway (Jamieson pp. 77–79), the yard was to remain active throughout the period reviewed.

As seen from OSO, the labour mix at British shipyards was not appropriate for offshore work, with an excess of finishing trades relative to steel workers. Labour relations were difficult and management resources thin. Though there were some good managers, many were poor. They were frequently intimidated by unions determined, or so it sometimes seemed, to remain in control and prepared to use demarcation disputes to that end. Technical resources, including the British Shipbuilding Research Association (BSRA), were good but their efforts not always well directed. As the offshore fabrication yards (where new investment, including government grants, was available) came into being, much labour migrated to them for better pay and conditions.

Ministers responsible for OSO took only spasmodic interest in BS. Dickson Mabon (Labour) had considerable interest in the Scott Lithgow yard, which was in his constituency. Conservative Ministers became interested when government policy was to sell off those parts it could (including Scott Lithgow) and to close the remainder.

Until the arrival of (Sir) Robert Atkinson in 1981 as BS Chairman, the individual yards continued to be managed by the pre-BS incumbents with limited central direction. Atkinson centralised the organisation, with an HQ first at Newcastle and then at Gourock (at Scott Lithgow). He identified offshore as an opportunity, setting up a heavily resourced Offshore Division, with its own semi-submersible design which failed to sell. The Scott Lithgow and Cammell Laird yards were to be the main focus.

With venture manager help, Scott Lithgow had managed to break into the DP drill-ship business early with the *Ben Ocean Lancer*, but delivered late. Drill-ship demand was insufficient as semi-submersibles came to the fore and the company had only partial control of the drill-ship design. However, the yard had the benefit of aggressive management and a local MP who was Minister in charge of OSO, as well as developing offshore facilities based on wide, open berths, good cranes and open water. It entered the semi-submersible business with BP's MSV, *Iolair*.

A later semi-submersible, the *Ocean Alliance* drilling rig, for use by BNOC, proved a commercial disaster and a major contributor to a BS loss of nearly £161 million in 1984, about £376 million in 2008 terms (Jamieson p. 76). The drilling contractor had negotiated a high charter rate of $100,000/day (about $182,000/day in 2008 terms) but by the time the vessel was delivered several years late, the market had collapsed. TH acquired the yard in 1984 but as its reputation made it impossible to obtain new orders in a depressed market, had to run it down, eventually closing it in 1990.

The situation at Cammell Laird was, if anything, worse. It had disasters with both rigs it built – a jack-up for BGC (where there was a sit-in) and a semi-submersible, the *Sovereign Explorer*, which was delivered extremely late.

Private yards did better. The Clydebank yard established a good record for the construction of jack-ups. Appledore, Cochrane, Ferguson, Hall Russell and Richards were fairly successful in the market for small ships like supply boats until demand collapsed.

Turning to BSC, the steel industry absorbed more OSO time than shipbuilding and its nature became well known to senior OSO personnel. The nationalisation, privatisation, nationalisation, privatisation cycle had a destabilising effect on planning and investment, discouraging a purely commercial outlook. Prior to the appointment of (Sir) Ian McGregor early in the Thatcher administration, there was a high degree of local autonomy under an ineffective top management. At operating level, the calibre of management was very variable. From the outside, it often seemed that the technical people failed to pay attention to market needs, while the marketing people seemed to have no influence with operations and technical staff. Despite investment in the 1970s, there continued to be failures in piles and pipelines due to limited capacity and quality control problems at Hartlepool where there was difficulty in keeping mill sheet tolerances to specification. Price was also a problem.

BSC did not originally think North Sea demand would be long-lived and, in OSO's view, management did not seriously address the needs of the market until forced to do so by (Sir) Ian McGregor who was trying to come to terms with the collapse of shipbuilding steel demand. McGregor made largely beneficial changes, such as the move of the corporate HQ to Glasgow from London and the introduction of a more centralised approach to the high volume but low margin 'tonnage' business.

The first North Sea platforms were constructed from standard mild steel as supplied to shipyards. Oil companies at the time bought steel in bulk, supplying it as 'free issue' to fabricators, and designers had to adapt their designs to the specification, contributing to the over-designed and massively heavy nature of early northern basin platforms. BSC was 'tonnage' and not 'value added' oriented and believed it could replace tonnage of rolled product demand lost in shipbuilding from the offshore market.

Generally, oil companies were happy to take 'standard' materials like steel from the UK to give themselves greater freedom with more specialised items.

BSC was accepted on the whole as a reliable supplier but was seen as 'pricey' when compared to its Japanese, German and Dutch competition. BSC relied on OSO to negotiate for it, which led to a tendency towards over-pricing. Late delivery was not a problem due to oil company bulk purchasing. Except where the closures of old over-manned plants were concerned, industrial relations were generally satisfactory.

However, designers began to conceive lighter-weight platforms based on high tensile structural steels, with plate requirements beyond BSC's capability. The Corporation lacked the incentive to respond to these new 'value added' requirements. When it eventually did so, it chose an inferior method of producing high tensile steel to that employed by the Japanese competition.

Foreign competitors could usually 'out-flank' BSC, whose investment levels were in any case long constrained by the national economic difficulties. Foreign companies were more market-responsive and had easier access to capital. BSC's attitude to them was pragmatic. For instance, where specification was a problem, BSC was willing to act as merchant for the Germans in order to keep control of its market.

Despite its problems, BSC also had a brighter side. Shell confirmed its strength in tubular goods, stating that in 1981 BCS had exported some 100,000 tonnes of such products to the USA (Shell UK 1982, p. 12). Stewarts &Lloyds based at Corby was the source of this success, being strongly estab-lished (see p. 42) in seamless tubulars and casings where it had developed an innovative 'stretching' technique. It also had a licence for French designed Vallourec tapered threaded pipe-end joints, for which it built a new plant at Bellshill (Glasgow). Not all oil companies would accept the Vallourec joint and the additional set-up costs for API threads to satisfy such customers was a problem. Oil companies sometimes (unsuccessfully) requested a government indemnity if they were to use Vallourec. Corby was largely left alone by central management.

6.4.4 Creation of New British Enterprises

With 'boom' conditions a thing of the past and many disappointing precedents, there was no longer an obvious reason for British companies to consider expanding in the offshore supplies industry. Nevertheless, some did, often those already with 'fringe' interests. A case in point was Simon Engineering. Failing to achieve a significant position in engineering and contracting, despite its SimChem subsidiary's early start with the Teesside terminal (see p. 120), it attempted in the mid to late 1980s to develop a strong subsurface group. It acquired the previously independent geological and reservoir engineering consultants, Robertson-ERC, and the former NCB marine seismic business, Horizon Geophysical. Horizon's high fixed costs and vessel under-utilisation eventually forced Simon to divest to a foreign buyer.

A larger company with ambitions to expand in the offshore industry was British Telecommunications (BT). It had a long association with the industry through the provision of communication systems. After privatisation in the early 1980s, its Marine Services Division (responsible for submarine telephone cables) began to compete aggressively in the offshore sector (Drury 1986). The Division had ambitious expansion plans calling for large-scale investment in vessel(s) directed at the offshore market. In the wake of the 1986 oil price slump, it failed to receive backing from the BT Group Board.

Marconi Avionics also had unfulfilled ambitions. It outgrew its original subsea controls joint venture with U.S. company Koomey and attempted to diversify into ROV manufacture. On two occasions during the 1980s it wished to acquire leading U.S. subsea completions manufacturer, Vetco, but was unable to gain the support of its parent.

In 1986, process designer and contractor Babcock-Woodall Duckham was poised to acquire Global Engineering, which would have given the company a leading position in the UK's 'front-end' engineering business. The deal collapsed when the oil price slumped.

Joint ventures with British partners were still seen as desirable as a marketing tool by foreign companies seeking access to the UKCS, some market segments of which remained relatively attractive on a global prospective. A short-lived example was a mobile drilling joint venture between ship owner James Fisher and U.S. drilling contractor, Zapata Offshore. ASCo-Smit was longer lasting. ASCo supplied British labour to work from two small Dutch owned and operated semi-submersibles. These joint ventures were fairly typical of the field generally, with the foreign partner providing the capital assets and the British, unable or unwilling to commit capital, providing only labour. This reluctance to invest led some to describe the British as the 'coolies' of the North Sea. Such a view failed to recognise the growing number of UK personnel in management positions with foreign-owned companies, some achieving very senior posts.

One joint venture with an unusual origin was Bechtel-Laing, which came into being as part of the settlement of the dispute with Sun Oil over the development of the Balmoral field (see pp. 181, 182). The formation of Brown & Root Vickers (BRV) may be seen as part of a phased withdrawal from the industry by a major British company. Even after the events of 1977–1979 (see pp. 165–166), Vickers had retained important interests in the offshore business. After a few years of independent operation, its design business was placed into a joint venture with B&R, which eventually absorbed it.

Although difficult trading conditions existed for many activities for long periods, there were also growth points related to the increasing concentration on incremental subsea production facilities (and to a lesser extent FPVs), the gradual substitution of ROVs for divers and the well service demands resulting from the growing maturity of the early giant fields. Subsurface technologies developed at a rapid rate.

New opportunities therefore arose for entrepreneurial companies and individuals. The lack of venture capital remained a problem. Eventually, the flow of new funds increased after 3i developed a successful Aberdeen practice. This served as a 'marker' for others to try to emulate, having shown that despite the disappointing performance of NSA, it was possible to for an investment fund to make money in offshore goods and services. As a result, the number of independent British entrepreneurial companies rose with time and modest management buyouts became possible.

However, in late 1985 it was still possible for the author to write '. . . *in 1984 the average venture capital investment made in the UK was less than £300,000. Such a sum is an entire order of magnitude away from what is necessary in the offshore scene and even syndication is no answer with such small individual investment units*' (Smith 1985b, p. 1).

In 2008 terms, £300,000 was equivalent to about £700,000. Ventures requiring large up-front new investment were virtually impossible to finance and any start-up other than the smallest very difficult. Faced with this situation, entrepreneurs faced difficult choices. The less determined gave up. Some took their ideas or embryo businesses to established companies and sometimes got backing at the cost of a premature loss of control. Others attempted to trade with insufficient backing. The situation improved somewhat as the commercial banks developed more confidence in the venture capital managers and became willing to extend debt finance, but this principally benefited 'buy-out' situations with established cash-flows.

A few start-ups, like Mentor Engineering Consultants Limited (Mentor) turned to foreign finance. Mentor was a subsea design and systems engineering company, part of a community of niche engineering businesses specialising in subsea and general 'front-end' design which developed rapidly in the mid-1980s in the London area, particularly around Woking. From its foundation in 1984 until its sale to U.S. company McDermott in 1989, financial backing was Norwegian, with ownership shared with the UK founders.

Most firms in the Woking cluster had at least part British ownership and all were almost entirely British staffed, including a subsidiary of the U.S. well head equipment firm Cameron. Other members of the group included Global Engineering, J P Kenny, Mentor Project Engineering, Baker Jardine, Furness Underwater Engineering (FUEL), Shearwater Engineering and Granherne. The creation of this specialists group met a long-standing OSO objective.

Costain eventually acquired FUEL and the Wood Group Kenny, but most of the other firms became foreign-owned. Indeed by the early 1990s, the UK had lost its pre-eminence as the international centre of subsea engineering design as the activity (and some UK designers) migrated to Houston due to the growth of deepwater activity in the GoM and the then lower cost of U.S. engineering man hours.

If the UK presence in subsea design expanded in the mid-1980s, the reverse was true for subsea services. Among the casualties was LOG, a group of companies put together by Lyle Shipping. The original main elements were Kestrel Marine (a fabrication business with a pipeline 'bottom tow' joint venture with the Dutch tug and supply boat operator, Smit-Lloyd) and Osprey Electronics (underwater cameras). The subsequent acquisition of Global Diving, a small British diving company, led to LOG's downfall. It appears that a condition of sale was that LOG would acquire the building contracts for two high specification diving support vessels under construction at a Dutch yard. Apart from a payment towards the cost from the Dutch government in return for an undertaking to sail under the Dutch flag, the vessels were effectively 100% debt financed (in Dutch florins), quite possibly contrary to international agreements. Following the 1986 oil price slump, LOG was unable to earn sufficient to service the debt, a situation exacerbated by the depreciation of sterling against the foreign currency liability. With conditions in the offshore industry generally difficult and Lyle's original shipping business not prospering, the entire Lyle Shipping group collapsed in 1987.

A niche business, developed on the back of the growth of well work-over, was SAI Tubular Services Ltd. (later Scotoil). SAI (Scottish Agricultural Industries) was an ICI affiliate with a fertiliser business adjacent to Aberdeen Harbour. In the early 1980s, alternative employment was sought for the site. Thanks to its possession of a marine discharge facility licensed for the disposal of low level radioactive waste arising from fertiliser manufacture, it became the location for the decontamination of mildly radioactive well tubulars. Eventually, the business and its associated oilfield chemicals activities were the subject of a management buy-out.

More highly technical subsurface businesses came into the limelight in the late 1980s and early 1990s, showing that as the stock of experienced British petroleum engineers grew, a proportion had both the ability to innovate in this 'core' area, previously a largely U.S. preserve, and the drive to become entrepreneurs. They were assisted by changing R&D funding priorities (see pp. 176–180) usually based their businesses on proprietary technology embodied in a product or in software. If successful, such ventures likely to be highly profitable, at least for a time. Like subsea design specialists, they were skill, rather than capital, intensive businesses. A ready exit route – a trade sale to an American or Norwegian company – could be presumed following commercial success. They there fore presented an attractive type of investment opportunity to venture capitalists.

Many of these newcomers were located around Aberdeen, examples being Anderguage, Petroleum Engineering Services (PES), Petroline and Geolink. Aberdeen also captured the few new independent service companies established, such as coiled tubing operator, Progenerative Services Ltd. (PSL).

However, a surprising number were located elsewhere. Edinburgh Petroleum Services (EPS) and Concept Systems, both software companies, were in Edinburgh, where two universities engaged with the offshore industry and the PSTI all were located. A band stretched between the southern English university towns of Reading and Southampton, mainly in Hampshire. Among its members were Sondex, Sensor Highway and TSL, all product or product development focused, and software company Oilfield Mapping Systems – additions to the long established head offices of Expro (oilfield services), Gaffney Cline (petroleum engineering consulting) and Sonardyne (underwater acoustics) already in the area. Access to the London airports was an important location factor, most Hampshire companies almost from their inception being focused on world rather than the North Sea markets. Indeed some, like Gaffney Cline, had little North Sea business. Availability of highly qualified manpower from the many military, aerospace and nuclear research establishments in the area and presence of Schlumberger's Europe Africa and Middle East headquarters at Gatwick may also have added to the attractions of the area.

Assessing OSO

OSO was the focus of almost continuous interest throughout its life, although this interest was on a long-term declining trend subject to occasional 'spikes' around some particular high-profile issue that reached the public domain. Much published comment was essentially journalistic and superficial in nature, although some was more painstaking and even extended to serious academic studies. It is this second category which is considered in this chapter.

7.1 THIRD-PARTY COMMENTARY

The first serious study was by Jenkin. Published in 1981, it dealt mainly with the period 1973–1976, although Jenkin's interest extended to 1978. It, therefore, dealt with an OSO that was something of a 'work in progress'. Jenkin's view was that OSO's FFO policy and the associated exhortation and auditing were important factors in the increase in British content. Not only did it draw British firms to the attention of foreign oil companies entering the North Sea but also all oil companies were aware of the adverse effect on their exploration licence prospects and general government relations of a poor UK content.

However, FFO was insufficient to address foreign competition in specialist sectors where British industry was commercially weak (or, he might have added, absent). Here, he found no effective OSO policy tool. Jenkin was very conscious of the particular difficulties faced by the venture managers, noting they lacked high-level commercial discretion, ready access to substantial City funding and faced bureaucratic difficulties in mobilising even modest Section Eight (selective) Industry Act finance. As far as the OETB was concerned, he appeared sceptical of the extent to which it would be commercially effective, though noting that this '. . .remains to be seen'.

He could not decide whether the oil companies or their suppliers were the main beneficiaries of the OETB programme, but formed the opinion that the IRG scheme had been of more benefit to the oil companies than their suppliers.

Norman J. Smith, The Sea of Lost Opportunity.

203

Jenkin (see also p. 112) canvassed the views of offshore supplies companies, although the statistical basis of his main 1976 survey results could be open to criticism. It was based on the results of one hundred questionnaires returned from a mailing to three hundred and eight firms –the members of one of the many trade associations involved in the industry. Most of the responses turned out to be from SMEs having little or no involvement in the offshore supplies industry. He, nevertheless, concluded from his survey that OSO had been in contact with a large proportion of firms in the industry, offering appropriate services in an ineffective manner due to lack of experience and expertise. He found that supply companies valued a government role in providing market intelligence and introductions to oil companies and in preventing discrimination against British suppliers, although there were calls for improvement. He found a lack of enthusiasm for the extension of government R&D support or financial assistance and only limited interest in help with joint ventures (Jenkin, pp. 136–138, 144–145).

Jenkin was very conscious of the constraints within which OSO functioned. These included the traditional British commitment to open markets, UK membership of the EEC and the over-riding government objective of maximum speed of North Sea development. He recognised that the last point alone precluded Britain adopting a policy comparable to that of Norway. By virtue of a more restrained rate of offshore development arising from the lack of an urgent need of the revenues and by vesting commercial control of its offshore sector in the state oil company, Statoil, Norway was free to promote the interests of its supply sector in ways not open to OSO.

The period covered by Cook and Surrey's 1983 work extended until 1982 and was an international comparative study of British, Norwegian and French policies towards the offshore industry, with a few additional observations in respect of the Dutch and Italian experiences. They noted that policies available to governments reflected not only the differing resource situations of their respective countries, but also differing economic structures, philosophies and administrative methods.

In the case of France, there was no domestic offshore resource base but offshore R&D started early, before this was even known. Recognising the strategic importance of the oil service industry in the immediate post Second World War period, the government, the state-controlled oil companies and national research institutes worked closely with industrial firms to develop a co-operative network involving users, contractors, sub-contractors and suppliers, in which the government had confidence. A highly nationalistic policy resulted.

Norway also pursued a nationalistic policy, but based on an entirely different set of circumstances. The offshore resource base was very large relative to the energy and financial needs of a small country and the priority was to restrain development in order to prevent the destabilisation of the non-oil economy and Norwegian society. The number of large industrial firms was small so that it was easy for the government to communicate with them and to co-ordinate their activities. Some of the firms in the marine sector exhibited entrepreneurial characteristics. 'Norwegianisation' was pursued initially through administrative means (i.e. licensing policy from 1968 to 1969), requiring preference for Norwegian suppliers, with the obligation to carry out R&D in Norway added 10 years later. However, it was recognised that to develop 'Norwegianisation' to its full extent, the state oil company, Statoil, had to exercise a leading role in offshore procurement decisions, even where higher costs resulted. The government also became directly involved in R&D and encouraged the partially state-owned Norsk Hydro and the Norwegian private company Saga Petroleum to complement Statoil's activities.

Cook and Surrey pointed out that whereas the UK shared with Norway a large resource base, it was more like France in having a large and diverse industrial base, with a good research infrastructure. However, it differed from both countries in a number of respects. The British government was late in recognising the need to become involved in the offshore supplies sector, while British industry seemed risk-averse and slow to learn. The life of the state oil company, BNOC, was too short to have much effect in supporting British industry. Unlike Norway, Britain faced an acute balance of payments problem and rising unemployment in its old industrial areas, arguing for reliance on the international oil companies for rapid resource development. Once in being, the offshore supplies industry became an additional voice favouring continued rapid development.

National industrial policy was already based on equality of treatment between local and foreign-owned firms, making it difficult to discriminate against the latter in the offshore sector, particularly as their experience and expertise could bring early benefit. While some government R&D support was seen to be important from 1973 to 1974, it took until 1983 before an attempt was made to involve UK companies in R&D through the licensing system. Overall, the British policy was based on free market principles rather than nationalism.

In considering OSO's effectiveness, Cook and Surrey believed the outcome was much as was to be expected, given the policies followed. A high proportion of the UK fabrication market was won and considerable strength built up by firms that adapted existing equipment for offshore use, but no leading firms developed where large-scale R&D and learning were required and few British firms achieved a strong position in international markets. Vickers was the closest that the UK came to having a lead firm. The assumption was that local firms did not need significant assistance to become competitive, unlike the view in France and Norway. They concluded that the '...*large scale opportunities were transitory*' and that even if radical nationalistic policies had been successful in the early years (which they considered doubtful, given British industrial problems), they would not have been effective in international markets (Cook and Surrey 1983, pp. 72, 81, 95). The following adapted and expanded Table 7.1 summarises their work. 'Export promotion' as a policy, at least as far as the UK was concerned, would probably not have been seen as a particularly important policy by Cook and Surrey as it only really became so after their work was completed. It has, however, been added.

Although this has not been added to Table 7.1, it is also now clear that the single major beneficiary from Norwegian policies has been its subsea sector, parts of which now have world leadership positions, which would also not have been apparent to Cook and Surrey.

TABLE 7.1 Policies Employed in the UK, Norway, and France and Their Beneficiaries

Policies Used	UK	Norway	France
Assisting early entry	Negligible	Slight	Substantial
Long-term major project R&D	Negligible	Increasing with time	Substantial
National oil company purchasing	Slight	Major	Major
Full and fair opportunity	Major	Major	Policy not available
Open door to foreign firms	Major	Controlled joint ventures	Strong restrictions
Selective sponsorship of major firms	Negligible	Significant	Major
Export promotion	Major	Significant	Major
Policy Beneficiaries			
Sectors assisted to obtain work	All, but especially fabrication and equipment	All, but especially civil engineering and fabrication	All, but especially exploration, rigs, subsea, pipelaying, and installation
Sectors in which national firms benefited most	Fabrication and adapted equipment	All, but especially civil engineering and fabrication	All, but especially exploration, rigs, subsea, pipelaying, and installation
Sectors in which foreign firms benefited most	Design & management, oil specific equipment; pipelaying and installation	None directly but long-term results of joint ventures unclear	None

Adapted from Cook, L. and Surrey, J. (1983), *Government Policy for the Offshore Industry: Britain compared with Norway and France.*

Hallwood based his work published in 1990 mainly on survey data collected in 1984 from the Aberdeen area. He drew forceful and essentially negative conclusions on the effectiveness of the UK government policies applied by OSO. He thought '. . . *successive British governments have failed to take into account global dynamics, barriers to entry and foreign protectionism*' with the result that British companies were '. . . *largely confined as providers of locationally determined inputs.*' (Hallwood p. 160).

He particularly noted that British companies had failed to enter the industry's *'technological core'*, in part due to a late start and low R&D. Though he does not define *'technological core'* with any precision, the implication is that it mainly comprises well related and other subsurface products and services largely predating the oil industry's move offshore (Hallwood p. 90).

Cameron had already given another indictment of the effects of British government policy on the development of the local offshore supplies industry in 1986. He considered that despite government initiatives '. . . *the performance of UK supply firms has been on the whole a poor one, given the scale of activity on their doorsteps'*. He attributed this to be due, at least in part, to the disadvantages suffered by new local suppliers competition with long-established foreign firms resulting from government determination to develop the UK's oil and gas reserves as rapidly as possible. He went on to say: *'The fate of the UK offshore supplies industry'* '. . . *could reasonably be described as the other price of North Sea oil'* (Cameron p. 28).

Kashani (2005) sought to assess the effect of OSO's policies in the entirely different terms of whether or not it had raised oil company costs on the UKCS. With data supplied by a reputable consulting firm on 133 oil fields (dry gas fields were excluded) brought on stream before 2000, Kashani considered the efficiency of UKCS development relative to what he saw as the level of OSO intervention in procurement, dividing the fields into seven groups by date.

He concluded that fields developed and brought on stream in the period of *'heavy intervention'* (1975–1984 inclusive) were less efficient than those brought on stream in periods of *'no intervention'* (pre-1975 and post-1992). The present author is not competent to judge the validity of the statistical analysis used to reach this conclusion. In any case, the result would have been heavily dependent on the quality and weighting of the data inputs. Kashani appeared to have gone to considerable lengths to allow for variations in physical parameters, scale and 'learning', but acknowledged that he had not allowed for such factors as concern for time, tax rates and the oil price. To the author's knowledge, the effect of the first of these alone would have been very considerable during the period of 'heavy intervention' and indeed was one of the reasons for it. Similarly, he rated 1989 as the least single efficient year, suggesting it resulted from the all-time peak in reported UK content (87%) having occurred two years previously. A more plausible explanation is that in 1989, the offshore oil and gas industry was suffering from the 'knock-on' effects of the Piper Alpha disaster of the previous year.

There were also other factors that Kashani seemed not to have recognised, such as the exchange rate, growth in the sharing of infrastructure, the 'step-change' nature of some technological innovations and institutionalised cost-reduction. He also failed to consider the potentially cost-reducing initiatives of OSO in encouraging supply capacity to grow, addressing 'bottlenecks' and sponsoring R&D.

Although OSO's FFO activities undeniably had some cost to the oil companies, there is little evidence that it was large enough to seriously concern them. It would, in the author's opinion, have normally been 'swamped' by influences such as those mentioned in the preceding paragraph. OSO did not intervene in the great majority of contracts and in the relatively few cases where it did, the price and delivery margin between the best British and the best foreign offer were usually small.

According to Jenkin, OSO would simply allow an order to go to a foreign firm if a British bid were 10–15% higher (Jenkin p. 181). In the author's view, a differential of 10% or less would have been more typical of the point at which OSO would have made a detailed review, although a few exceptions may have existed where a contract was particularly sensitive. It must also be kept in mind that price was not necessarily the sole determining factor. Delivery dates and sometimes other issues would also have to be considered. Finally, British bids were by no means invariably the higher and nor would a review necessarily result in a British award.

A small number of retired oil industry executives had an opportunity to express their views of OSO when they gathered together in 1999. Despite some criticisms (none on the grounds of increasing costs), on the whole they endorsed the opinion of a former engineering director of Shell Expro. He was not prepared to criticise OSO. His view was "...OSO on the whole was reasonable. They had a difficult job. I think that what they did was fairly well balanced". He added that, although OSO could be a source of delay, it "...promoted British industry in a proper way" and did not support non-performing firms (Cook 1999).

7.2 ONE INSIDER'S VIEW

The writer joined OSO as Industrial Director on a 3-year secondment in March 1977. He had already been heavily involved with the North Sea and had some knowledge of OSO. His initial primary responsibilities were to administer the venture managers and the export effort, as well as sometimes to deputise for the DG on other matters. In late 1978 he was appointed as DG, a role he had already been performing 'de facto' for some time, and held the position until spring 1980. On returning to the private sector, he continued to be involved with the offshore industry and hence broadly aware of OSO policies, but did not qualify as an 'insider' as he had done during 1977–1980.

With offices in both the DEn's London headquarters and in Glasgow, office time was divided roughly equally between the two, with the need to combine participation in departmental decision-making and to support Ministers usually taking precedence over managing OSO. Unlike most civil servants, OSO's DG was required to spend a great deal of time visiting company offices and sites, and occasionally to go offshore. In addition there was foreign travel involved in export promotion and international negotiations, primarily with Norway. Not only was this work pattern very demanding, it also inhibited the development of the close personal relationships with other members of DEn's oil and gas management team which would have speeded up OSO becoming a more integral part of the whole.

In order to avoid the benefit of hindsight, in part the result of writing this book, the author has tried to base his inside view on his feelings at the time. Fortunately, this was made relatively easy by the fact that in the last few weeks of his role as DG of OSO, the writer gave both a major address to London's Oil Industry Club and a number of press interviews. The former – Smith (1980b) – was somewhat anodyne and diplomatically structured, having been prepared in advance and 'vetted', whereas the latter, being unstructured and not 'vetted' were less inhibited.

The address to the Oil Industries Club was devoted in part to OSO's relations with the oil companies, claiming that these were generally good, with company top management, for strategic reasons, reconciled to the government's aims but recognising that difficulties did arise with middle management. This was at the level charged with the prime responsibility for the success of individual projects, where wider issues were of less immediate concern and where, with career development in view, individual mobility was high. There was also a discussion of the unusually diverse and multidisciplinary nature of OSO's work, which had to be carried out under the scrutiny of outside interests involving business, politics, the unions and the media. The problem of managing OSO from Glasgow as a decentralised unit from a Whitehall department, with the further complication of OSO itself having Aberdeen and London locations as well as Glasgow, was discussed at some length. The risks of over-selling the export prospects of the UK's still immature offshore industry were mentioned. For the future, the author thought there was still a job for OSO to do, particularly in using FFO to ensure offshore operators did consider competent British suppliers and in enhancing UK industrial capability. A contribution towards this could be made through the closer integration of OSO's R&D branch with its Engineering branch, something the author had tried to set in hand.

However, the bulk of the address was devoted to the successes and failures of the British offshore supplies industry. It was noted that, although considerable progress had been made towards meeting the IMEG 70% target, creating offshore related jobs and developing a specialist labour force, such success had rested mainly upon the fabrication sector, the adaptation of standardised industrial equipment and USA-owned subsidiaries. The author made it clear that in a truly international industry, such as oil and gas, it was to be expected that the UK supplies component would consist of a mix of domestically owned firms, foreign subsidiaries and joint ventures.

It was further noted that '... *the UK's worst record is probably in marine activities requiring very large capital investment for example drilling rigs, DP diving support vessels, heavy lift and pipelay vessels*'. He attributed these and other deficiencies in part to structural factors affecting the UK industrial and commercial scene generally, such as a lack of risk capital for large projects in volatile markets. However, he also identified four issues relating specifically to British entrants to the offshore supplies industry – the lack of a customer 'comfort-inducing' track record, activities by conglomerate companies unable to adjust to offshore market conditions, a dearth of well financed and managed medium sized companies dedicated to the industry and an excess of undercapitalised ventures.

While some press comment confined itself to reporting the talk, two publications had already sought and been granted 'valedictory' interviews. In one of these, the author was quoted as saying: "*No part of OSO role is to obtain preferential treatment for British industry*" (Algar 1980, p. 26). There was also a public admission that the heavy lift market was now considered by OSO as impenetrable to new entrants.

The second interview (*Financial Times North Sea Letter* 1980, pp. 2–3) was couched in more outspoken terms and can be taken as a reasonable representation of the author's views at the time. With the benefit of hindsight, there is relatively little the author would wish to change. However, more could have been made of the over-riding priority given to early oil production under the conditions of economic crisis prevailing at the time of OSO's foundation and the implausibility of achieving a higher level of indigenous British ownership given the scarcity of 'new equity' and the risk aversion of 'old equity'. The main points made were:

i.   Failure on the part of British business to take advantage of available opportunities was partly responsible for the UK's share of the North Sea market not being higher;

ii. Things would have been worse without OSO, but the "... *British system*" had in some areas frustrated its efforts, drilling being an example. It was not within OSO's remit to address the structural problems of the UK economy, but it would continue to seek FFO for British firms and promote exports;

iii. OSO had assisted in restructuring British companies, preventing control passing overseas. Examples were Vickers Offshore and Atlantic Drilling;

iv. Running an organisation like OSO within the confines of the civil service was difficult;

v. OSO had been over-staffed when established in Glasgow and needed to be
pared back to its core functions, from which IRG administration could be
gradually removed. It was intended to merge the R&D and Industrial Capability Sections.

The aim of integrating the R&D Branch with the Industrial Capability Section
was never achieved. Had it been, in the author's opinion, the chances of
OSO developing and implementing a long-term strategy of creating strong
and technically advanced British-controlled companies in the industry would
have been much improved. Nowhere would this have been more the case than
with underwater products and services. The combination of Bondi's Underwater Initiative (see p. 138) with the scale and growth of underwater activity
on the UKCS offered a chance for the UK to overtake the foreign competition
in some parts of this key area, provided appropriate strategies and structures had
also been put in place. The writer did not then foresee that it would be the
Norwegians rather than the British who would achieve this.

The British and Norwegians were both late starters in modern underwater services compared to the Americans and the French. The USA had the advantage of
an early start in underwater tasks for the offshore oil industry and the French of
heavy pre-investment. Both had early experience of manned and un-manned
underwater vehicles for military and oceanographic application, technology readily transferable to oil industry applications. The UK had no underwater vehicle
experience prior to development of the North Sea. Thanks in part to its stock
of Royal Navy trained divers, it rapidly reached parity in underwater oil field
expertise with the USA and France and was to show itself as fully capable of technical innovation. It was to fail in respect of finance and strategy.

Although the Norwegians were also late starters, they were to adopt a much
more holistic approach to catching up than the British. Norway not only had a
vested interest in specialist shipbuilding but also offered the financing mechanisms appropriate to speculative risks, such as the construction of sophisticated
subsea support vessels. Far more decisively than the British, Norwegians turned
their backs on diving in the late 1970s in favour of subsea robotics. By supporting locally owned firms in the field of subsea equipment, Statoil and Norsk
Hydro successfully strengthened the domestic supply chain in this increasingly
important activity.

At the time the writer left OSO, the Norwegian underwater strategy was not
yet fully apparent. He still believed that subject to sufficient funding, the former
Vickers interests, resuscitated and expanded through BUE, could form the key-
stone of a successful British underwater sector. As it was, foreign-controlled
firms came to occupy the upper tiers of the British underwater supply chain.
UK-based subsidiaries of USA wellhead equipment companies such as Vetco,
McEvoy, Cameron and National Supply served world markets in subsea completions from Britain. Britain soon lost its always limited position in subsea support and construction vessels, though Aberdeen became the principal global

operating base for the worldwide leaders in this field. The UK held on to a presence in the field of underwater services, especially ROVs, longer, but foreign ownership again eventually prevailed. There was little that OSO could have done about this with the tools available to it, but given the high growth potential of this particular sector and its technical sophistication, the relegation of domestically owned companies to relatively lowly positions in the subsea supply chain should be viewed as a policy failure.

7.3 OSO'S STATISTICS

With respect to OSO's UK order content statistics, these were based on the quarterly returns submitted by the Operators, which did their own procurement and supplied their own estimates of the UK content. The returns were not systematically checked, but there was some case-by-case checking/querying to 'police' the process (e.g. in the differentiation of 'front' companies for foreign suppliers from genuine inward investors). The UK content figures could be regarded therefore as somewhat subjective in character and were certainly always 'political'. Although OSO was often accused of inflating the UK content, prior to the arrival of EPIC-type contracts, it could well have understated it because the figures were based on initial contract values rather than the final value. Large cost over-runs were common, especially in the early days, and usually largely represented British labour costs.

Until 1981, the publication of the UK content figures in the annual 'Brown Book' was accompanied by extensive footnotes as an aid to their interpretation. Later, the practice varied; some of the same qualifications were included in the main body of the text while others might reappear as footnotes.

In 1981, the 'Brown Book' pointed out, for example, that data was collated form individual returns submitted by each Operator designed, to the extent possible, to fit in with normal corporate recording methods and therefore unable to '. . . deal in detail with what is imported and what is home produced'. This was a cautionary note to set against a preceding statement about UK content representing '. . . the value of the contracts placed with companies, which through employment, manufacturing or sub-contracting, makes a substantial contribution to the UK economy'. Further, order and expenditure figures could not be directly compared for a number of reasons (Department of Energy 1981, p. 21).

Except in the very early years, UKCS expenditure was invariably higher than new orders, as commonsense suggested that it should be, but not only because of order execution times. Another reason was contract value 'creep' (whether due to over-runs, specification changes or inflation adjustments), although orders could also be reduced or on occasions cancelled. Possibly most important of all was the exclusion of oil company in-house expenditure devoted to the UKCS. In the early days of FFO, BP and Shell in particular maintained massive engineering departments with large labour and overhead costs and also provided some offshore services in-house. Though these expenses were by then in decline, attention does not seem to have been drawn to them until late in the history of FFO – Department of Energy (1989, p. 89). Unrecorded small orders would also enter the expenditure figures and might have affected UK content. Though these small orders would be expected to contain a large element of domestically acquired 'house-keeping' items, this could have been outweighed by foreign purchases of 'niche' products and services.

Particularly during the period after the move to Glasgow, when OSO employed a professional economist and a professional statistician, there was unease within OSO over the high profile given to the order statistics, particularly by politicians. There was a temptation to present the figures as representing the contribution of the offshore supplies industry to the British economy, a role for which they were ill-suited. Most UK orders would have had some foreign content even if not large enough to be identified by the Operator or OSO; the converse would also have been true of many non-UK orders.

Although OSO did not make any attempt to assess the foreign content of gross domestic orders, the Norwegians do appear to have done so. Cook and Surrey (p. 46) stated that in 1979, the gross Norwegian share of 67% fell to 52% on a net basis, suggesting the foreign input to domestic orders was about 22%. Given the much larger and more diverse nature of the British economy, it is possible to speculate that the gross to net differential in the UK may have been less.

To assess the real contribution of the offshore supplies industry to the British economy would have required the calculation of its domestic value added, which to the author's knowledge was also never attempted, presumably because it was seen as 'too difficult' and might have led to increased problems in collecting quarterly returns. Such difficulties may have been more apparent than real. After the introduction of value added tax (VAT) in 1973 the value added concept soon became a familiar one to UK companies and had been so even earlier for companies in most other EEC countries. For companies from countries where the concept was not familiar, a simple OSO guidance note should have sufficed, subject to sample checking against calculated or pre-existing coefficients.

Most external criticism of the figures focused on the less complex issue of the failure of OSO to distinguish between the orders gained by British-owned companies from those gained by the subsidiaries of foreign-owned companies operating in the UK. An Aberdeen MP asked the MoS for Energy to confirm that, despite OSO figures showing UK content in excess of 70% "... *the true figure for wholly-owned British companies is only about 40 per cent and that does not tend to be in high technology matters?*" (Malone 1984)

In his reply, the Minister of State described the 40% figure as "*conjectural*" and reminded his questioner that OSO figures related to firms contributing directly to the British economy – Buchanan Smith (1984a).

Malone gave no source for his figure but it seems likely that it related to research work then being undertaken in Aberdeen, but not formally published until several years later by Hallwood (1990) who concluded that indigenously controlled companies accounted for no more than 39% of orders placed in 1984 and perhaps for as little as 36%. With OSO claiming a UK order content of 74% for the same year, the implication was that foreign companies operating in the UK accounted for 35–38% (Hallwood 1990, p. 157). However, in the author's opinion, it is probable that Hallwood overestimated the share of foreign affiliates. The data from which it was derived was drawn from the Aberdeen area, which, as the main UKCS service centre, contained the local operational headquarters of almost all the large foreign-owned oilfield and underwater service companies working on the UKCS; they dominated their market segments. The fabrication, general manufacturing and design engineering components of the offshore supplies industry where British ownership was much higher were widely distributed across other regions of the UK.

The most consistent critic of OSO's UK content figures was Mackay Consultants, which published estimates of the share of total UKCS expenditure taken by UK-owned firms. It claimed that OSO's figures gave only a partial and therefore misleading view of UK industrial performance in the North Sea.

Its estimate of 48% for 1985 led the Select Committee (1987) to request that the Minister of State provide it with a written note on the differences between the MacKay and OSO figures, although since OSO did not produce figures for the UK content of expenditure, a proper comparison could not be made. In response, the Minister argued – Select Committee (1987, pp. 204–205) – that on the basis that 1985 expenditure figures would have largely reflected orders placed in 1983 and 1984, the two measures were broadly consistent. Mackay's overseas content of 24% of expenditure was similar to OSO's 2-year average of 26%. The difference was that Mackay sub-divided the UK element (76%) between British-owned (48%) and foreign-owned companies (28%), whereas OSO did not do so for its UK element (74%).

The Minister took advantage of his submission *"...to demonstrate OSO approach"* with two examples. One concerned a foreign-owned platform yard where it was assumed that apart from a 5% profit margin passed to the parent, all other expenditure was in the UK, resulting in a UK attribution of up to 95% *"subject to evidence"*. The other was an underwater construction contract awarded to a British diving company using a foreign-owned support vessel costing 50% of the contract value. In this case OSO, *"subject to evidence"*, would record a 50% UK content. The author is doubtful whether this methodology was always consistently and rigorously applied. Even if it were, it could still be open to criticism. For example, the overseas owner of the platform yard had many ways of extracting 'value' from its UK subsidiary in addition to its profit margin, by such means as transfer pricing, management charges or interest on inter-company loans.

The author does not have access to Mackay's methodology, but his opinion is that Mackay's local content figures are probably more representative

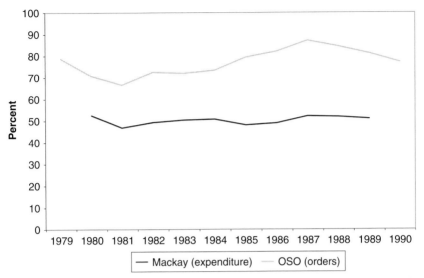

CHART 7.1   UK Content Comparison. Data Sources: DEn/DTI (1980 –1991) 'Brown Books', Mackay Consultants (1990) *North Sea Oil & Gas Commentary*.

than Hallwood's, although Cameron (p. 69) reported that BRIT claimed that UK-owned and -controlled companies obtained "less than 30% of North Sea contracts".

Chart 7.1 shows a decade's comparison between OSO's UK order content series and Mackay's UK-owned firms' expenditure share.

From 1988, successive 'Brown Books' provided a commentary on the performance of British-owned firms (see p. 187), although without any statistical support. The charitable view of its absence was that to provide the information would be difficult and prohibitively expensive. The uncharitable view is that the data would have embarrassed the government by showing the British offshore supplies industry in a less flattering light than 'up-beat' ministerial statements suggested, particularly had the frequent foreign take-overs of British-owned firms also been recorded.

7.4 A SUMMING-UP

Any attempt to assess the extent of OSO's impact on the offshore supplies industry in the UK is essentially doomed to failure since it is impossible to know what would have happened without its coming into existence. By the time OSO was fully functioning, many British firms had already decided to enter the market and local content would have risen substantially even without OSO intervention. During the early 1970s, there was not only widespread interest in entering the offshore market from existing businesses but also some significant institutionally financed start-ups, both features being largely absent in subsequent periods.

The rate of new business entry was further stimulated by the activities of OSO, notably in highly specialised areas where it sought to fill 'capability gaps' by encouraging joint ventures between British and overseas firms. Direct inward investment was also encouraged and throughout its existence, OSO was unable to discriminate in favour of British-owned firms. Not only would such discrimination have been contrary to general government policy, but also creating new British-owned capacity would have risked development delays. Given that the overwhelming early government priorities were security of supply and the balance of payments, avoidable delays were not countenanced.

OSO had few powers of its own. To influence procurement, it had to rely mainly on the threat of government sanctions in licensing and field development approval. In addition, it had to contend with the structural problems of British industry, British aversion to risky capital-intensive projects, early hostility from the USA and periodic interventions by the EC. There were also barriers to entry for British firms to overcome, the most serious being established client–supplier relationships.

Nevertheless, in its own terms, from 1979 onwards, OSO consistently outperformed its local content target of 70%, peaking at 87.2% in 1987, the end of its most interventionist period. The figures were widely criticised (often rightly so) and many in OSO were all too aware of their deficiencies and misuse. Unfortunately, once the system was established and had become recognised as a useful political tool, there were strong vested interests that would have opposed changing it, even if OSO had had the resources and the oil and gas companies the willingness to do so. However, at the very least the figures provided some sort of a benchmark against which OSO and the oil companies could judge the extent to which government aspirations were being met.

Apart from the adaptation of standard industrial products, British success was greatest where activities were locationally determined and/or where labour rather than heavy capital investment or complex technology was required, particularly fabrication. Most such activities proved transitory in nature, with overcapacity soon emerging, and they offered few export opportunities. Topside design, commissioning, inspection and maintenance were also areas of some success, with better long-term and export prospects.

British engineers had the competence to design North Sea platforms, but it was a long time before they had much opportunity to do so. One important North Sea sector, offshore installation and pipelay, was never penetrated. Others, particularly 'core' subsurface activities, such as mobile drilling and drilling fluids, were penetrated but entry was not sustained. In the promising area of underwater engineering products and services, British-controlled companies were relegated to 'bit part player' status. At one time or another, OSO had devoted a considerable part of its limited resources to such areas. The inadequate capital resources of the British contenders and their initial lack of credibility in the face of established client–supplier relationships usually frustrated its efforts.

Many business exits, both voluntary and involuntary, occurred in a long period of difficult trading conditions in the 1980s, particularly after the 1986 oil price crash. Foreign buyers, mainly American and Norwegian, acquired many 'high technology' or otherwise strategic British interests. The lack of British buyers was matched by a virtual absence of start-up capital to back a new generation of innovators and entrepreneurs then emerging. There was little OSO could have done about this.

Despite the difficulties, there were examples of UK-controlled companies that survived, prospered and internationalised. Some, such as the well tubular business of BSC and KCA Drilling, had already done this long before OSO had been created. Others such as Expro were able to do so without much need of OSO support. However, it can plausibly be argued that for this group to have been larger would have required the government to have intervened earlier and to have created a different business climate, providing *inter alia* limited discrimination in favour of British-owned firms, easier access to risk capital and a more substantive and focussed R&D programme. OSO had no powers to deliver any of these things.

In short, OSO did what it could with the limited means at its disposal. Other than its role in focusing oil and gas company minds on the issue of procurement, which almost certainly had a beneficial effect on UK-based suppliers, perhaps its greatest success was the extent to which USA-owned service and supply companies were cajoled into doing more in the UK than otherwise would have been the case. To have expected more of OSO would have demanded a change in the mind-set of its political masters, themselves constrained by circumstances and probably in thrall to Treasury ideology.

Chapter 8

Case Studies and Expert Testimony

In Chapters 5–7, British industrial performance has been largely viewed through the prism of government policy and commentaries upon it. This chapter adopts a different methodology, seeking first to examine directly examples of British business performance, within clearly defined market segments. Lack of readily usable data has limited the extent to which this could be attempted to only two segments. Considering the development of individual British companies can provide an alternative perspective and this has been the second approach. The problem here is the selection of appropriate companies from the large number potentially available. In the event, four were selected. Their common feature is that all of them have been important within particular parts of the industry and, in the author's opinion, are also otherwise in some way representative.

The chapter closes with opinions on British industrial performance and that of OSO from a group of former senior government officials and executives from the offshore sector.

8.1 MARKET SEGMENT CASE STUDIES

8.1.1 Drilling

Overall, drilling is probably the most important single market segment in the industry, as well as usually one of the largest. Information on the market position of British contenders in the UKCS market is available for 1982 and 1989 from two reports in the 'Offshore business' series. Conveniently, these 'snapshots' allow a comparison between periods before and after the 1986 oil price slump.

It needs noting that that during most of this period the UK market, particularly for E&A drilling, was growing (see Chart 4.2, p. 102). This reflected the addition to the need for companies to drill mandatory licence obligation wells of a considerable financial incentive through a special PRT tax relief, which operated from 1983 to 1993. The proportion of development wells drilled from mobile rigs was also increasing. The resultant buoyancy of the UK market was in marked contrast to the situation in most other parts of the world where high specification mobile drilling rigs were employed, which were mostly characterised by poor demand and excess capacity. As a result, the UKCS acted as a 'magnet' to the international drilling industry, which prevented the contract hire rates for mobile rigs there from reflecting favourable local demand conditions rather than the generally poor rates prevailing globally. However, in terms of offshore wells drilled it was a relatively small element of the world market. In 1982, its 249 wells represented about 7% of the global total. With UKCS wells increasing to 331 by 1989 against a background of reduced numbers elsewhere, its share would have grown substantially. Moreover, since drilling costs were higher than the global average on the UKCS, where conditions demanded the use of higher specification equipment than was normal elsewhere, in value terms it might have represented at times as much as 30% of the world total during the 1980s.

Smith Rea Energy/Hoare Govett (1983b) provides data from which it can be estimated that the UKCS market for mobile drilling in 1982 was worth over £500 million (over £1.3 billion in 2008 terms) and that for fixed platform drilling over £90 million (nearly £240 million in 2008 terms). Total well costs would have been substantially higher in both cases.

Companies the report considered as 'British', excluding oil and gas Operators, were reckoned to occupy about 23% of the mobile drilling market and about 13% of the fixed platform drilling market. U.S. companies occupied about 53% of the mobile market, the balance being shared between the Operators and French, Danish and Norwegian contractors. In the fixed platform market, North American firms held about 63% of the market, a German firm about 8% and the Operators about 16%.

The report noted that British-owned Operators on the UKCS retained in-house major drilling operations. BP still owned and/or operated semi-submersible and platform drilling rigs, and also had joint ownership of an advanced drill-ship with the U.S. contractor SEDCO and a purely financial interest in a heavy-duty jack-up working on the UKCS. Shell continued to own and operate a semi-submersible and also owned the rigs on some of its UKCS fixed platforms operated by contractors, one a Shell/SEDCO joint venture. There were also financial interests in mobiles operated by foreign contractors. Although neither operated the rigs, both BGC and Britoil had financial interests in rigs still under construction in 1982. Taking account of Operator interests and treating Shell as a British company, would have increased the British shares to about 29% for the mobile sector and about 28% for the fixed platform sector.

The report went to some length in considering which contractors should be regarded as British, not always a straightforward issue. Two of the companies operating British-flag mobile rigs (Norway's Fred Olsen group and Sonat, formerly the Offshore Company, of the USA) were eliminated, leaving five that were considered to qualify as meeting at least '. . .some of the following criteria: British origin, British-managed, substantial British equity ownership' (Smith Rea Energy/Hoare Govett 1983b, p. 16).

The five companies then operated eight semi-submersibles, five drill-ships and two jack-ups, with an additional three semi-submersibles and two jack-ups under construction. With the benefit of hindsight, it is possible to realise that this was probably the largest mobile fleet that the British contractors were to have.

Of the five, the KCA Drilling Group plc pre-dated the North Sea industry and was British by origin, control and management. However, in 1982 less than half its turnover arose from the UKCS, where it was confined to two fixed platform drilling contracts. Its sole mobile unit – a sophisticated DP drill-ship – was operating off Spain, with the balance of turnover arising from land drilling and work-over rigs in a number of countries.

The nature and origin of Ben ODECO (50% British) has already been described (see p. 146). In 1982 it owned and/or operated four mobile rigs, comprising an old moored drill-ship, a modern DP drill-ship, a jack-up and a semi-submersible. Only the last was in the North Sea, although in 1982 Ben ODECO entered an agreement with Britoil's St. Vincent Drilling subsidiary to co-own and manage the latter's advanced semi-submersible, expected to be delivered by Scott Lithgow in 1984. In addition to its interest in Ben ODECO, Ben Line had acquired full ownership of Atlantic Drilling Ltd. (see p. 145). In 1982, this company had two semi-submersibles, a third under construction and a conventional drill-ship, with capacity largely committed outside the UKCS.

If Ben Line was Britain's first force in mobile drilling, Houlder Marine Drilling Ltd. was its second. Its shipping company parent, Furness Withy, had come under Hong Kong (then still a British colony) control in 1980 but management and personnel remained overwhelmingly British. It held a 60% interest in the two Kingsnorth Marine Drilling Ltd. semi-submersibles (see p. 145) with Norwegian investors holding the balance. Allowing also for rigs operated but not owned, the Houlder fleet in 1982 totalled three semi-submersibles (one under construction) and three jack-ups (two under construction), all intended for UKCS work, predominantly for BGC. Houlder also provided onshore drilling services to BGC.

The other mobile rig owning contractor, the report considered to be British, was Jebsens Drilling plc, a company founded in 1979 by the transfer of two semi-submersibles from the Norwegian to the British flag. In 1981, an introduction to the Unlisted Securities Market (USM) resulted in British financial institutions acquiring interests of about 45%. The balance and management control remained in Norwegian hands. In 1982, the company owned three semi-submersibles, all working on the UKCS, together with a Scott Lithgow built DP drill-ship working off Canada. At position three among the 31 contractors contesting the UKCS market in 1982, it was the highest placed of the British contractors, with an 8.7% share of mobile rig-months (Smith Rea Energy/Hoare Govett 1983b, p. 13).

The fifth and final company treated as British in the report was Dan Smedvig Ltd, a 50:50 joint venture between Norway's Smedvig Drilling and Britain's

Davies and Newman, a shipbroker and airline owner. The company's operations were confined to oil company-owned fixed platform rigs.

The picture of the British offshore drilling business presented by the report is of a small but steadily developing industry faced – in its dominant mobile drilling segment – with fluctuating demand, high capital intensity and strong competition. It was clearly earning market credibility while struggling to develop fleet sizes and mixes by mobile rig type and ages that would enable its members to operate commercially in terms of a portfolio of contracts. This was a facility already open to its larger foreign competitors and offered some protection in market downturns. It was the conventional wisdom of the time that the minimum fleet size to achieve a balanced contact portfolio was six, so that this was the minimum economic scale for a successful operation in a competitive environment. If this was accepted, in 1982, Houlder was on the verge of achieving minimum economic scale and Ben Line had already attained it, if Atlantic Drilling and Ben ODECO could appropriately be treated as a unit. Even so, according to the report, with eight rigs, it would not make the 'top twenty' mobile drilling contractors on a world scale. Companies operating nine mobile rigs occupied positions 18–20. Companies with between 39 and 20 mobile rigs occupied positions one to five (Smith Rea Energy/Hoare Govett 1983b, pp. 8, 22).

Jebsen and particularly KCA were far from approaching minimum economic scale. Fortunately for KCA, its main activity was the altogether commercially less demanding fixed platform drilling market, where its clients (BP and Mobil) owned the rigs. Nevertheless, the activity was small in size and vulnerable to substitution from mobiles as pre-drilling development wells became the norm for fixed production platforms and subsea and floating production became more popular.

Evidence to the Select Committee (1987, p. 102) showed that following a collapse in offshore exploration in the wake of the price slump, drilling contractors were in a financially desperate situation. Mobiles operating on the UKCS fell from 57 to 24 during 1986, with the industry making heavy losses.

The second 'snapshot' showed the state of the British drilling industry after the price collapse. In 1988 and 1989, the UKCS market for mobile drilling contracting was estimated to have shrunk to about £260 million per annum in value (about £460 million in 2008 terms). Excess capacity reflected in low day-rates was the primary cause. At about £70 million (about £125 million in 2008 terms), fixed platform drilling had been squeezed less (Smith Rea Energy 1990b, p. 1).

The British drilling contractors had suffered badly from these difficult market conditions. Houlder had ceased to exist as a separate entity, following its sale in 1989 to Sweden's Stena Drilling. Jebsens disposed of its drill-ship prior to being absorbed into the fully British company Midland & Scottish Resources (MSR). Early in 1990, MSR sold two of its three semi-submersibles into foreign ownership, retaining the third for conversion to a FPV. By the end of 1989, KCA Drilling was no longer engaged in the mobile rig sector (though it continued in platform drilling) and 51% of its shares were then in Norwegian hands. In the same year, Smedvig acquired its former British partner's interest in platform driller Dan Smedvig.

Only Ben Line appeared to be largely unaffected by this radical restructuring, though it too experienced some changes in equipment ownership and operating arrangements. At the end of 1989, Atlantic Drilling was operating three semi-submersibles (one owned by Ben ODECO), a jack-up (Norwegian owned) and a conventional drill-ship unsuitable for UKCS operations. Ben ODECO continued to operate a DP drill-ship and a jack-up; it was also equal partner with Britoil's St. Vincent Drilling subsidiary in the venture set up to operate the dp semi-submersible *Ocean Alliance*, delivered from Scott Lithgow 4 years late in 1988. The Ben Line interest in this combined fleet of eight vessels was all that remained of the British mobile drilling contracting industry.

BP and BGC continued to maintain some ownership and operating interests in UKCS mobile rigs, although Shell had withdrawn by 1989. There was no evidence of continued fixed platform drilling activity by British Operators.

Given the many mid-year changes, it was difficult for the report to establish 1989 market share figures reflecting nationality of ownership. None was available for platform drilling. However, in the mobile drilling sector, companies still British at the year-end, including Operators, appear to have held about 15–16%, of which Atlantic Drilling/St. Vincent Drilling accounted for about 9–10% and Operators for most of the remainder (Smith Rea Energy 1990b, p. 20).

Although Ben Line retained its interest in the mobile drilling business longer than the other British contractors, it too was to withdraw. Having first dissolved the Ben ODECO joint venture, it next disposed of its interests in the joint venture rigs and of its shipping interests so that it became almost synonymous with Atlantic Drilling, which was absorbed by Stena Drilling in 1996.

It had become clear by the end of the 1980s that, having missed out on the initial northern North Sea drilling boom, the persistent attempts to establish a viable UK mobile drilling contracting capability had been unsuccessful. Given the circumstances, this is not entirely unsurprising. Not only were the technical and management requirements exacting, making it difficult for new entrants to acquire a credible track record, but the business was also both capital intensive and subject to demand volatility.

The scarcity of large-scale equity capital for high-risk investments relative to the much easier access to debt and lease finance has been mentioned at several points in this study. It resulted in most British participants being heavily indebted from the outset, a problem compounded by the fact that much of their equipment was delivered after the steep decline in day-rates that began in 1981. The inevitable result was pressure on cash flow and a weak negotiating position. Such support as OSO could offer may have increased the ability of British companies to win contracts but not the rates that could be earned.

IMEG had recognised the difficulties likely to be involved in British entry to this key activity and had made one of their strongest and most interventionist recommendations with respect to it (see p. 100). Rightly or wrongly, these were ignored and the 'drilling initiatives' attempted by OSO (e.g. p. 145) did not prove to be an adequate alternative. However, the late entry to a market that became over-supplied and experienced long periods of depression could probably not have been overcome under either scenario without an overtly protectionist policy.

Given the importance (however measured) of drilling contracting to all phases of offshore operations, British lack of success in the field was a serious strategic failing, indeed sufficiently so to undermine any pretext that a well-rounded independent UK offshore capability had been developed.

8.1.2 Drilling Fluids

Muds, brines and associated chemicals and services are indispensable to all phases of drilling operations, forming a small but important part of the drilling and well services sector. Information on the market position of British contenders in the UKCS market is available for 1984 and 1990 from the 'Offshore business' series. As in the case of drilling, these 'snapshots' allow a comparison between the period prior to the 1986 oil price slump and that after, which reflected a significant restructuring.

According to Smith Rea Energy/Hoare Govett 1985 (pp. 7, 16), the UKCS market for drilling fluids in 1984 was estimated at just under £74 million (about £173 million in 2008 terms), shared between nine suppliers. It probably represented only 2–3% of a world market, the scale of which was largely determined by the very great number of (inexpensive) onshore wells rather than the small number of (expensive) offshore wells.

Companies the report considered as 'British' were reckoned to occupy about 29% of the UK market. U.S. companies mainly parts of U.S. oilfield service majors occupied about 67%, the balance being shared between French and Norwegian firms. British firms held the second and fifth positions in the market. International Drilling Fluids (IDF), number two with a 17% market share, was a wholly owned subsidiary of English China Clays, a public company. It had started as a joint venture with a Canadian oilfield chemicals supplier. BW Mud held the fifth position with a 10% share. Originating as a joint venture between Berry Wiggins (later KCA International) and a U.S. mud company, it had by this time become independent following a management buy-out (Smith Rea Energy/Hoare Govett 1985, pp. 9, 10, 11,16,17).

The drilling fluids business had a difficult time following the 1986 oil price crash and considerable changes followed. Although the number of wells drilled recovered from its trough (see Chart 4.2, p. 102), the mud companies had to deal with fierce competition stimulated by depressed conditions internationally. Environmental concerns were driving a switch from oil-based to water-based muds, while the effects of a major restructuring of the U.S. oilfield service industry were becoming apparent. Also the oil companies were close to a major shift in the way they dealt with their suppliers. By placing greater responsibility and risk on them, this would favour consolidation of the supply chain and thus the larger, more integrated suppliers.

For 1989, the UKCS market was estimated at about £80 million (or about £140 million in 2008 terms), suggesting that in real value terms the market had contracted significantly. The number of suppliers had shrunk to six. BW Muds had grown strongly to become market leader with an estimated share of over 22%. By contrast, IDF had fallen to fifth place with a 16% market share. By this time, IDF had run losses for several years, becoming a problem to its parent, which sought buyers. It was not successful until 1992 when Dowell Drilling Fluids, then part of Schlumberger, acquired IDF (Smith Rea Energy 1990c, pp. 19, 34).

Though BW Muds was a reasonable size in the context of the UKCS market, by the standards of the American owned companies with their global reach with which it competed, it was a small business. Moreover, the U.S. firms were in the main parts of diversified oilfield service businesses and thus better equipped to offer the type of integrated drilling and well service contracts that came into vogue.

BW Muds continued as an independent company for the remainder of the period covered in detail in this book. In 1998, it was acquired first by the UK's Abbot Group, reuniting it with its erstwhile sibling, KCA Drilling, which 3 years later sold it to U.S. mud company M-I Drilling Fluids, jointly owned by Smith International (60%) and Schlumberger (40%). The latter had acquired its stake in exchange for its Dowell Drilling Fluids unit. Hence the only two serious British contenders in the business eventually both came under common U.S. ownership. The demise of UK-ownership mainly reflected an inability to adjust to changes in the way business was conducted, which favoured the major foreign owned diversified oilfield service businesses.

8.2 CORPORATE CASE STUDIES

8.2.1 British Underwater Engineering Limited (BUE)

This company has been included for three reasons. Firstly, it was more directly a product of state intervention than any other British offshore service and supply company. Secondly, the author was for a time personally associated with it. Thirdly, it was engaged in advanced underwater services and the UKCS was the largest part of the world market at the time of existence.

The origins of BUE have already been described (see p. 166). By early 1980, the NEB had bought out Wharton Williams and established a dedicated management team with an NEB employee as managing director and the author as chairman. Later in 1980, two ex-Shell employees joined the board, one as technical, and the other as non-executive, director. A financial director was not in place until early 1981 (Smith 1982).

The new team faced considerable problems. BUE's largest subsidiary, manned submersible operator British Oceanics (BOL), had become inactive in 1979, resulting in a consolidated loss for a 1979 10-month trading period of over £2.2 million (over £8.1 million in 2008 terms) on a turnover of nearly £4.9 million (or about £18.1 million in 2008 terms). The new management's first move was to acquire Sub-Sea Surveys (or SSS), originally founded as a consultancy by VOL's former survey manager, but with the longer-term objective of operating ROV's. SSS had leased the 'Consub 2' ROV (Figure 8.1) built by BAC with part government funding, the lease guaranteed by the NEB and a private sector partner (the Peckston Group) in return for equity stakes, and had shown that ROVs could undertake North Sea underwater survey tasks, such as pipeline surveys, more cost-effectively than manned submersibles. A substitution risk was thus hedged.

At much the same time, Intersub, BOL's largest remaining competitor, collapsed. Some of its assets – particularly its support ships – were better than those of BOL. To avoid the risk of competing in a shrinking market with a party that had acquired them at a 'distress-sale' price, BUE acquired the assets for £4 million (about £12.3 million in 2002) terms. This was financed from shareholders funds, with 50% being provided by a new private sector investor in BOL, NSA, seen at the time as the first stage in the re-privatisation of the group. An NSA director joined the BUE board. BUE turnover in 1980 reached nearly £14.5 million (over £38 million in 2002 terms), with trading close to breakeven. The Intersub acquisition increased the number of ships owned

FIGURE 8.1 Consub 2 ROV. Courtesy Mr. J. D. Westwood.

by BUE to eight, leading in 1981 to the establishment of BUE Ships (Smith 1982).

During this period, the management took stock of BUE's existing position and their perceptions of the future direction of the offshore industry. They concluded that there was scope in both domestic and international markets for a specialist British company offering a wide range of underwater services and their marine support, backed by specialist engineering design and manufacturing. Slingsby Engineering offered a satisfactory manufacturing base, while it was decided engineering design could be developed internally. This was broadly the strategy successfully adopted by the foreign firms that came to dominate these growing segments of the North Sea (and world) market. Existing commitments to part OETB funded R&D were reviewed. After prototype testing, the Intertek encapsulated 'neutrabaric' subsea production system was abandoned in light of growing industry acceptance of non-encapsulated systems. The pipeline repair project was re-focussed on underwater structural connections, such as piles to platform jackets and to subsea production structures. Renamed BUE Hydra-Lok, development continued with OETB support and considerable commercial success was eventually achieved.

A disappointing feature of 1980 and 1981 was the disappearance of two major pipeline projects from BUE's prospective workload. At this time, the company was the market leader for pipeline route and 'as laid' surveys. First, a proposed central North Sea gas gathering pipeline system failed to receive British government approval. This was followed by BUE's loss of Norway's Statpipe project. 'Inside information' from Statoil's technical advisers (U.S. engineering firm Fluor) indicated that BUE's bid had been recommended as the preferred service provider. Operator Statoil had allegedly over-ruled this in order to create a Norwegian capability, awarding the contract to a Norwegian newcomer to the field, Stolt Nielsen Seaway. If correct, this account provides a good illustration of the way Statoil used its privileged position to fill gaps in Norwegian capability.

By the end of 1981, the generally supportive NEB was replaced by the British Technology Group (BTG), which made it clear that no further investment funds would be made available to BUE, which would be sold as quickly as possible. Merchant bankers were appointed to bring this about but failed to identify purchasers acceptable to the BUE management, which wished to develop the company independently. BUE was permitted to continue to pursue its corporate strategy but forbidden to seek additional finance from new external shareholders.

During 1982, BUE acquired HMB Subwork, founded by former VOL employees and operator of a type of underwater vehicle capable of either a manned or unmanned role. The acquisition was financed by the sale of Slingsby's aviation interests. A serious deficiency in underwater services remained the lack of a diving arm, where acquisition prospects were few and start-up prospects daunting. The following year saw entry to the diving sector through the acquisition of

KD Marine, financed partly from internally generated funds and partly by borrowing.

By now, stresses arising from the integration of newly acquired subsidiaries (which had created a group stretching from London to Aberdeen) and differences of opinion between executive management and the controlling shareholder were serious. Divisions emerging within the BUE management itself further complicated matters. In particular, one view was to continue a policy of rapid debt-financed expansion whilst another believed the company should sacrifice investment opportunities rather than incur excessive borrowing. This issue became focused on whether to order a new ship for a support contract for the Argyll field. A decision to go-ahead was made shortly after BTG had removed the author as chairman and director of BUE in December 1983.

Notwithstanding these upheavals, 1982 and 1983 were good years financially, with pre-tax profits in 1982 of £1.25 million and £1 million in 1983 (respectively, about £3.2 million and £2.4 million in 2008 terms). The deterioration in 1983 was partly due to the cost of opening a Singapore office as an entry point to Far Eastern markets.

By the end of 1984, BUE's original technical and finance directors and the ex-Shell non-executive director had all left and defections among senior managers in the operating units were by now also in progress. During 1984–1986, BUE continued to expand its shipping interests, entering the supply boat field.

A pre-tax loss of £5.25 million (about £12.3 million in 2008 terms) was recorded in 1984 on a turnover £31.5 million (about £73.6 million in 2008 terms). Turnover rose sharply in the following year to £38.5 million (about £84.8 million in 2008 terms), with a return to profitability – £0.8 million pre-tax (or nearly £1.8 million in 2008 terms). The respite was short-lived and the loss (including asset write-downs) for the oil price slump year of 1986 was nearly £10 million on a turnover of £26.4 million (respectively, about £21 million and £56 million in 2008 terms), leading to 'negative equity' for BUE and qualified accounts for both it and its parent, now NSA. The loss would have been £1.25 million (nearly £2.7 million in 2008 terms) greater had the company not taken a credit to income of that amount from liquidated damages totalling £3.85 million (about £8.2 million in 2008 terms) from BS for late delivery of the field support vessel *British Argyll*, originally scheduled for May 1985 but achieved in May 1986 (North Sea Assets PLC 1985 and 1986).

During this period, NSA (whose portfolio was generally performing badly) and BUE had become increasingly intertwined. NSA both made and guaranteed loans to BUE and allegedly co-invested with certain BUE directors in a shipping venture. NSA acquired sufficient BUE shares from BTG in 1984 for BUE to cease to be a BTG subsidiary. At the start of 1986, NSA purchased further shares with the result that BUE became an NSA subsidiary (fully owned from 1987), an event leading to the loss of NSA's investment company status and the eventual re-naming of BUE as NSA. Not only did the performance of BUE in 1986 demand action from NSA, it also forced other creditors to try to salvage what they could. The result was an eventual total 'clear-out' of the boards of both NSA and BUE. Among the new appointees of 1987 was David James, a 'company doctor' and 'crisis manager' nominated at the insistence of the principal lenders to BUE. He became executive chairman of BUE and chairman of NSA's executive committee, which took over the functions of Ivory and Sime, previously its investment manager and secretary. He subsequently became NSA's deputy-chairman and main organiser of the 'rescue' of both companies.

The reconstructions of the two companies involved Court approved changes to capital structures as well as new capital. Success depended on the willingness of BUE's creditors to write-off indebtedness and, in some cases, to make cash injections, assume responsibility for debt repayment, or convert short-term liabilities to long-term ones. The result was an 'extraordinary credit' to the 1987 profit and loss account of just over £12 million (over £24 million in 2008 terms), a pre-tax profit of £10.6 million (about £21.5 million in 2008 terms), a fall in bank indebtedness of £13.4 million (over £27 million in 2002 terms) and a return to positive equity. The principal contributors were BS and the Clydesdale Bank. Turnover in 1987 was marginally down on 1986. During 1988 and 1989, BUE was rationalised into three trading units – BUE Ships, HMB Subwork (ROVs) and Hydra-Lok. This involved two major disposals. Diving Services was sold to the Swedish Stena Group and Slingsby to a UK purchaser but soon passed on into French ownership. Both were considered too capital intensive for the reconstructed company. The non-BUE holdings in NSA's portfolio, seen as having limited potential, were all sold or wound up by early 1990, allowing James to hand over responsibility to a new permanent executive management. He left behind a business with a sound balance sheet, a stagnant turnover and a low level of trading profits – North Sea Assets PLC (1987 and 1989).

The new management inherited an NSA that acted as a direct holding company for the remaining BUE subsidiaries, although references to BUE were soon rare. With the disposal of the BUE shipping interests in 1991 and 1992, the use of the term BUE was entirely discontinued. Some shipping interests passed directly into foreign, mainly Middle Eastern hands, although after a management buy-out, BUE Ships traded successfully as an independent British supply boat company for a many years before also becoming foreign owned.

The final years of NSA/BUE were relatively tranquil and the company remained in profit. The period 1990–1993 saw acquisitions as well as disposals. The largest were Huntly Equipment Rental (hire of rig mooring systems and wire rope product supply) and SeaMark Systems (underwater protection systems). Low margin businesses, they nevertheless made positive profit contributions. HMB and Hydra-Lok enjoyed unspectacular but steady growth. After the acquisition of Scandive in Norway in 1993, HMB became the largest British operator of ROVs. Between 1990 and 1993, group turnover grew from £21.6 million to £29.3 million (respectively, about £36.6 million and £42.5 million in 2008 terms), with pre-tax profits rising from £1.57 million to £1.95 million, respectively, nearly £2.6 million and over £2.8 million in 2008 terms (North Sea Assets PLC 1990–1993).

This performance failed to excite the share price in the way needed for the company to raise new capital for expansion, leading to the 1993 chairman's statement in the annual report saying '... that other ways of demonstrating the underlying worth of the Group will also be considered'. The statement was a clear indication of what was to come. Hydra-Lok was sold in 1994 and the remainder of the business a year later, in both cases to U.S. companies.

8.2.2 Expro International Group PLC (Expro)

Expro's inclusion among the corporate case studies can be justified on two grounds. It is a rare example of a company that received sufficient venture capital to grow to a size 'to go public' without succumbing to foreign take-over and has flourished in the drilling and well services sector, part of the industry's old 'core', where successful British firms have been few. The early history of Expro is inseparable from John Trewhella, a Cornish mining engineer who contributed, to an oral history project that provided an invaluable input to this account (Manson 2002).

After qualifying in 1953, Trewhella worked in gold mining, but soon realised that petroleum engineering offered a more remunerative career, joining Kuwait Oil in 1956. After training, he worked on well completions and work-overs before taking charge of the wireline and well test section, with UK expatriate staff. Recognising a trend to outsourcing, he conceived the idea of buying his employer's equipment and winning a service contract. Having secured local financial backing, he resigned in 1964 in order to submit a proposal.

He was unsuccessful and joined a 10-year-old French well service company, Flopetrol, in Paris, where he noted that French oil companies were expected to employ French service companies. After spells in Algeria and Nigeria, he moved to the Netherlands in 1965, soon taking management control of Flopetrol's Great Yarmouth base. UKCS clients in the late 1960s included Amoco and Conoco, with production crews in demand. Concerned at the low level of UK content, he began tracking down British engineers from Kuwait.

Eventually, aware of the limited competition faced by Flopetrol from U.S. firms like Baker Oil Tools, he again crystallised the idea of having his own company and in 1973 called on OSO's DG. On learning no U.S. partner was needed, OSO advised Trewhella to prepare a business plan for placing before potential investors. To provide support and help develop the plan, Trewhella brought in another Cornish-trained engineer, Humphrey Green. Meanwhile, Trewhella had parted amicably with Flopetrol, which doubted he could compete successfully.

Ten potential backers were approached. New Court Natural Resources responded very quickly, investing £100,000 (about £900,000 in 2008 terms) for a 75% interest in a new company, Exploration and Production Services (North Sea) Limited, soon known as Expro. Trewhella held 15% and

Green 10%. Headquarters and laboratories were established in Reading, with operational bases in Great Yarmouth and Aberdeen. Services to be offered were well testing, wireline operations, bottom-hole pressure tests, PVT analysis and production crews, a range offering worldwide potential.

Expro soon won contracts and grew fast, finding the oil companies keen to have a UK competitor to the French and the Americans. Ten operatives defected from Flopetrol to Expro, which was also able to attract other good people. These included former managers from Kuwait Oil, such as Jimmy Ross who replaced Green as Trewhella's confidant and co-management shareholder.

The first (well test) contract was from BP, followed by others from Amoco, Shell, Phillips, and Burmah. The latter, following an unexpected southern North Sea oil discovery (see p. 67) had to improvise a well test burner. At Burmah's request, Expro developed an efficient burner, helping it consolidate its market position. Provision of field production crews soon emerged as the second key activity, rapidly taking Expro personnel numbers into the hundreds. Fields included Argyll (with Hamilton as Operator), Montrose (Amoco), Tartan (Texaco) and Thistle (BNOC).

In 1979, New Court had accepted an offer for its Expro shares from KCA, a deal strongly opposed by Expro's management. Trewhella and Ross succeeded in raising the £300,000 (about £940,000 in 2008 terms) necessary to exercise their pre-emption rights from three other venture capital firms led by Flextech.

By 1987, bases had been set up in the Netherlands, Norway, China, Egypt, Libya, Malaysia, Thailand and Tunisia. Despite difficulties following the 1986 oil price collapse, the company now employed 500–600 and had a turnover in excess of £30 million (about £61 million in 2008 terms). However, Trewhella left the company in 1987. By 1991, Flextech was sole owner and in 1992 sold the company to a management buy-out (not involving Trewhella). The lead financier was another venture capitalist, CINven Limited.

A Stock Exchange listing in 1995 revealed that turnover in the year to 31st March 1994 had risen to £65.4 million (about £93 million in 2008 terms) generating an operating profit of £10.4 million (over £14 million in 2008 terms), of which 60% came from the original group and nearly all the rest from two acquisitions made between 1987 and 1992. At the listing, the company was valued at £103.2 million (about £143 million in 2008 terms) and employed 832 people. The Group was described as providing 'high technology services', with an increasing focus on development and production, then about two-thirds of turnover. It was trading in 30 countries. Only 46% of sales arose from the UKCS, where it was believed to have about 50% of the well test market, over 30% of wireline operations and about 50% of the provision of downhole pressure surveys and data acquisition services. The Group's development work concentrated on the adaptation of existing technologies, with key developments protected by patents (Expro International Group PLC 1995, pp. 10, 13,14,17,18, 20).

Expro has continued to expand and, at the time of writing, remained an independent British-based company, although it has reverted to private (private equity) ownership. Expro showed that with competent management, a consistent strategy and adequate finance, it was possible for a British 'start-up' to compete successfully in a 'core' oilfield service segment.

8.2.3 Trafalgar House plc (TH)

Many of the large British companies that sought to enter the North Sea market could be classified as either construction groups or as conglomerates. Starting in the first category, TH moved into the second and in some ways can be treated as a representative of both, which is why it has been included.

The company originated in 1957, undertaking property development in London. In 1962, it became involved in house building and the following year was first quoted on the Stock Exchange. It grew rapidly, in large measure from acquisitions, moving from residential to commercial property, from house building to general construction, then to hotels, shipping and (for a few years) newspaper and magazine publishing (The Monopolies and Mergers Commission 1984).

A by-product of acquisitions in the 1969–1971 period was to bring TH into contact with the oil and gas industry. Cleveland Bridge and Engineering, one of the UK's principal bridge builders and heavy steel fabricators, was acquired in 1969. With 'offshore' antecedents going back to the Thames forts (see p. 45), it seemed almost pre-ordained to play a part in North Sea development. In the words of TH's driving force at the time the company '... *had ideas for their own design of steel production platforms*' (Broakes 1979, p. 217).

Despite considerable marketing effort and attempts to establish a joint venture with McDermott (see p. 124) as well as some initial interest in its design from BP for the later Forties field platforms recorded in the BP Archive, the company did not succeed in establishing itself in the platform area. BP was concerned with its inexperience and its unwillingness to quote against a BP/B&R specification. However, thanks in part to substantial investment, it did become an important module fabricator, trading as Cleveland Offshore.

The acquisition in 1970 of specialist construction and engineering contractor The Cementation Company, brought with it some servicing of oil wells and drilling equipment in the Middle East and Geoprosco International, a small geophysical survey business (Broakes, p. 217). A larger exposure to oil and gas followed in 1971 when the acquisition of the Cunard shipping company included a fleet of 28 offshore support vessels trading as Offshore Marine, sold to the U.S. firm Zapata in 1978. At the time of the Monopolies and Mergers Commission Report (p. 78), 16 of the vessels remained in TH's ownership.

These almost chance engagements with the offshore oil and gas industry and the emergence of substantial demand following the oil discoveries in the northern North Sea led TH to decide to enhance the existing interests with new ventures. In mid-1973, on his retirement as managing director of Shell Expro, George Williams was appointed to pursue this strategy. In the same year,

TH joined a consortium to construct concrete platforms at Portavadie (Broakes p. 219). The only activity was to develop the site under a government contract, which Broakes regarded as a 'lucky escape', given the over-estimation of platform demand.

Williams's remit was to expand rapidly in the service and supply sector, direct E&P involvement being ruled out. As joint ventures seemed to offer a way forward, discussions were held with Heerema, who soon realised that TH had little to offer. Better progress was made in the USA, particularly with the acquisition in August 1973 of a 40% interest in Dearborn-Storm, a Chicago-based company with interests in computer leasing, supply boats and jack-up drilling rigs.

In October 1973 came the oil price hike, followed by offers for the Dearborn-Storm shares. The first was rejected, but a higher offer was accepted from ODECO. Broakes (p. 218) rationalised this decision as a reaction to U.S. flag restrictions and rig oversupply. Williams, it seems, saw it rather as evidence that the TH management was a short-term rather than a long-term investor, probably contributing to his decision to leave at the end of 1974, after a stay of only 18 months.

The Monopolies and Mergers Commission Report showed TH had a turnover in 1982 of over £1.05 billion (over £2.7 billion in 2008 terms) and a pre-profit of £65.6 million (about £169 million in 2008 terms). Nearly 70% of turnover and over 54% of trading profit arose from 'Contracting, civil engineering, etc.' By then, continuing offshore business was effectively confined to the Teesside module yard of Cleveland Offshore, which escaped mention.

With timing triggered in part by opportunities arising from privatisation, 1982 saw the beginning of a new drive into the offshore sector. TH bought RDL from BSC, adding a second Teesside module yard to its fabrication interests. The acquisition did not include the Methil platform yard, although this followed in 1984 when 'a right of first refusal' was exercised, bringing with it Syd Fudge who was appointed managing director of TH Offshore. By this time, Methil's link with de Groot (see p. 156) had been terminated and replaced by an 'informal' joint venture with UIE. The Group's yards earned a generally good reputation. However, by the time of the new acquisitions, the longer-term outlook for their products for the UKCS was already questionable and export prospects poor.

As well as consolidating its position in fixed structures, TH turned its attention to floating structures, buying Scott Lithgow from BS in 1984. The move was intended to allow it to compete in the FPV market. Although the yard had experience in the construction of both semi-submersibles and drill-ships, it was not commercially successful and was accumulating large losses from late delivery penalties in respect of a semi-submersible drilling rig contract (see p. 196). The 1986 oil price crash damaged demand for production facilities, with the market for drilling vessels already depressed, Unable to make a success of Scott Lithgow with its by now poor reputation, in this negative environment, TH felt compelled to close the yard in 1990.

Also in 1984, TH took a 29.9% interest the John Brown Group, moving to full control in 1986. It included gas turbine manufacture at Clydebank and design and project management services in London, both successful businesses, and made TH a 'force to be reckoned with' within the industry. When in the early 1990s it became necessary to bid for the integrated risk/reward sharing contracts by then in vogue, the design and fabrication businesses were combined, thereafter trading as Trafalgar John Brown Offshore (TJBO). Among TH's other new ventures in the 1980s was the acquisition of North Sea E&P interests, after a few years floated on the Stock Market as Hardy Oil and Gas.

In 1991, TH purchased the Davy Corporation, a long-established British process plant contractor, with particular strengths in the metallurgical sector, which, it is said, TH wished to enter. Davy had made a late entry to the North Sea construction industry, taking a fixed price contract to convert an exploration semi-submersible into an FPV for the Emerald field, using LOG's former Kestrel Marine facility at Dundee.

The financial losses and commercial complexities of this contract (which led TH acting for a time as floating production contractor) was one of a number of factors that undermined the viability of the Group. As at Scott Lithgow, the difficulties seem to have been more commercial than technical, though in both cases design work was still being carried out during construction. Labour productivity was low at both sites and there was also labour militancy at Dundee. However, poor management was almost certainly more to blame for the problems than the work force; in particular lack of pre-acquisition due diligence exposing TH to the consequences of badly drafted contracts.

In 1992, control of TH passed to Hong Kong Land, followed in 1996 by an agreed take-over bid of £904 million (about £1.35 billion in 2008 terms) from Norwegian offshore group Kvaerner. TH had followed an opportunistic strategy, which was undermined by management failures in matters such as due diligence, timing misjudgements and short time horizons.

8.2.4 John Wood Group PLC

The company has been included as the most successful example of an Aberdeen firm becoming a significant member of the international energy support industry. Whilst a few other local firms grew into significant businesses with some international operations, only Wood achieved both a global reach and listed public company status.

The force behind the John Wood Group's drive into the oil and gas business was (Sir) Ian Wood. He joined the family business in 1964 when turnover was £550,000 and pre-tax profits £20,000, respectively, about £8.5 million and £300,000 in 2008 terms, generated from fishing, fish processing and marine engineering, (Mackie 2001, p. 383).

In 1969–1970, the company's engineering workshops first worked for the oil industry. In 1972, it acquired a small shipyard, giving it space to establish new oil-related activities such as a supply base. About this time, Wood hired two American oilmen, who *"knew the business"*. By 1975, the group had a turnover in excess of £18 million (about £111 million in 2008 terms), now mostly from the oil and gas sector, and had entered into the first of the joint ventures that were to figure prominently in its early expansion. This was the formation of Weir-Wood, aimed at servicing industrial equipment employed offshore (Mackie, pp. 383, 385, 388*)*. The arrangement helped develop expertise that allowed Wood to expand into the North Sea modifications and maintenance market, for long its most important activity.

At the time of its listing on the London Stock Exchange in 2002, the Group's prospectus referred the extension of its activities in the 1980's and 1990's ... *'into infield engineering, well support and gas turbine overhaul activities'*, with growth and International diversification stemming from '... *strong organic development, a series of targeted acquisitions and long-term joint venture*s' (John Wood Group 2002, p. 4).

By 1982, the Group had de-merged its original fishing related activities into a separate family-controlled holding company. The Group had an energy sector turnover of £70 million (about £180 million in 2008 terms) in 1982 and began seriously to internationalise, investing in the Middle East – gas turbine maintenance and repair – and the USA – oilfield services (Mackie p. 389). The latter was a new activity, but a small acquisition had already been made in turbine maintenance and repair in Aberdeen.

Acquisitions of varying sizes followed, mainly in the UK and USA, consolidating existing activities and entering new ones such as manufacture of well-related hardware and engineering design, including subsea production and pipelines as well as topside facilities. Joint ventures also continued to feature, particularly in the field of aero-engine derivative gas turbines. Part of the aim in growing the turbine service business, which for a time extended from aero-engines to heavy industrial gas and steam turbines, was to lessen dependence on the oil and gas industry.

The difficulties faced by the upstream industry during the mid-1980s to the early 1990s slowed development and led to divestments, such as the supply base. Despite large acquisitions, inflation-adjusted revenues in the mid-1990s were not much more than double the 1982 level. Growth accelerated again from the end of the decade, with revenues reaching about £1.05 billion and operating profits of £71.5 million in 2001, the year before listing, respectively, about £1.27 billion and £86.5 million in 2008 terms. At the offer price, the company's ordinary share capital was valued at £943 million (over £1.1 billion in 2008 terms). Financial reporting was by then in U.S. dollars, reflecting an international oil and gas business operating in more than 30 countries and employing over 10,000 people (John Wood Group pp. 4, 8). It had proved adept at adapting to the changing contracting climate in the oil and gas industry, with its increasing emphasis on out-sourcing, risk/reward sharing, broader contract scopes, and margin pressure.

Group structure at the time of listing reflected its evolution over the previous 30 years, with three divisions. Engineering and Production (50% of revenues) began with maintenance and production support and eventually extended to design services. Well Support (29% of revenues) included an electric submersible pump (ESP) unit designing and manufacturing equipment for production enhancement, a Pressure Control unit designing and manufacturing surface wellheads, valves, chokes and actuators and a Logging Services unit providing wireline services and permanent well monitoring. Well Support accounted for most of the Group's R&D activities and hence for most of the many patents held, though no single patent on its own was considered '*material*' to the business (John Wood Group, pp. 19–22, 28). The third division, Gas Turbine Services (21% of revenues), needs no further elaboration.

As the business grew, Ian Wood came to be seen as a spokesman for the indigenous service and supply sector and was prominent in BRIT during its short existence. He later became increasingly involved in official policy-making and implementation. Such diversions necessarily reduced his day-to-day management involvement in the Group, a loss compensated for by the development of a professional management team. Nevertheless, he remained chief executive and in control of the Group's strategic direction.

The success of the Wood Group began with the recognition of how the location and assets of an existing business at risk of decline could be used to facilitate entry to a new local market, then of an unknown scale. The necessary initial 'know-how' was gained from a combination of staff recruitment and joint ventures. Funding requirements and risk exposure were limited by avoiding capital-intensive activities, generally the most cyclical elements of the oil and gas industry. With this self-imposed constraint, financing could be met from a combination of retained profits and prudent borrowing, although in due course there was some dilution of the family shareholding by the issue of small equity interests to Scottish financial institutions and ultimately a move to the full public market. Acquisitions, where great care was taken to avoid over-payment, were used both to diversify the Group away from oil and gas and also to take it into the 'core' areas of oil and gas production, though not of exploration. Consistent long-term adherence to such policies protected the Wood Group from the worst effects of the large demand fluctuations with which the service and supply industry had to contend.

8.2.5 Generalisations from the Corporate Case Studies

The main question that arises is the extent to which the subject companies faced barriers to entry and how they addressed them. The question of barriers to new entrants is mainly discussed elsewhere (especially pp. 81–83). The most directly relevant work identified was that undertaken in Aberdeen by Hallwood (1990) during the mid-1980s.

Hallwood concluded that Aberdeen-based new entrants to what he considered to be the 'core' area of drilling and well services, faced severe barriers to entry from incumbents possessing technical 'know-how'/patents and operational credibility with an established client base. He could have added that most were adequately financed as members of long-established groups that had earned good profit margins during periods of excess demand in preceding periods. Most Aberdeen companies were engaged in locationally determined non-'core' activities. Hallwood's concentration on service companies and his narrow focus on Aberdeen limit the extent to which his work is appropriate to the four companies reviewed in above.

Only one of the four – the Wood Group – had its origins in Aberdeen. Its initial entry points were undoubtedly geographically determined, but management recognised that better margins and a long-term future with a large international component depended on breaking this link. It succeeded in doing this so, moving closer to, and eventually entering Hallwood's 'core', through joint ventures and acquisitions, overcoming its lack of 'know-how' and patents, of which it became a considerable holder. It preserved a sound financial base by 'pacing' its rate of expansion and avoiding capital intensive and highly cyclical activities. Though it used OSO services, nothing suggests that it depended on them.

The other company whose offshore activities were largely geographically determined was TH. It made acquisitions of established fabricators on the Tees, Clyde, and Forth estuaries, whose initial credibility had already been established with considerable OSO support and – in the case of platform construction – joint ventures. Since as contractors working to third-party designs, the TH yards were largely immune to the need for patents and employed only limited amounts of other specific 'know-how', available on the open market. Methil's joint ventures were mainly concerned with the provision of project management and 'buying-off' foreign competition. Since the fabrication sites were in locations that qualified for extensive regional aid and the parent was (at least ostensibly) financially strong, funding was not an issue. However, there was generally over-capacity in the fabrication business, producing fierce competition. Poor strategic management produced a business portfolio with too many risky large contracts and too few 'cash cows', which appear to have been the main reasons for its eventual demise.

Expro did not owe its origin and success in penetrating the 'core' of the service business to geographical chance or inherited corporate credibility. Though operating from bases in Aberdeen and other industry service centres, it established its head office in Reading. The reasons for this choice have not been investigated, but probably rested on easy access to London its airports and – possibly – to Cornwall. The necessary technical 'know-how' and client credibility was so clearly embodied in the founder himself and his personal associates – all initially British – that any idea of needing a foreign partner was positively rejected and the company soon developed its own proprietary technology. Venture capital was available to meet the fairly modest start-up costs, possibly because it was arranged before the true cyclical, technical and other risks faced by new entrants to the sector had been recognised in British financial circles. Competent management and successful trading in well-selected segments sufficed to retain venture capital interest for the 'long-haul' to a public listing. Support from OSO seems to have been important only at the very outset.

The activities of Vickers as precursor to BUE cannot be categorised as having been geographically determined, starting as they did in Barrow-in-Furness. From the start, Vickers embarked on an internal development course of building technological 'know-how', sometimes leading to patents, in what was to be a new 'core' area, underwater services. Since it was confined to offshore oil and gas, this new 'core' was smaller than Hallwood's, which covered both onshore and offshore oil and gas, but it was also less mature and therefore should have been easier to enter. Vickers failure can be attributed to a combination of political interference, market contraction and managerial errors.

BUE itself concentrated more on acquisitions than internal development. As a late starter in the new 'core', it saw this approach as the most cost-effective means of catching up with the French and American 'front-runners' in its range of service provision, rather than as a means of gaining 'know-how'. With OETB support, it continued parts of the Vickers R&D programme, which led to the foundation of the successful patent-based Hydra-Lok business. The failure of BUE can be attributed in part to difficult market conditions, but more to lack of equity finance originating in a change in political ideology, compounded by management misjudgements.

This analysis suggests that lack of technical 'know-how' and patents were relatively unimportant as barriers to entry. When felt to be necessary, they were relatively easy to obtain via licences, joint ventures, acquisitions or internal development, a conclusion consistent with the market segment case studies. The author's own recollection is that difficult to circumvent patents were important in only isolated cases of manufactured products such as drill bits (Hughes, USA), anchors (Bruce, UK), jack-up mechanisms (Le Tourneau, USA) and tubular joints (Vallourec, France and Hydra-Lok, UK). Client preference for working with 'tried and trusted' suppliers presented a more difficult barrier, though OSO could provide some help.

OSO was unable to help much in respect to a late start, adverse market conditions, technological and contractual changes, lack of equity finance and management failings. Both the corporate and market segment studies suggest that these areas were where the main responsibility for British failures lay. It is certainly no coincidence that of the four companies studied, the two successes were both characterised by consistently strong management.

8.3 EXPERT TESTIMONY

From Table 3.2 (see p. 79), it can be seen that the responses of five senior public servants and politicians suggested that support for British industry rated fourth out of the five main UKCS policy drivers. The same group was also asked to address questions relating to problems faced by British industry in the offshore sector, particularly by new entrants, but only two felt able to respond.

It was decided to include these two individuals with three former senior OSO officers (the author not included) and to compare the views of this public sector group of five with those of a private sector group of nine business figures, six British and three foreign, all with considerable experience of the UK offshore industry. Twelve of these fourteen individuals were also asked a number of other questions on the effectiveness of OSO policies and all engaged in a more unstructured discussion of issues where they were believed to have special knowledge, though the extent was limited with the foreign nationals who were consulted by email rather than in face-to-face.

The analysis of the questionnaire results and pertinent generalised points the respondents made are presented on a non-attributable basis. Although the groups are small and clearly cannot be presented as a statistically valid sample from which reliable general conclusions can be drawn, the individuals represent long experience across a diverse range of offshore activities. The opinions expressed were sometimes contradictory and are not necessarily shared by the author.

A few minor adjustments were made to the numerical assessments submitted because not all respondents had, as requested, 'scored' to a base of 100 and in one case a respondent had given a large weighting under 'Other' to 'Fear and ignorance'. The author combined this with 'Risk aversion', one of the potential constraints suggested and from which he did not think it differed materially.

Although both private and public sector groups agreed the performance of British industry was adversely affected by multiple problems, there were considerable differences between them in the importance attached to individual factors. Table 8.1 overleaf deals with the public sector group.

The public sector group ranked 'Poor management' (1) and 'Shortage of funds' (2) as the most serious constraints, followed by 'Existing client–supplier relationships' (3=) and client 'Perception of superior overseas management' (3=), with 'Poor labour relations and productivity' (5) a little behind. The ideas that there was 'Insufficient government support' (9) or a 'Lack of time to acquire experience and/or technology' (10) found little support. The remaining three factors were held to be of middling importance. One of these was 'Some others' (6=) where respondents were invited to remedy what they saw as any important omissions from the list. Additional suggestions volunteered were (i) the short life anticipated for North Sea production and (ii) risk aversion on the part of the oil companies (as opposed to their potential suppliers).

The private sector group's responses are summarised in Table 8.2 (see p. 241). The private sector group believed that 'Risk aversion' (1) and 'Existing client–supplier relationships' (2) were seen as the main constraints The next highest ranked were 'Insufficient government support' (3) and 'Shortage of funds' (4), although the latter would have rated as the most severe constraint of all if only responses from the four with entrepreneurial experience were taken

TABLE 8.1 Public Sector Group's View of the Main Constraints Faced by British New Entrants to the Offshore Service and Supply Sector

Factor	Average Score	Range	Rank
1. Lack of time to acquire experience and/or technology	5.4	0–14	10
2. Risk aversion	9.6	0–16	6=
3. Shortage of funds	13.4	5–23	2
4. Poor labour relations and productivity	11	5–15	5
5. Poor management	13.8	10–20	1
6. Insufficient government support	5.8	0–15	9
7. Existing client–supplier relationships	12	5–20	3=
8. Perception of superior overseas management by clients	12	0–18	3=
9. Perception of superior overseas management by government	7.4	0–18	8
10. Some other(s)	9.6	0–18	6=
11. Total	100	n.a.	n.a

into account. At the other extreme were 'Some other(s)' (9) and 'Perception of superior overseas management by government' (10). The private sector group 'Some others' constraint suggestions were lack of a national champion, lack of training and education and protectionism in Norway, the USA, and Brazil.

Although both groups agreed that 'Perception of superior overseas management by clients', 'Existing client–supplier relationships' and 'Shortage of funds' were important obstacles for British new entrants, there was otherwise a general lack of correspondence between their opinions. While it is perhaps not surprising that the private sector group should place 'Insufficient government support' high among the constraints or that the public sector group should do the same for 'Poor management', the fact that 'Risk aversion' was seen by the private sector group as the greatest single constraint when the public sector considered it of only middling significance is more remarkable. No member of either group volunteered patents (or rather lack of access to them) as a constraint, which supports the view already expressed (see p. 238) that this was not seen as a problem.

Ten of the twelve respondents (two public sector and eight private sector) also responded to set questions about British government support policies,

TABLE 8.2 Private Sector Group's View of the Main Constraints Faced by British New Entrants to the Offshore Service and Supply Sector

Factor		Average Score	Range	Rank
1.	Lack of time to acquire experience and/or technology	9.1	0–30	7
2.	Risk aversion	18.0	5–50	1
3.	Shortage of funds	10.6	0–32	4
4.	Poor labour relations and productivity	7.3	0–20	8
5.	Poor management	9.6	0–20	6
6.	Insufficient government support	11.3	0–35	3
7.	Existing client–supplier relationships	15.1	0–35	2
8.	Perception of superior overseas management by clients	10.6	0–20	5
9.	Perception of superior overseas management by government	3.1	2–10.5	10
10.	Some other(s)	5.1	0–20	9
11.	Total	100	n.a.	n.a.

whether or not an internationally competitive British offshore industry had emerged by the mid-1990s, whether the outcome could have been improved by different policies (and, if so, which) and what became the strength and weaknesses of the industry. Table 8.3 overleaf below deals with the support policies. In view of the smaller number of respondents, they were considered as a single group.

Although the results shown in Table 8.3 seem narrowly to confirm the effectiveness of British government policies in the fields of FFO and the use of licensing powers, they hardly amount to a ringing endorsement overall. This ambivalence is scarcely surprising as on the key question of whether or not an internationally competitive British offshore industry had come into being by the mid-1990s after 20 years of government support, the group was equally divided, although two of those responding positively did so only with heavy qualification.

Nine out of ten believed that, from a British perspective, the outcome could have been improved upon, with one of the foreign respondents as the only dissenter. Only one of the nine put the main responsibility for underperformance upon industry and only two upon government, with the remainder holding both to blame.

TABLE 8.3 Opinions on British Government Support Policies for the Offshore Industry

Policy	Viewed As
'Full and Fair' Opportunity Policy	Generally effective – 6 Generally ineffective – 4
Use of Licensing and other Regulatory Powers over oil companies to favour of local suppliers	Generally ineffective – 6 Generally effective – 3 Don't know – 1
R&D Support Programmes such as OETB and MTD	Inadequate in scale – 4 Adequate in scale but ineffective – 3 Don't know – 3
Parity of treatment between British-based suppliers irrespective of ownership	Desirable in principle – 5 Undesirable in principle but unavoidable – 4 Don't know – 1
Inability to nurture key technologically or otherwise strategically important British owned companies or to protect to them from foreign takeover	Undesirable in principle but unavoidable – 5 Undesirable in principle & avoidable – 3 Don't know – 1 Unusable answer – 1
OSO Export Promotion Efforts	Generally ineffective – 5 Generally effective – 4 Don't know – 1

The principal criticisms of government policies included the failure of British governments to persuade leaders of industry that the prospect of the UK becoming a world leader in offshore service and supply was real, its failure to get state scientific and engineering bodies to give a high enough priority to E&P technology and its failure to persuade the City to provide the venture capital for its commercial exploitation. Moreover, the government needed to have better understood the importance of national ownership of companies and technologies and to have had a policy/framework for continued support of UK-owned businesses, some adding that equal treatment of foreign owned companies operating in the UK actually created confusion. Further, in the early years, the British government should have provided more and better focused R&D support, more 'coaching' or guidance of young companies and more 'teeth' to OSO. A large, well-funded research institute (like IFP) would have provided a significant catalyst for the development of British companies, which after the demise of BNOC were placed at a major disadvantage by the lack of a state oil company compared to those in Norway and France where it proved critical to their development. Criticism of a different nature came from one contributor who believed government should have exerted greater pressure for improvement on inefficient companies benefiting from the UK content requirement.

As far as British industry was concerned, criticisms included the view that in the early days, companies that might have become major players failed to recognise the long-term opportunities, were reluctant to take risks and preferred to pursue low-risk defence and other public sector contracts. Managements were often poor in terms of ability to focus on a corporate aim. This was compounded by the big company culture of the day, such as rigid pay scales and stratified management hierarchies, which made appropriate offshore terms and conditions of employment difficult to implement. Such companies also sometimes became over-reliant on OSO to do their marketing. More broadly, there was too much focus on finance and too little on technical entrepreneurship; concentration on the home market also continued for too long.

Finally, the respondents were asked to give their views on what became the 'Strengths' and 'Weaknesses' of the British offshore industry. There was widespread agreement (six out ten respondents) that the main strength had been the development of a large, skilled and internationally mobile labour force. The only other strength mentioned by more than one respondent was the breadth of small, specialised companies created. With respect to British-owned companies, one respondent went so far as to say there were none with major strengths, but ignoring current ownership, there were or had been strengths in deepwater and floating technology and subsea equipment and services.

There was less of a consensus on weaknesses. Three respondents referred to the inability of the British offshore industry to handle complete turnkey projects from within its own resources, principally due to a failure to invest in offshore drilling and installation equipment. Four others referred to corporate structure issues; a lack of consolidation had led to a lack of companies with the critical mass to compete effectively internationally or to provide the number of quoted companies necessary to give the London Stock Exchange a worthwhile 'Oil Equipment and Services' listings group. It was held that the latter would have benefited the industry generally by making it better known and understood in the financial community, thereby facilitating the raising of finance. One respondent saw the British industry as fragmented (impacting on margins), failing to have developed UK technology base and insufficiently invested overseas.

Some of the other noteworthy observations made by the respondents during more general discussion are worth repeating, though many have already been alluded to elsewhere. Few saw labour militancy as a problem, though low productivity was, particularly in former shipyards where unions were reluctant to abandon demarcation. The real 'human relations' problem was poor management, particularly short sightedness and lack of leadership at Board level. There was no lack of entrepreneurial skills or innovations, but domestic equity backing was lacking. The contribution of venture capitalists was minimal because the offshore service sector did not fit their model. This made the creation of new, dedicated businesses difficult, forcing the oil companies to use the existing industrial structure with all its problems. The commercial banks acted within their normal business model; the Scottish clearers were better than the English but offers were almost invariably in the form of overdrafts. This led to over-geared companies, exposing both companies and banks during market downturns. The Treasury's 'addiction' to inward investment as always being in the UK's best interest prevented the government giving greater support to UK owned companies. Support should have been more concentrated on 'high-tech' activities and less on fabrication, that is the government was too concerned with short-term employment. This led to foreign-owned companies being able to 'steal' 'high-tech', more value added activities, leaving lower margin roles to the British. This problem was compounded by a lack of an integrated R&D strategy. Too high a proportion of R&D results were placed in the public domain. The MTD programmes in particular were mainly driven by the commercial needs of the oil and service companies but resulted largely in free 'knowledge transfer' (e.g. publication of codes and guidance) rather than the creation of proprietary products and services. Moreover, British universities that did technically clever things were inept at exploiting them commercially.

As far as OSO itself was concerned, given 'lack of teeth' and the constraints under which operated, it was widely seen as a success. Politicians of both major parties were supportive. Constituency interests were not abused. However, OSO suffered from its inability totally to escape from the traditional civil service culture or to co-ordinate fully with the PED on the areas of expertise required for the future. OSO also sometimes lacked support from departmental top management, leading to over-reliance on Ministerial backing. This itself fluctuated under the Thatcher administration, affecting the strength of resistance to periodic pressures from the EC. OSO was also adversely affected by many changes in government administrative structures and from the flawed decision not to create and maintain a separate 'Ministry of Petroleum'.

OSO was not without internal problems. In particular, it failed to develop a consistent internal view of which supply sector companies to see as 'champions', partly due to internal differences between the Engineering and R&D branches. This inhibited OSO's ability to develop its own 'champions' within the oil companies, who would have taken a special interest in companies 'flagged' by OSO. On the international front, most of the time the EC took little interest in OSO, despite periodic protests from the Dutch and Germans. The USA was at times actively hostile to OSO and Norwegian and French chauvinism were always present.

In general, the expert testimony reinforces the findings from the other research undertaken, adding relatively little that was new, which in itself is reassuring to the author. Unfortunately, although there were some useful suggestions on how British industrial performance might have been improved, no clear set of practical alternative policies amounting to a different strategy to the one adopted by the British government emerged.

Chapter 9

Looking Back on a 30-Year Journey

The point has now been reached where it is possible to try to assess the response to the North Sea from British industry and the way this was influenced by government priorities and policies. Perhaps the easiest way of starting to do this is to devise and examine a number of propositions.

9.1 SOME PROPOSITIONS

Based on what this book has exposed and/or premises that have circulated within the British offshore industry, there are nine such ideas to be investigated. They are:

i. *The scale of North Sea demand and the requirement for technical innovation were such that they offered the opportunity to create a major new industrial sector, with export potential.*

The new technical demands placed on the offshore industry by the opening up of the North Sea province were clear and are discussed on pp. 56–59. For many years, the UKCS had been the world's largest theatre of offshore operations and still accounted for 20% of global offshore expenditure as late as 1983 (see p. 174). It was also prominent in the development of new 'core' activities specific to offshore operations, particularly in the underwater field (see pp. 174–175), offering the opportunity of an export base.

There can be no doubt of the scale of the demand or that simple extrapolation of existing offshore technology was an inadequate means of addressing much of it. The evidence would seem to support this proposition.

ii. *The UK had failed to establish an internationally competitive offshore service and supply industry by 1993, 30 years after its inception.*

During the mid-1980s, considerable misgivings were expressed about the export performance of the British offshore supplies industry (see pp. 187–188). Despite substantial export promotion efforts by OSO over the following decade, half of the industry participants consulted did not believe that an internationally competitive industry had been established (see p. 241). Moreover, of those that did, some made their answer subject to serious qualification.

Support for the proposition is therefore quite strong. Moreover, it would be difficult to avoid the conclusion that had the question been framed to exclude the foreign-owned component of the British industry, it would have been much more so. The weak positions in mobile drilling and in construction alone would have demanded as much. The remaining propositions concentrate on why this was the case.

iii. *The opening conditions of the British economy and the general policy priorities of British governments did not create a favourable environment for the development of major new industries.*

As shown on pp. 1–4, during the 1960s and 1970s British economic policy became determined by a succession of balance of payments crises, leading to a series of 'stop–go' economic cycles, devaluation, a currency float, and rising trends in inflation and interest rates. It was widely perceived that British industry was suffering from a range of ills, including poor management, poor labour relations, low productivity, under-investment, insufficient innovation and general lack of enterprise. Such issues are discussed on pp. 10–19. It is clear that, although some of the perceived difficulties may have been exaggerated, the British economy was in long-term relative decline, as well as suffering from severe short-term problems. This environment discouraged the establishment of new businesses requiring substantial new investment and long payback periods. Entrepreneurs were accordingly attracted to short payback projects in the property and financial sectors, where they were readily accommodated by the banking system (see pp. 88–89).

The evidence clearly supports this proposition, which provided the drivers for the government policy of maximum speed in early North Sea oil extraction, regardless of the consequences to UK industry. It was also probably a partial explanation of the 'addiction' of successive British governments to inward investment.

iv. *Within government oil and gas policy, support for the domestic service and supply industry came late and was always a subsidiary aim rather than one of the primary aims, to which it was subordinated.*

The industry was already nearly 10 years old when OSO, and with it a formal policy of support, came into existence in 1973 (see p. 109). By contrast, the primary policy aim of the rapid exploitation of the hydrocarbon resources of the UKCS had been in force from the outset (see p. 70). In the view of a group of former policy makers, 'Encouraging British industry' ranked only fourth among government objectives behind 'Security of Supply', 'Improving the Balance of Payments', and 'Supporting British E&P companies' (see p. 79).

The poor position of industrial support among policy objectives extended to the R&D area. Industrial support accounted for only about a fifth of the

OETB's early annual budgets (see pp. 137–138) and remained small throughout the period reviewed (see pp. 177–180).

The evidence supports this very proposition very strongly and it was the inevitable corollary of proposition (iii) above. It can be seen as a defence from some of the less well-informed criticisms made of OSO.

v. *The exploration and production companies were uninterested in the development of an indigenous British service and supply sector per se.*

In the initial phases of North Sea activity, British companies continued drilling their own wells (and occasionally even hiring their rigs to third-parties) and committed large sums to the construction of new drilling rigs in British yards before becoming disillusioned by instances of poor response from British industry (see pp. 94–95). The arrival of a significant number of U.S. companies totally altered both the demand and supply sides of the industry. U.S. Operators in many cases had longer offshore experience but on the whole were smaller than their UK peers. Their U.S. Operations had depended heavily on the use of the contracting and service company community that had developed to serve them, a community large enough to offer a measure of internal competition. Finding no similar community in the UK, such companies encouraged their existing suppliers to establish themselves in the UK. The arrival of the U.S. contracting and service companies was generally welcomed by the British government and freed the British oil companies from the need to foster the growth of the domestic alternatives otherwise required as rapid growth in demand outran their abilities to source in-house.

The evidence seems support this proposition to the extent that the British Operators were relieved from having to address the issue.

vi. *The cyclical pattern of demand and uncertainty about the potential scale and lifespan of UKCS activity worried business decision makers, particularly those outside the oil companies.*

As is apparent from Chart 4.1 (see p. 94), a cyclical pattern of demand for goods and services came to characterise the UKCS, particularly in the development expenditure area where British-owned firms were concentrated. This led to alternating 'boom and bust', with periods of excess capacity and low prices, especially in fabrication where a change in the nature of demand away from large fixed installations towards lighter structures, FPVs and subsea installations created additional difficulties.

Always in the background, particularly so in the 1970s when most new entry decisions were made and when the briefness of the initial period of capital investment in the southern North Sea gas basin was still fresh in people's minds, as was the uncertainty surrounding the ultimate scale of UKCS reserves and the timescale for their exploitation (see pp. 53–56). Though proof is hard to come by, it seems certain that these considerations provided support for those who argued in favour of a cautious approach to the offshore service and supply industry.

The evidence supports this proposition. Its impact could have been lessened and a steadier pace of development achieved by the adoption of a depletion policy. This might also have lessened the very high proportion of UKCS oil production occurring under conditions of low oil prices.

vii. *By the time the North Sea opened up, U.S. service and supply companies had already achieved an unassailable lead in offshore oil and gas technology based on reputation, patents and scale economies, presenting potential British competitors with insurmountable entry barriers.*

Since the USA saw the origins of the offshore oil and gas industry, this inevitably gave U.S. companies a 'head start' in the provision of specialised services and the equipment, such as mobile rigs and offshore construction and pipelay equipment (see pp. 32–34). Similarly, the large number of wells drilled and in operation (albeit mostly shallow and low productivity) onshore provided economies of scale for firms involved in the provision of products and services related to the 'core' areas of drilling and well services, to a large extent transferable to the offshore scene. The large number and often modest size of U.S. E&P companies favoured reliance on specialised contractors, already well-experienced when the development of the North Sea began. Though patents undoubtedly existed, the protection they offered seems to have been important in only a few cases. Market position was based mainly on reputation and an abundance of experienced manpower, with economies of scale and ownership of specialised equipment also relevant in some cases.

By the time the development of the North Sea began, European competition to U.S. contractors had already begun to emerge – from the Netherlands, France and Italy (see pp. 50–53). In the first case, it was in offshore construction and offshore marine equipment and essentially a private sector initiative. In the other two cases, the scope was wider, with government support for the aim of the national oil and gas companies being able to operate without the need to use U.S. contractors. France had been strong in drilling and well services from much earlier (see pp. 27, 50) and had state sponsored R&D programmes from the end of the Second World War.

The position in the UK was different. BP and Shell were among the largest oil companies in the world and their extensive in-house resources had left little need for the use of contractors, though this began to change as they moved offshore. UK offshore contractors did not exist prior to Wimpey's entry in the late 1950s. British contenders suffered from a numerically limited client base compared to the Americans and the absence of the government support available to the Europeans. Although it offered a domestic market with a good client spread for the first time, British firms generally reacted sluggishly to the development of the southern North Sea.

Nevertheless, despite the scale and urgency of the situation, 'sucking in' a very large part of the world's existing stock of specialised equipment and experienced personnel, the initial development of the northern North Sea still allowed British-owned firms with only modest government support to enter almost all sectors of the offshore contracting and oilfield services businesses, with the notable exception of offshore construction. With structural design and specialised well hardware progress was less, thanks to the early implantation of U.S. subsidiaries, though for a time British manufactures emerged as leaders in subsea well controls.

The evidence does not support this proposition. The widely held perception that American companies held unassailable positions when North Sea activity began did not always reflect the reality. Indeed in some areas, such as Heerema (Dutch) in heavy lift and Schlumberger (French) in well-logging, European rather than U.S. companies were already the leaders.

viii.   *the British offshore service and supply industry failed in general either to produce entrepreneurs or technical innovations;*

The presence of entrepreneurship is confirmed by Tables 5.2 (see p. 154) and 5.3 (see p. 168). Although less systematically presented, the study presents sufficient cases of British technical innovations during the first 30 years of North Sea activity to show that this too was present. Notable examples are the Bruce anchor (see p. 150), the PORES reservoir simulation software (see p. 141), the Vortoil separator (see p. 137), the Hydra-Lok tubular steel bonding system (see p. 176), the Gall Thomson marine breakaway coupling (see p. 139), subsea pipeline survey by unmanned vehicle (see p. 226) and Expro's efficient well test burner (see p. 231). Others are mentioned briefly in the context of the OETB (see pp. 177–180).

Additionally, British engineers played the leading part in the North Sea's role as the proving ground for subsea production technology from the time of the Cormorant Central scheme (see pp. 130–131) onwards, although the critical components such as flexible pipelines, umbilicals, wellheads and controls remained (or became) almost entirely foreign-controlled, despite various British initiatives.

The evidence does not support the proposition, even though few significant and long-lasting new British-owned businesses resulted and many of the innovations passed quickly into foreign hands before their full commercial potential had been realised.

ix.   *Lack of venture capital/private equity inhibited the formation and adequate finance of new enterprises, resulting in a reliance on the existing industrial structure.*

The case that venture capital/private equity was in short supply in the 1960s and 1970s and that the scarcity of external start-up finance constrained the initial capital of many new ventures is outlined on pages 90–92. Shortage of start-up funds and the lack of an effective 'junior' market for trading in shares was a recipe for a static industrial structure in the unquoted sector, sustained by the banks and development capital institutions. This necessarily impacted most heavily on activities where technical risks and/or capital requirements were high, such as many segments of offshore activity, which remained true even after venture capital/private equity became more readily available late in OSO's life. British entry to such segments was, therefore, mainly dependent on the decisions of existing large companies, few of which made any attempt to raise new equity for this purpose.

There is no formal statistical data on the provision of institutional funding for the service and supply sector. From the anecdotal evidence, it seems fair to say that start-up capital was most readily accessible in the early 1970s, with Seaforth Maritime founded in 1972 (see p. 147), Expro in 1973 (see p. 151) and Star Offshore Services in 1974 (see p. 147). No latter start-ups of a comparable scale followed. The early 1970s also saw the creation of specialist investment institutions targeting the industry, particularly Flextech (see p. 150) and NSA (see p. 112). At about £194 million (in 2008 terms), the latter raised a significant amount but its poor performance did not encourage imitators.

Throughout the late 1970s and most of the 1980s the provision of equity finance for the specialist British service and supply sector was particularly weak, not surprisingly given the often poor offshore market conditions. Thereafter, the situation began to improve. State equity providers such as the NEB and the SDA/SEn did not materially alter the situation, the former being too short-lived and the latter too small a scale investor. Government grants of selective financial assistance were too limited to have much overall effect (see pp. 151–152). Established companies and new inward investors were the main beneficiaries of mandatory regional financial assistance. The clearing banks, particularly the Scottish, reacted positively to the new business opportunities created in the service and supply sector, but did this predominantly within their existing business model based on debt rather than equity finance, although in the early 1970s the Bank of Scotland in particular took a number of minority equity stakes. The fairly ready availability of bank finance and the absence of alternatives contributed to many unsound financial structures.

Despite the obstacles, numerous established companies did enter the sector, albeit often at inadequate scale and for a short period. They usually did so without raising new equity, though there exceptions such as Robb Caledon/BEFL (see p. 147) and Ben Line/Ben ODECO (see p. 148). Many already faced difficulties in their existing businesses and were ill informed of the nature of offshore opportunities; caution and the need for positive short-term financial results were commonplace.

There can be little doubt that the evidence supports strongly this proposition, or that it paid a major role in the disappointing performance of many British firms trying to enter the market in competition with longer established and better financed foreign competitors. However, it is probably not too difficult to understand the attitude of financiers after the initial euphoria of the early 1970s gave way to a long period of difficult trading for many – though not all – parts of the North Sea service and supply industry and the difficulties they long had in understanding it. These difficulties were for all aggravated by the extremely few service and supply companies with public markets in their shares and hence of interest to investment analysts.

9.2 SOME CONCLUSIONS

British industry's failure to use the North Sea opportunity to establish a prominent position among the leaders in the global offshore service and supply market, or even to have a significant share of many important sectors of its domestic market, was clearly a disappointment to the many British nationals who sought a different outcome.

This negative outcome would probably have come as a surprise to an informed observer in the period immediately prior to the start of North Sea exploration. Such an observer would have known that the UK was 'home' to three of the world's oldest and largest oil companies, rich in experience, technical expertise and funds Their far-flung activities already included offshore operations, albeit in benign environments, since there were no others at the time. He or she would have known that Britain already had a long tradition as a supplier of goods to the oil and gas industry, with a supply chain comprising both domestically and foreign-owned firms. A new breed of British contractor was also emerging, with the likes of KCA (drilling contracting) and Wimpey (offshore contracting) already active in the Middle East and North Africa.

Given the country's large heavy engineering, shipbuilding and marine industries and the reputation of its structural engineering designers, he or she would not have doubted its ability to design, construct, install and man the structures needed to support offshore drilling and production. The Maunsell structures remained intact after 20 or so years standing in shallow waters some miles from the British coast and the early appearance of Cleveland Bridge's deepwater 'Colossus' platform design was to provide evidence that such design skills remained. Even in the emerging discipline of offshore pipelines, the informed observer would probably not have doubted that these too could be provided domestically. After all, Britain produced line pipe and had invented and first employed the reel lay technique that was eventually to become widespread. In short, it was reasonable to believe that initial gaps in capability could soon be rectified.

For this promising prospectus to be unfulfilled, there must have been powerful countervailing forces at work, as indeed there were. Except, presumably, within the oil companies themselves, there appears to have been little or no anticipation at senior levels in British government and business of the extent to which oil and gas production would move on to the continental shelves. There is nothing to suggest that there was any serious early thought given to the potential export market opportunities, whether or not there were discoveries off the British coasts.

Similarly, there appears to have been little or no appreciation outside the oil companies of the speed with which a 'cluster' of contractors with a distinct and rapidly evolving technological base was developing in the GoM in the 1950s and early 1960s. As a consequence, there were no British 'special measures' such as long-term R&D programmes, education and training initiatives or encouragement of investment in specialised equipment put in place in advance of UKCS operations. There were such measures elsewhere in Western Europe, particularly France.

To achieve rapid exploration and exploitation of the North Sea, the UK encouraged participation by foreign, mainly USA, E&P companies as well as British ones. For the initial southern basin activity, government limited its support for British suppliers to exhortation to the oil companies. The latter responded to the extent that British capacity existed but were soon discouraged by instances of poor performance, particularly in rig construction, or disinterest, as in platform construction. A tendency for early British new entrants to give up when difficulties arose rather than to persist and move along the 'learning curve' must have been a disappointment for their customers.

Meanwhile, generally smaller and less technically self-sufficient than Shell and BP, U.S. Operators responded to the lack of a British oilfield service and offshore contracting industry by encouraging companies known to them from their U.S. operations to establish themselves in the UK. One such company, B&R, working with its Dutch partner Heerema, had installed fixed platforms in the Dutch and German sectors of the North Sea in 1964, the year of the first British licensing round, and was already well known to BP. B&R was awarded the first field development contract for the UKCS (BP's West Sole gas field), beginning its long period of dominance of the UKCS as an offshore contractor.

The reaction of British industry to the opportunities offered by the southern North Sea market can at best be described as sluggish. There were a few initiatives from outside the oil sector, such as supply boat and helicopter operations and small-scale fabrication and telecommunication businesses. There was also an expansion of the B&R/Wimpey Middle Eastern association into the UK supply base and supply boat areas. The most significant developments were probably those undertaken by the three main British oil companies themselves in offshore drilling.

The weak response by British business and consequent penetration by foreign firms resulted in part from the factors described immediately above and in part from the relatively limited scale and short life of the initial southern basin opportunity. Other important underlying causes lay in the structural problems of the British economy, such as bad management, low productivity and poor project control. Moreover, government policies and the financial system were biased in favour of maintaining existing industries and firms, often by encouraging mergers, rather than creating new ones. Finally, for some companies it made more sense to employ resources on low risk contracts then available from government-funded defence and infrastructure programmes. In any event, the experience of the southern North Sea did little to prepare British business for the much larger opportunities that were to arise in the northern North Sea.

By 1971, it had become apparent that the northern North Sea contained large commercially viable oil fields, development of which would require more than a further simple 'stretching' of existing offshore technology such as had occurred with the southern basin, as well as extremely large expenditures; the extent of both factors was underestimated. Underlying economic conditions were deteriorating so that by 1973–1974, the UK was in the throes of a serious economic crisis from which it would not fully recover for many years, overlain by a pervasive fear for the security of energy supply.

Whether Conservative or Labour, governments inevitably saw the most rapid development possible of the North Sea as a means of escaping these problems and gave it maximum priority, safe in the knowledge that the oil companies were also committed to a similar policy of rapid development. The result was that UKCS expenditure multiplied in real terms more than fivefold between 1971 and 1976, when it exceeded £13.4 billion in 2008 prices, which resulted in a scramble to mobilise resources of near wartime proportions. Almost irrespective of the condition of British industry, it would have been impossible for such explosive growth in demand to have been met without a heavy reliance on foreign companies. Allowing domestic capability the time to 'catch up' with the burgeoning market was not an option; it was clear that a very large part of the global offshore industry's existing resources would have to be mobilised to match demand. As this new demand was in effect incremental to them, it allowed established foreign firms to achieve a combination of high margins and high capacity utilisation, giving them the cash to continue to invest heavily, retire borrowings and/or build up liquid assets. Such benefits were unavailable to British firms recently entering the industry, which were therefore less well-placed to cope with the downturn that was shortly to occur.

Despite its unresolved problems, British industry, including the many foreign-owned firms by now established, reacted more vigorously to the northern North Sea oil 'boom' than it had done to the opening-up of the southern North Sea, establishing in particular fabrication facilities and a supply base infrastructure, with government support mainly from regional development grants. Nevertheless, unless the supply response from British industry was further 'catalysed', there was likely to be deterioration in the balance of trade as equipment and service imports surged in advance of oil production, as well as the loss of opportunities to mitigate the unemployment arising from the run-down of shipbuilding and heavy engineering.

The result was the commissioning of the *IMEG Report* in 1972 and the establishment of OSO at the start of the following year. Its responsibilities included helping to speed development as well as increasing UK content, objectives not always mutually compatible. Its powers were weak and, beyond exhortation and the provision of information, depended mainly upon leverage arising from the DEn's licensing role, which could not come even partially into play prior to the Fifth Licensing round in 1976/1977, by when many potential opportunities for British-owned businesses had already been lost. The formation of the OETB in 1975 initially did little to strengthen OSO's role. BNOC, established in the same year, was too short-lived to have very much effect. The government's policy of encouraging inward investment denied OSO the ability to favour British-owned companies or to prevent foreign acquisition of 'strategic' British-owned companies. Its successes were largely confined to encouraging investment in facilities in the UK by overseas suppliers, facilitating joint ventures as a means of technology transfer to local firms and concentrating the minds of oil companies on achieving high UK content as a means of doing well in licence awards. The last was exploited as a means of ensuring that labour-intensive fabrication work was largely undertaken at UK sites, a matter of great interest to politicians given the employment implications and high public visibility of the contracts. Being measured in turnover terms, it also encouraged domestic procurement of 'commodity' type low-value added supplies. The net effect was to give the Operators considerable freedom to procure high-value added items abroad.

After its 1976 expenditure peak the market became cyclical for the remainder of the period studied, with troughs in 1979 and 1987 and peaks in 1984 and 1992. The periods of declining demand exposed over-capacity in areas like platform and module construction, leading to closures. Elsewhere, they revealed the presence of companies of sub-economic scale, with parents having problems elsewhere in their businesses, or simply looking to capitalise on their market or technology position. In most such cases, the result was sale to a foreign purchaser. As a consequence, footholds established by British-owned industry in key areas such as underwater technology were lost, despite having been in part publicly funded. Another issue was lack of access to long-term risk capital (i.e. equity), with the resultant undercapitalisation of companies all too common. Such difficulties undermined much of OSO's early work. Later OSO successes such as the 'supply boat initiative' and the greater engagement of oil companies in R&D provided inadequate compensation.

9.3 COULD IT HAVE BEEN OTHERWISE?

Inevitably, the question arises of whether the British industrial response could have been improved, and if so how. This is not easy to answer, partly since it is almost impossible to eliminate the effect of hindsight. Even more important is to keep any suggested modifications of government policy consistent with what would have been possible at the time. Thus, it is pointless to suggest that the most obvious way to have improved performance would have been to adopt the Norwegian policy of relatively slow depletion, in part determined by a desire to match demand to industrial capability.

Though such a policy would have been an option for an independent Scotland, it was never possible for Britain whose circumstances demanded a policy of rapid development and maximum output at least until the early 1980s, by which time the main elements of the industrial response had already been determined. Thereafter, a more measured rate of depletion would have probably made little difference beyond some possible smoothing of demand after the initial North Sea investment surge. Even this is doubtful since oil price movements and tax changes could still have introduced cyclicality, whilst pressure from the bloated fabrication sector, which had arisen from the events of the early 1970s, would have continued to favour unconstrained development.

Again, it would not make sense to suggest policy changes other than those likely to be acceptable to both Labour and Conservative administrations. To do otherwise would to be to invite across the electoral cycle more of the disruptive policy changes like the establishment and dismemberment of BNOC. Thus, the successful Norwegian approach of endowing a state oil company with powers to ensure that licensees co-operated in the development of industrial capability over a protracted period must be ruled out. Similarly, it would be unrealistic to propose policies involving massively increased levels of government direct investment or public funding of the private sector. Although these would have been ideologically acceptable to the Labour party of the time, administrations of neither party were often in a budgetary position to undertake them. Conversely, industrial policies favoured by both parties, particularly freedom for inward direct investment and equal treatment for all companies irrespective of ownership, would probably have had to be maintained for the offshore sector in order to remain consistent with broader policies.

Collectively, these constraints significantly reduce the additional government policy options available. Nevertheless, the author still believes that much could have been done to improve the industrial response. To this, the question of timing remains critical. Unfortunately, short of an early lobbying campaign by the British oil companies directed at bringing the likely future opportunities in international offshore oil and gas exploitation to the attention of British government and industry, there is probably little hope that much more could have been achieved prior to the move to the northern North Sea, leaving the few southern basin supply sector pioneers in their exposed positions in the interim.

Even so, government action should have started earlier. The commissioning of the *IMEG Report* and the establishment of OSO could, and should, have been brought forward by a year at the very least. Given the speed with which events were then moving, this alone could have had a major beneficial impact. By bringing the government earlier into direct and objectively critical contact with the oil industry's likely rather than stated procurement needs, it may even have helped prevent over-commitment of public funds to concrete platform yards and the creation of excess fabrication capacity. It took time for OSO to learn that oil company forecasts were better viewed probabilistically than accepted at face value.

It must be asked whether the government's general reaction to the *IMEG Report* was correct and what feasible alternatives existed to important recommendations that were not implemented, or only partially so. On this, the author believes that the government was right to reject IMEG's proposal for an independent PSIB and to keep what became OSO within the ministry responsible for oil and gas policy. In this way, its leverage on the oil and gas companies could be maximised, despite the downside of making it impossible for OSO to escape fully the restrictions inherent in civil service 'culture'. However, OSO would have been better called the Petroleum Supplies Office (PSO). This might have helped it focus earlier on the critically important 'core' area of drilling and well services, to which IMEG devoted little attention, being over-focused on 'big ticket' items. Failure to catalyse more British-owned corporate activity and R&D in this area early on adversely impacted on the long-term opportunities at home and abroad.

Similarly, the manner in which OSO set up the FFO and quarterly return system as recommended by IMEG, with the addition of the MoU as the means of governance of the policy, was essentially the correct means of providing a cornerstone to relations with the oil companies. However, the 'areas of special interest' mechanism should have been in place from the very outset, covering from then at least two areas – underwater services and products and drilling and well services. Moreover, a means of measuring UK and foreign content in terms of value added, even approximately, should have been established. This would not only have provided a better measure of the contribution the offshore supply and service industry was making to the UK economy than simply adding up orders, but would also have revealed whether what many believed, that is that British suppliers were concentrated in 'low-tech' low margin commodity-type items, leaving foreign suppliers a near-free hand in 'high-tech' higher margin items, was actually true. A value added dimension might also have provided a new perspective for assessing UK content in licence awards.

IMEG was right to point out that the heavy investment required to establish British-owned drilling and offshore construction contractors would require government support. However, the methods it proposed of insurance against inadequate returns plus direct government investment and/or purchase of equipment for lease to British contractors were probably both too radical and the total potential costs to the Exchequer too difficult to quantify to find favour with the Treasury.

It was correct to follow up IMEG's identification of cheap foreign export credit as a problem, but the IRG scheme might not have been the most appropriate response. As it resulted in cash payments to oil companies as customers, it was perhaps easier to categorise as a direct subsidy than the provision of supplier credit support would have been. In any case, it would have been preferable for the scheme to be introduced with a provision for it to end on the implementation by OECD members of a unanimous agreement to cease providing cheap credit for the offshore industry in member states.

IMEG also was correct to suggest joint ventures with foreign companies as a means of technology transfer and the right of the joint venture or the UK joint venture partner to trade freely around the world. However, it should have placed greater emphasis on the case for British majority shareholdings in joint venture companies and also put the case for British businesses obtaining the required 'know-how' by means other than by the relinquishment of equity, such as licensing, hiring individuals with the necessary expertise and contacts or outright purchase.

With respect to R&D and education and training, the IMEG recommendations were a useful starting point but probably insufficiently developed from a strategic viewpoint. Again, while IMEG correctly drew attention to the importance of Industry Act finance to the development of a British offshore supplies industry, it failed to develop a strategy as such, particularly with respect to the use of selective financial assistance.

Given its deep-seated problems, IMEG made a mistake in devoting so much attention to the traditional shipbuilding industry. Rather than seeking within it an economic solution to the construction of, in particular, semi-submersibles, a better approach would probably have been to offer Industry Act aid to any suitable platform construction yard willing to make the appropriate modifications to its facilities. The overlap in skills, facilities and equipment between semi-submersible and steel jacket construction were probably greater than that between traditional shipyards and semi-submersible construction.

There are five additional measures that the author believes could have been employed to give greater support to British industry within the constraints of the time. The emphasis is very much on what could have taken effect in the period 1972–1976, that is from the presumed start of a PSO to the end of the first investment upswing.

The question of the compatibility of the measures suggested with European competition law as it then stood immediately arises. It should be noted that the UK did not become a member of the EEC until 1973, which would have allowed the possibility of any enabling legislation required to have been put in place prior to UK accession. Two of the measures would have required changes in licensing conditions, whilst the other three would have required a new 'Strategic Industries Act' and/or a radical overhaul of the existing Industry Act. The concept that some industries, including but not confined to defence, are 'strategic' in character and justify special treatment by national governments continues to be recognised internationally.

In essence, it is believed that the measures suggested would not have excited any greater or earlier interest from the EC than those that actually came into force. This assumes that they had continuous strong ministerial support preferably extending to a willingness to form a tacit alliance with the French, themselves pursuing strong 'national interest' policies in the petroleum supplies business. The possibility of a successful challenge to the measures by the USA or some other country is also considered to be remote, given that they would not impinge upon access to exploration rights or impose import restrictions and would continue to treat foreign-controlled suppliers based in the UK as British companies.

Two of the measures would have related mainly to government inter-action with the oil companies through the offshore licensing regime. There would have been immediate changes in the licensing regulations. Most importantly, there would have been a requirement for a non-voting government representative to sit on all joint operating committees or their equivalent in single licensee situations. His or her main role would be to ensure that companies fully understood government policies and aspirations and to make it apparent to them that wilful non-compliance could lead to difficulties in respect of drilling consents, Annex B approvals and future licence awards. He or she would also be available for consultation by the government, particularly prior to new licensing awards. Additionally, a requirement for licensees to support R&D in the UK and to offer prototype-testing opportunities for UK innovations would have joined FFO/UK content as a licensing criterion at the outset (see also below).

A much more aggressive approach towards R&D and education and training should have been adopted than IMEG suggested, centred on the establishment of a British Institute for Petroleum Science and Technology (BIPST). Government would have had to provide initial launch funding (say for 2 years) and continue to fund on a 100% basis R&D needed to support statutory requirements in such fields as safety and environmental protection. BIPST would otherwise have been largely financed by a tax-deductible levy on oil company expenditure relating to the exploration, development, decommissioning and abandonment of UKCS licences, imposed as a licensing condition. The suggested initial rate would have been 0.75%. This would have produced a large but fluctuating income, requiring BIPST to set up a stabilisation reserve so that it could maintain, or indeed possibly increase, its expenditure during recessionary periods.

BIPST would have been an agency outside the civil service. Beyond offices, it would have had no facilities of its own, acting through the existing infrastructure as far as possible and sponsoring by the award of long-term contracts the creation of any new facilities needed. In its first 3 years, a large part of its activities would have been involved in training. A good example would have been the charter of a marine drilling rig (rather than the purchase proposed by IMEG) from a British-owned drilling contractor for use as a training facility, probably under a management contract awarded to a U.S. organisation. As a secondary benefit the charter would have provided 'launch aid' for a nascent drilling contractor.

On and from the rig people could have been trained in several specialist disciplines in short supply, such as drillers, mud, wire-line and logging engineers, divers and ROV operators. In the main, trainees would have been existing employees of British companies, trained at below cost. If places were available, individual British nationals could have been trained, with fees funded by a loan to be written off provided they were employed by a British-controlled company for 3 of the 5 years following the completion of training. Non-nationals would have been charged commercial rates. After 3 years, the programme would either have been ended or handed over to a commercial trainer. Other industry training schemes and relevant university courses would also have been encouraged, though all subject to BISPT accreditation. As an interim measure, BISPT would have been able to subsidise attendance on overseas courses.

In the R&D field, BIPST would have commenced by taking over responsibility for all existing upstream projects where government finance was involved. European funding would have been channelled through it. The formation of MTD/PSTI and the OETB would have been unnecessary. BIPST would have differed from them by virtue of a higher level of spend, earlier attention to drilling and well services and more direct supervision of projects. Crucially, BIPST would have been a research, development and demonstration (RD&D) body, through time increasingly focused on development and demonstration of prototypes intended to give British industry a range of up-to-date products and services. While the majority of projects would have been on a shared cost basis with oil company and supply sector partners, with no provision for repayments to BIPST, it would also have been able to undertake 100% funded 'commercial pathfinder' projects in challenging subject areas like multiphase flow and subsea separation. This would have allowed it to develop its own intellectual property portfolio to be made available on 'soft' terms to British industry and to trigger direct oil company involvement in 'follow-on' projects. Generally, the results of such BISTI-funded research projects with commercial potential would not have been released into the public domain until after suitable arrangements for their commercial exploitation were in place.

The management of BIPST would have been responsible to a board comprised of representatives from government and universities, British-controlled oil and gas companies, the British supply sector, financial institutions and foreign-controlled oil and gas companies. The first three groups would have been appointed on 3-year terms but the fourth would have rotated annually in order to disseminate knowledge of BSPTI as rapidly as possible within the industry internationally. Board responsibilities would have included identifying prototype test opportunities and possible commercial 'spin-outs' arising from academic research.

The three other possible measures would have related to direct government inter-action with the service and supply sector. Firstly, the petroleum supplies business would have been declared an industry of national strategic importance, allowing the government to treat UK-owned firms differently from foreign-controlled businesses. This would include the power of veto over the foreign acquisition of companies or intellectual property deemed as being of special importance, the power to deny regional aid to any unwelcome (because, for instance, excess capacity would result) new inward investment as well as the exclusion of foreign-controlled businesses not already established in the UK from eligibility for selective financial assistance other than in exceptional circumstances. There would, however, have been no discrimination in favour

of British-controlled businesses under the FFO policy and no import restrictions, maintaining the interest of foreign firms in the British market.
Ministers would have exercised powers through and on the advice of the
PSO. In principle, the powers would have been permanent and their exercise
reported upon annually to Parliament.

Secondly, in order to encourage entry to capital intensive and/or high risk
areas, for British-owned and flagged mobile offshore assets, including supply
boats, anchor handling tugs and underwater support vessels as well as mobile
drilling rigs and heavy lift and pipelay vessels, there should have been a special
'launch aid' package for a maximum period of 5 years. Its key features would
have been a first year capital allowance of 125% of investment, ship finance
credit terms for new builds (sometimes previously refused on definitional
grounds), plus selective financial assistance grants of up to a maximum of
20% of the capital expenditure involved. During the late 1960s, a scheme
of a broadly similar nature had substantially stimulated investment in conventional British shipping (Institut Européen des Affaires, or INSEAD, 1976),

Thirdly, the 'venture' arm of PSO would have been more important and
better resourced than it was within OSO, with the clear aim of creating more
genuinely British-controlled enterprises and a correspondingly smaller role
for joint ventures and subsidiaries of foreign-owned firms. This would have
been achieved in a number of ways. Within government, a close working association with PED would have been put in place from the outset, meaning that
PSO would always have been involved with the Annex B process and PED
always actively helping PSO to identify future needs. PSO would then have
been charged with ensuring these could be met from British-controlled sources,
as by alerting existing suppliers, persuading BIPST to arrange the necessary
RD&D programmes and occasionally 'catalysing' the establishment of a new
enterprise.

The creation of new enterprises without the focus on short-term financial
results inherent in the conglomerates, or the endemic structural problems of
traditional heavy engineering and shipbuilding, would have been a PSO priority
second only to FFO. To achieve this, PSO would have had to actively seek out
potential entrepreneurs and investors – unlike OSO, which normally only
responded to approaches – and to use selective financial assistance aggressively
in a way that OSO could never do. Selective aid funds could only be deployed in
circumstances where PSO was both satisfied on the basis on research and analysis of the potential to create a viable enterprise and had established that the
commercial sector was willing to take the greater part of the risk. Swift, if
necessarily conditional, commitment to selective financial assistance would
have provided a basis on which to approach financial institutions or existing
controlling shareholders. Without their financial support, the venture would
not proceed. With it, the selective aid would have been intended to act as a
'safety margin'. In this way, the chronic under-capitalisation of new ventures
might have been mitigated, although the underlying scarcity of new long-term

equity capital would have remained a problem pending changes in the wider economy.

Though start-ups would figure prominently, existing businesses committed to 'desirable' specialisations would also qualify for selective aid as a means of accelerating their development. In all cases, the emphasis would be on creating long-term businesses, for the international as well as the domestic market, rather than simply filling UK supply gaps, which was for long OSO's starting point. The need for rapid access to funds would have required PSO to have had delegated authority over an initially large annual tranche of selective aid funds, with responsibility to the IDE restricted to accounting matters. The funding requirement would decline with time as the 'launch aid' period for the mobile assets sector came to an end and other capacity gaps filled. Further PSO venture support would have included helping to source technical assistance, licences or key personnel as possible alternatives to joint ventures.

While no attempt has been made to compare the cost of these proposals with what actually happened, they should not have involved any large overall increase in government expenditure. Some measures like the declaration that the offshore service and supply industry was strategic, the closer relations with PED or the placing of government representatives on operating committees would have had no budgetary implications. There would have been a small net loss of tax revenue from the oil companies as well as probably some loss and/or deferment of tax receipts from mobile asset owners.

The 'front-loading' and earlier start of industrial support would have brought supply sector expenditure forward. Encouraging investment in areas like heavy lift and pipelay vessels and mobile drilling rigs would undoubtedly have run the risk of large losses, particularly if the investments had missed the initial market surge. However, the risk would have been reduced if one or even two 'equipment generations' had been 'skipped' without losing the benefit of an early start. The willingness to back start-ups would of itself have ensured many failures along with, it is to be assumed, some outstanding successes of the like of Expro.

The BIPST mechanism would have almost totally relieved the public purse from funding R&D other than that required for its start-up and for regulatory purposes and any net expenditure on training would have been short-lived. BIPST should not have needed as large an R&D staff as OSO/MaTSU/OTU, MTD and PSTI combined. A little less would probably have been spent on regional development grants and concrete platform sites but much more on selective aid. Regional aid savings would have become much larger had legislation excluded oil and gas terminals on the grounds they offered no 'additionality' (i.e. construction was unavoidable, with or without assistance). Indeed, it is possible that this change would have released more than enough money than that needed for the additional funding of the service and supply industry, although the front-end loading of the service and supply measures would have caused a timing mismatch.

The aim of the enhanced policy mix would have been to increase the British-controlled content of the service and supply sector, widen its scope and improve its technology level over what was actually achieved by 1993. The confidence of those involved would have improved as they saw the interests of British industry brought closer to parity in importance with security of supply and the balance of payments. A reduction in inward investment might have been the cost. The risk of delays and possible failures as more new British firms 'cut their teeth' would also have had to be faced, but these were far from unknown with supposedly experienced foreign suppliers.

Whether or not the additional policies would have succeeded in creating a truly internationally competitive industry can never be known. Much would have hinged on how much earlier than OSO the PSO and the accompanying policy mix could have been launched. They certainly could not have resolved all the 'background' problems of British industry, which manifested themselves in offshore service and supply as elsewhere, but they might have mitigated their impact there.

Chapter 10

Postscript

Much has changed over the 17 or so years since the main narrative ended, but much else has stayed the same. To explore fully the current scene, let alone to describe how it has arisen, would require another book, which the author does not propose to combine with this present one. Rather what is attempted in this chapter is something of a 'canter across the scene' in order to try to answer some of the questions which the reader might otherwise feel are left 'hanging in the air'.

The UK economy experienced a long period of growth after the end of the recession of the early 1990s, eventually brought to an end by the 'credit crunch' of 2008. The growth was accompanied by a continued decline in the importance of manufacturing in favour of the growing service and public sectors, the latter's expansion little affected by the decline in defence spending that followed the end of the 'Cold War'. Trade union power and membership continued the decline set in train by the Thatcher administration, with strikes becoming rare. British labour relations were no longer seen as chaotic, or British management as incompetent. However, strikes still occur and low productivity relative to other countries is still sometimes perceived in certain industries. Many employers complain of skilled labour shortages. All three of these issues have been observed at one time or another on large engineering construction sites, though the traumas of the 1970s (see pp. 14–19) have not returned.

The UK became an enthusiastic adopter of the globalisation trend and widely seen as an economic success story. Much of the apparent prosperity was generated by rising debt, both private and public, and by the continued transfer of companies, many 'household names', from British to foreign ownership. Balance of payments deficits were regular features, but kept to reasonable proportions, thanks to a continued large positive contribution from the UKCS and readily financed by a globalised banking system and by the asset sales already referred to. Until the 'credit crunch', venture capital and private equity firms continued to advance. Nevertheless, government recognised that start-up and early stage development capital remained scarce and committed limited public funds to various initiatives in this area. The traditional encouragement of inward investment has continued, with emphasis now on services. Selective regional aid is still available, though within EU-wide constraints. It is mainly in the form of discretionary grants.

Significant public funds are still invested in university research and some in non-defence industrial R&D, often through newly created mechanisms such as the Energy Technologies Institute and the Technology Strategy Board. The proliferation of overlapping research institutions is in stark contrast to the much more limited attention paid to commercialisation issues. Governments paid little or no attention to industrial strategy until the last few months of the Labour administration that lost office in May 2010. During 2009 – early 2010, there were a number of favourable ministerial references to German and, particularly, French industrial policies and the issue of discretionary financial assistance to individual companies was reopened, both, it seems, prompted by a recognition that the economy needed to be re-balanced towards private sector investment and exports, to which manufacturing is widely seen as holding the key. Up to the time of writing (late 2010), the new Conservative-Liberal government has, beyond rhetoric, done little more than agree to fund greater university-industry collaboration while cancelling some specific company aid and hinting that take-overs might be more tightly regulated. Nevertheless, all political parties continue to stress that the economy must be re-balanced. There is little evidence they recognise the difficulty in achieving this if UKCS output and investment continue to decline. According to UKOOA's successor, Oil and Gas UK (2010, pp. 10, 11), in 2009 oil and gas output still made a positive contribution of £27 billion to the balance of trade and in 2008 the industry also accounted for about a third of industrial investment.

10.1 THE UKCS OIL AND GAS INDUSTRY AND ITS SUPPLY SECTOR TODAY

The UKCS, though still providing the largest element in UK primary energy supply and much business for many companies, is now only a modest and declining part of the global offshore oil and gas business and no longer the most technically advanced. Hydrocarbon production, which has already continued a decade or two beyond what was originally envisaged, seems likely to continue for at least a decade or two more. However, to the extent that the British government has an energy policy, its main objective is to 'decarbonise' the economy, primarily by switching electricity generation away from fossil fuel towards renewable sources, although their limitations have also left the nuclear industry as a 'guest' (if an unwelcome one) at the table.

UK oil production peaked in 1999 at about 2.9 million barrels per day and gas in the following year at about 108 billion cubic metres (BP). The country is no longer self-sufficient in either and imports are fast increasing, particularly in the case of gas. Gas production doubled during the 1990s, facilitating a large-scale replacement of coal in generation and in the process substantially reducing carbon emissions. The price was paid first in terms of rapid reserves depletion and subsequently in steeply falling production. Reference to Chart 10.1 shows

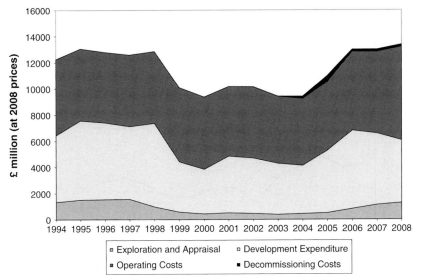

CHART 10.1   UKCS Expenditure (2008 prices) 1994–2008. Data source: Department of Energy and Climate Change.

that falling oil and gas production has not resulted in falling operating costs, contributing to the UKCS's reputation for high unit costs.

High costs also follow from the nature of continuing production. Most of it passes through an ageing infrastructure, whether sourced from the redevelopment of old or 'brown field' reservoirs or from the tie-ing back of small new developments, usually themselves subject to rapid depletion. To maintain the infrastructure is vital to these production sources but is becoming more difficult and expensive, despite the fact that over-designed structures of the 1970s and 1980s have proved remarkably resilient. Nevertheless, decommissioning expenditure, already in evidence on Chart 10.1 above, will rise sharply in the years ahead as uneconomic facilities are shut-down and removed.

Both short life new fields (usually exploited as subsea satellites or by chartered FPVs) and 'brown field' redevelopments require new capital expenditure, which, as Chart 10.1 shows, has been remarkably stable over the last decade, although at a level much below its earlier peaks (see Chart 4.1, p. 94). By contrast, drilling activity, whether for E&A or development, has been declining to historically low levels, as can be seen by comparing Chart 10.2 overleaf with Chart 4.2 in Chapter 4 (see p. 102). After a buoyant beginning, this decline has been especially marked in the case of development drilling. E&A drilling seems to have become more sensitive to oil price changes than it was in the days of strong tax incentives. Despite a general perception of declining prospectivity, regular licensing rounds continue to be held, 2010 seeing the 26th.

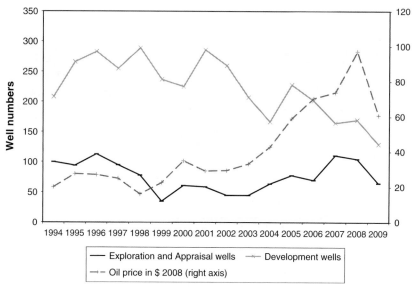

CHART 10.2 UKCS Drilling Activity 1994–2009. Sources: Department of Energy and Climate Change, BP (2010), *Statistical Review of World Energy*.

While large (but not 'giant') discoveries are still occasionally made, the general experience is that new discoveries are small and only marginally economic. Given that the early large discoveries are now approaching the end of their economic lives, these expectations have led the major oil companies to run down UKCS exposure, passing tail-end production to smaller independent companies whose concept of materiality is less demanding. These smaller companies are also responsible for much of the remaining exploration activity. Whereas the majors have chosen to continue to increase their reliance on out-sourcing, these smaller Operators have little option but to depend almost totally on it.

In the years immediately following the end of OSO, the level of political interest in the UKCS seemed to remain low. This was to change after an oil price decline and threatened tax increases in 1998, which were followed by a sharp decline in activity (see Chart 10.1, p. 267 and Chart 10.2 above).

Alarmed by the prospect of acceleration in the decline of North Sea production and the associated revenues, the government, with the support of the major E&P and contracting companies launched a plethora of collaborative initiatives, such as those set out in Table 10.1, under the umbrella of the Oil and Gas Industry Task Force (OGITF), which itself set the industry a number of challenging targets.

Whether as a result of these initiatives or, more likely, due to oil price developments, exploration and development expenditures stabilised until 2004 after which they generally rose. This was despite sharp tax increases in 2002 and 2006 (somewhat mitigated by subsequent changes).

TABLE 10.1 Some Post-1998 Initiatives

Name (Acronym)	Purpose	Comment
Oil and Gas Industry Task Force (OGTIF)	To set agendas, targets and take initiatives; joint government/industry body	Replaced by PILOT
PILOT	As above	Still active
Leading Oil and Gas Competitiveness (LOGIC)	To promote supply chain best practice	Still active
Digital Energy Atlas & Library (DEAL)	To provide an index for UKCS data through an interactive map	Still active
Licence Initiative for Trading (LIFT)	To promote web-based licence trading	Still active
Industry Technology Facilitator (ITF)	To facilitate development and adoption of new technology	Still active
Economic Advisory Group (EAG)	To carry out analysis and offer advice to industry side of PILOT	Inactive
NOVA Technology Fund	To invest in small, innovative supply and service companies	Oil company funded and now fully invested

Sources: Various.

Turning to the supply side, according to Oil and Gas UK (p. 33), the UKCS still directly supports 240,000 jobs, of which only 32,000 are with E&P and major contracting companies, with the balance lower down the supply chain. Furthermore, the same source estimates some 100,000 additional jobs are supported by supply chain export activity.

There are no longer any official figures on which to base estimates of the UK content of UKCS expenditure, However, Oil and Gas UK (p. 36) claimed that in 2009 '… *expenditure entering the predominantly British supply chain amounted to about £12 billion*'. To reduce the risk of timing vagaries this may be compared to an average UKCS expenditure (including decommissioning) over the 3 years ending in 2009 of about £13 billion (in 2008 prices). Ignoring the possible implications of the word '*predominantly*', this suggests that, at over 90%, UK content is now higher than anything ever achieved under OSO. This is quite plausible in that it is operating expenditure that has been growing in recent years, an area where there is a 'proximity premium', favouring the use of local labour and other resources, which could well have cancelled out the loss of the FFO policy.

This point illustrates how the effects of a declining domestic market are unevenly spread. With demand for fixed platforms and the associated topsides now spasmodic events at best, the once great fabrication industry, for long OSO's 'flagship' for UK content, is only a shadow of its former self. Unlike sectors offering specialist services or equipment, fabrication cannot find its salvation by increased export sales.

Oil and Gas UK (p. 36) also claimed that exports generate another £5 to £6 billion of supply chain turnover. To estimate what share of the global market outside the UK that this represents is by no means easy as it is difficult to establish consistent estimates of the global market's size, particularly in an era of floating exchange rates, nor to know with any precision how much of that market is freely open to international competition. However, one recent estimate of the total global spend for 2008 (including the in-house costs of E&P companies) available from a reputable consultancy (Douglas-Westwood Ltd, 2009) put the figure at $243 billion. Using an exchange rate of $1.85 to the pound (the average for 2008) and removing the UKCS element (seemingly nearly 10% of the total) would suggest a spend elsewhere of about £118 billion, with a UK share of 4–5%. Additional to direct exports, there are the sales of overseas subsidiaries of UK companies to be taken into account. No published figures for the UK as a whole have been found but Oil and Gas UK (p. 38) quotes a figure for 2008 Scottish companies of £4.1 billion, taken from survey by the Scottish Council for Development and Industry (SCDI). As Scotland accounts for slightly less than half of employment in the sector, this suggests the total for the UK of around £8 billion, about 6–7% of global spend outside the UK. How much of this is sourced from exports by the parent companies is impossible to estimate. It would be interesting to know how many of the local subsidiaries came about to meet local content requirements. OSO and Norwegian policies favouring local companies were used as models in many other parts of the world.

Taking all three sources of revenue together, it seems that UK-based firms might meet as much as 16–20% of total worldwide offshore oil and gas expenditure. If this were the case, it is a better outcome than most analysts would have predicted 20 or years ago. However, there is no information available on what proportion of this activity is by locally owned firms, but much clearly is not.

Moreover, the international success does not compare particularly well with that of the equivalent Norwegian industry. In 2010, INTSOK (Norwegian Oil and Gas Partners, a joint government industry body) revealed that direct exports of goods and services in 2009 were 80 billion Norwegian kroner (say nearly £8.2 billion at an average exchange rate of 9.8 kroner to the £). A further 38 billion kroner (say about £3.9 billion) of international sales arose through overseas subsidiaries. It is interesting to note that whereas the total international revenues of the two countries are very similar, the Norwegians are much more successful with direct exports (usually a much better contributor to domestic value added than the sales of overseas subsidiaries). This strongly suggests that the long-running Norwegian policy of fostering the development of technology intensive products under domestic control, thereby presenting barriers to entry by competitors, has been successful. It also probably accounts for what appears to be the higher labour productivity of the Norwegian sector. Again according to INTSOK, the Norwegian industry (whose domestic market is at least as large as the UK's) employs about 250,000 people compared to 340,000 in the UK, though it must be stressed that no attempt has to check the extent to which valid comparison can safely be made between the two figures.

The present structure of the British supply industry probably owes as much, perhaps more, to international developments as to domestic influences. The price weakness of 1998, like its predecessors, set in train extensive restructuring and mergers, acquisitions and divestments have continued apace. Nevertheless, it is probably best to start with the fate of the prominent domestically owned companies figuring in the main narrative.

Of the 'case study' companies, the John Wood Group and Expro both remain independent and UK headquartered. While Wood is still a listed company, Expro has been taken private by a private equity consortium. Abbot (KCA's parent) was a fully listed company for a number of years, during which it acquired German rival Deutag and re-entered the mobile drilling rig field (for benign environments). Abbot was acquired in 2008 by Cayman Island private equity vehicle and consequently delisted. Its operational subsidiary, KCA Deutag, continues to be managed from Aberdeen. Two other long-established British-owned companies in the sector retain full London listings, Hunting (of which the dominant component is oilfield products and services) and AMEC.

Two other fully London listed companies have very different backgrounds. Petrofac originated in the USA where it now has little business. Before entering the British market and relocating its corporate headquarters to London, it was already well established in the Middle East. With private equity support, it moved to a full London Stock Exchange listing in 2005. It is now the largest by capitalisation of the London listed oilfield service companies. Although in some ways analogous to AMEC and the John Wood Group, its activities are broader and it has out grown both in terms of market capitalisation. The other, Wellstream (a leading manufacture of flexible pipe) also originated in the USA, originally as a private company. It was first absorbed by Dresser, a large supplier to the energy sector, itself soon to be taken over by Halliburton. While part of Dresser, the company invested in a large new manufacturing facility on the Tyne, which opened in 1997, the prior to the Halliburton take-over. This made it Coflexip's major competitor. In 2003, as part of its strategy of exiting construction related activities, Halliburton sold the business to the management backed by British private equity interests. It was listed on the London Stock Exchange in 2007. In this roundabout manner, it could be argued, flexible pipe manufacture 'came home' to the UK (see p. 31).

Whereas as what are essentially large energy service conglomerates, AMEC, Petrofac, Wood and – to a lesser extent – Hunting are fairly unattractive to predators, Wellstream with its proprietary technology and strong presence in the rapidly growing Brazilian market is unlikely to remain independent for long, the purchaser almost certainly being a foreign company. A similar fate has already overtaken another fully listed company with strong proprietary technology (in part developed with UK public funding) – Sondex, a supplier of logging equipment. Fully listed in 2003, it was acquired by General Electric of the USA in 2007.

In addition to the small group of fully listed companies, there are a few other, much smaller, companies with shares traded on the AIM market. These include Offshore Hydrocarbon Mapping and Getech (both offering specialist geophysical exploration services) and KBC Advanced Technologies (providing consultancy, information and software). From 2005 until late 2009, Hallin Marine shares traded on AIM, prior to its acquisition by Superior Energy Services of the USA. Founded in Singapore in 1998 by a British 'expat', it had invested heavily in an attempt to become a full range underwater service company with its own support vessels, setting up operations in most important offshore oil and gas centres, including Aberdeen.

Apart from the small group of companies with public share markets, the British oilfield supply industry now consists of a few large companies, usually subsidiaries of foreign-owned groups, and many small locally owned units, some whose business is still purely local and locationally determined and others occupying product or service niches with international potential. Among the foreign-controlled subsidiaries is now Indian owned Corus Tubes, successor to Stewarts and Lloyds. It is still a force in the market for oilfield tubulars and line pipe, where the former difficulties with British produced submarine pipeline (see p. 196) now seem firmly in the past.

There are at least three private companies large and established enough to trade on a global scale. Two have organically grown from their roots in Aberdeen, the Craig Group (marine services and international distribution) and Balmoral Group (buoyancy, insulation and protection products). The third, Production Services Network, or PSN, has a different origin. Until 2006, when it was subject to a management buy-out backed by the Bank of Scotland, it was KBR Production Services. At that time, KBR (formed by a merger of B&R and Kellog) had not yet been spun-off by Halliburton. Although already with non-UKCS activities, it was in effect a lineal descendent of the original Aberdeen operations of B&R. Since independence, it has expanded rapidly, particularly outside the UK. Like Wood and AMEC, it operates to a favoured UK business model – high labour intensity rather than major investment in proprietary technology or fixed assets.

It seems clear that the largest remaining sector where local ownership predominates is production services with which topside maintenance, inspection and repair are closely associated. Since AMEC, Petrofac, Wood and, to an extent, PSN, along with a number of smaller locally controlled participants, all have design and project management skills to offer, these activities represent another area of British strength. A little UK-controlled fabrication capacity still exists, such as that at Burntisland Fabricators (a descendant of the original BEFL), although this is increasingly focussed on support structures for offshore wind turbines.

Provided KCA Deutag and Expro continue to be seen as British businesses, the country retains a good presence in drilling (except in high specification mobile rigs) in and drilling and well services (though not drilling fluids). Thanks to Hunting and a number of mainly small companies, a good UK-controlled representation remains in well hardware, as well as in less oil and gas specific product areas such as pumps, valves, turbines and controls.

In exploration services, apart from a number of small software firms, 'innovatory' geophysical data collectors and consultancies, there is little by way of UK-controlled capability, with no conventional seismic data collection and processing remaining. However, the fast growing Reservoir Group, now diversified beyond its origins as a core acquisition and processing company, is managed from Aberdeen, although controlled by a U.S. private equity firm.

Turning to marine support, despite a period in the early 1990s when it appeared that British-owned supply boat operators, particularly OIL, were set to be important long-term players in the market, few British significant marine contenders remain. The largest is the North Star Shipping division of the Craig Group, which not only continues to dominate the emergency response and rescue (formerly standby/safety) vessel market, but also has some capabilities in other fields, such as platform supply and ROV support. Provided it is considered to be British, Swire Pacific Offshore (a partly owned, Singapore headquartered affiliate of the British John Swire Group) is probably the only other UK company of significance in this area. Its fleet includes anchor handling tug supply boats, platform supply vessels, as well as specialised hydrographic and seismic survey vessels and vessels for diving and ROV support. Most of the fleet, including the seismic support vessels, are deployed outside of the North Sea. Bibby Offshore owns a small fleet of DSVs, operating in the North Sea and elsewhere.

As far as underwater services, the newer 'core' sector of the offshore oil and gas industry, are concerned, the UK lacks any presence in the 'top tier' contracting sector of the subsea supply chain. This is now dominated globally and in the North Sea by two companies. One is Paris-quoted Technip (which has absorbed the merged Coflexip Stena) and the other Subsea 7, a company quoted on the Oslo Stock Exchange but, following its recent merger with Acergy (once Stolt Comex Seaway) of mixed American, British, French and Norwegian lineage. However, lower down the chain, independent British subsea service companies still exist, notably Bibby Offshore, which in addition to its DSVs also offers a range of in-house underwater services, and Integrated Subsea Services (ISS).

Until the almost inevitable foreign take-over, Wellstream can be considered as the lead company of the British subsea product sector, which otherwise mostly consists of a large number of small companies, some with niche proprietary technologies and others with a limited contracting capability. However, JDR Cable Systems, manufacturers of subsea power cables and umbilicals is growing into a significantly sized business. It was formed by merger the of a British cable company (Jacques) with a Dutch competitor (De Regt).

Subsea UK, an industry body which absorbed the former Scottish Subsea Group (see p. 191) has some 200 members, many foreign owned. With a number of universities, it has sponsored the National Subsea Research Institute (NSRI), basically a co-ordinating body in the R&D field. Both NSRI and Subsea UK have had a small degree of government funding.

The UKCS continues to stimulate technological innovations, probably because it has employed a large part of the nation's engineering talent over more than a generation. The process of transferring ownership of British companies with proprietary technology or strategic market positions to foreign ownership has continued unabated, often following a fairly standard pattern. Once an innovation shows signs of market acceptance and long-term prospects, with sustainable barriers to entry, development capital is sought from venture capital/ private equity sources. After a short period of development under the new structure, the company is sold on, usually as a trade sale to a foreign owner. The fact that, subject, to proper initial due diligence, this process will almost always end with a handsome profit to the financier, has encouraged financiers actively to seek new deals. Table 10.2 overleaf shows a number of the deals consummated over the past decade and a half.

With the exception of Alpha Thames (then still at the development stage), all the above companies were well established and profitable at the time of acquisition, operating in markets with considerable growth potential.

10.2 WHAT OF THE FUTURE?

Prediction is a dangerous business, but short of some currently unforeseen circumstances, the domestic oil and gas supply chain market decline will continue to a point in the not too distant future when facility abandonment will lead to large-scale reductions in the manpower involved in offshore operations. It does not look likely that any increase in exploration and development will arise to compensate. For a time, the decline in current operations may partly be offset by the requirements of the decommissioning and removal process itself, though the lack of British ownership of the major floating assets (once used for installation) may reduce the share of work available.

The larger employers of operational manpower, such as AMEC, PSN and Wood may be able to replace the lost turnover in overseas markets, though mostly with local labour and thus little value added for the UK. The smaller employers are unlikely all to survive without finding alternative local markets. Firms offering niche products and services should be able to survive and grow through exporting, but it seems unlikely that this will demand sufficient new labour to prevent total employment from falling. Moreover, as local demand declines, some companies, particularly foreign-owned ones, will find it less and less viable to keep regional or international headquarters functions in the UK.

TABLE 10.2 Recent Foreign Takeovers of British Private Firms with Proprietary Technology and/or Strategic Market Positions

Subject Company	Area of Activity	Nationality of Acquirer
Alpha Thames	Subsea production equipment	Swedish
Brisco Engineering	Subsea and surface wellhead controls	American
Concept Systems	Marine seismic software	American
CRP Group	Buoyancy, insulation and protection products	Swedish
Ensign Geophysics	Seismic processing	American
Edinburgh Petroleum Services	Subsurface software	American
Gaffney Cline and Associates	Subsurface consultancy and software	American
Petroleum Engineering Services	Intelligent well control systems	American
Petroline	Speciality downhole technology	American
PII Group	Pipeline internal inspection	American
PSL Energy Services	Downhole and pipeline services	American
Reeves Oilfield Services	Downhole tools and services	Canadian
Rovtech	ROV services	Dutch
Seaye	ROV manufacture	Swedish
Sondex	Downhole tools	American

Encouraged by much political hyperbole, many firms believe their survival will be assured by the expected upsurge in energy investment in the UK as the country attempts to renew much of its energy infrastructure while 'decarbonising' the economy. In pursuit of the latter, recent governments have introduced a plethora of measures, some punitive, but mainly subsidies of one sort of another financed by levies collected by energy companies from their consumers through higher prices.

Most relevantly to the industry that forms the subject of this study, the government has committed to the EU that it will raise the renewable element in UK energy supply to 15% by 2020 from its current 2–3%. The main burden of this adjustment will fall upon electricity generation where the required increase is from 6% to 30%; offshore wind generation is seen as the chief contributor. There are also other areas of potential activity that might be seen as benefiting the offshore oil and gas supply chain; one is carbon capture and storage, necessary for any future investment in coal fuelled generation. It would involve transmitting 'captured' CO_2 for permanent storage in depleted oil and gas reservoirs, which can also be used for natural gas storage as the Rough field already has been for many years. Others revolve round marine energy sources, such as waves, tidal streams and tidal range through the use of barrages. Only the last of these is a proven technology but it is one that faces ecological as well as economic barriers.

Taking wind and marine resources together, there is no doubt that the renewable energy resources of the UKCS are very substantial. Indeed according to a recent report by the Offshore Valuation Group (2010), they would hypothetically allow the UK to become a net energy exporter by 2050, without requiring their full utilisation. Since they are renewable, such an outcome, the report argues, would make the UKCS's wind and marine resources more valuable than its oil and gas. Whether this eventuality will ever come about depends upon on a wind range of economic, technical and political uncertainties and in the author's view is improbable.

Be that as it may, the exploitation of the largest and most technically mature of the offshore renewables (wind) is already proceeding at a considerable pace, despite the fact that it remains an uneconomic source of electricity by comparison with fossil fuels. Investment viability depends on politically driven subsidy, making the industry totally different from North Sea oil and gas, which was commercially driven and profitable almost from the outset. Perhaps this critical difference may partly explain why an indigenous supply chain has been slow to develop, despite it having obvious for some years that the UKCS was, for a time at least, about to become the world's largest market for equipment and services for offshore wind generation. Since government has known for even longer that this market was sure to develop, its failure to apply the lessons to be learnt from the arrival offshore oil and gas and to stimulate supply chain development in advance of demand is an indictment of the costs arising to the UK economy from the lack of any industrial policy. The cynic may with good grounds argue that, carbon reduction apart (a universal benefit), subsidy-driven investment in offshore wind both increases the costs of British electricity and fattens the profits and payrolls of overseas rather than UK suppliers. He might also wonder why government failed to act to prevent such a state of affairs. Unlike offshore oil and perhaps also gas, it cannot be claimed that the arrival of the offshore wind industry was unexpected.

Nevertheless, it is already clear that very large sums are to be spent on offshore wind generation and transmission over the next 20 years. It is not realistic to quote hard numbers given uncertainties over timing and exchange rates, which are particularly significant given the current weakness in the UK supply chain, but a total annual average expenditure of £7–8 billion a year (including the offshore grid) is one insider's recent guess. Garrard Hassan Ltd (2010) reported to the British Wind Energy Association (BWEA) that the annual rate of offshore wind turbine installation between 2010 and 2015 would be between 600 and 900+. Thereafter, the number would increase sharply to a peak round about 2020. In the absence of further licensing rounds, the domestic market would shrink thereafter. A subsequent study by Douglas-Westwood Ltd (2010) for RenewableUK, successor to the BWEA, considered a number of different demand scenarios. It set out the case for a flattening of the potential unconstrained demand peaks of over 1300 turbines a year it foresaw for Round 3 licence award schemes alone in 2018 and 2020 in order to avoid a potential 'boom and bust' situation. To develop what it described as a *'healthy industry'*, it suggested maintaining overall turbine demand at between about 550 and 700 units per year from 2016 to 2030, though in 2017 demand might temporarily 'spike' at about 850 units. The danger of clashes with the offshore oil and gas industry over scarce resources, particularly the heavy lift vessels employed in both industries, needed also to be noted.

Comparisons between the offshore oil and gas industry and the offshore wind industry and the latter's UK job creation potential can be found in Smith (2009). Most plausible forecasts put the number of jobs arising from the offshore wind industry in the range 25,000–70,000, the higher figures mainly reflecting more manufacturing being undertaken in the UK. Employment potential under the Douglas-Westwood *'healthy industry'* scenario was estimated at about 45,000. There is nothing to suggest that the job creation potential of the offshore wind industry will ever approach that of the offshore oil and gas industry.

As in the offshore oil industry, many jobs will be locationally determined, favouring local procurement. Thus UK suppliers of environmental, legal, and some technical services, as well as operations and maintenance services, are well placed. Indeed the last mentioned could offer new opportunities for workers displaced from similar occupations by the decline of offshore oil and gas activity, although timing and skills mismatches may limit the scope for this.

For proprietary products, major component supply and offshore installation services where high order values and the scope for good profit margins may be available, the outlook is less favourable. The best UK product opportunities derived from offshore oil and gas experience are probably substructures, foundations and underwater electrical power cables and connectors. A few companies in these fields have not only recognised this but have also shown a willingness to invest to meet the demand. They include JDR Cable Systems, Harland and Wolff and Burntisland Fabricators.

Thanks to the North Sea oil and gas industry and to long experience in laying telecommunications cables, the UK is the base for a number of the specialist companies and vessels involved in offshore cable laying and burial; many are foreign owned, with the Subocean Group as a notable exception. Similarly, as the wind industry moves into deeper water, the demand for other sophisticated subsea services will increase, hopefully absorbing some of people and assets no longer required for oil and gas operations. OSO never succeeded in creating a UK-owned installation vessel capability and at the present all turbine installation vessels are foreign owned. One early (and undercapitalised) UK entrant has already failed and the vessel is now under Dutch ownership. However, adequately capitalised Swire Pacific Offshore has two such vessels under construction, though it is not clear where they will work.

The turbine itself is the most expensive element in an offshore wind power unit. It is a piece of sophisticated machinery, susceptible to continued improvement through research and development and its design and manufacture involves high value added (and hence potentially high wage) jobs. Component supply is an important element in the supply chain. No offshore turbines are currently produced in the UK. Until they are, it seems unlikely that the UK sourced proportion of an installation will rise above about 45%. What is more, as imports they must be purchased in foreign currency, which in a world of floating exchange rates complicates budgeting. A decline in the value of the £ sterling has already had an adverse economic effect on some schemes.

By the time the full realisation of this state of affairs finally dawned upon the UK government, the technology was already fairly mature and overseas companies well established. By then, it was probably too risky commercially to launch a new all UK venture. The government reacted in its time-honoured default mode by seeking inward investment. In total, five foreign companies have expressed an interest in establishing UK manufacturing facilities – Clipper Windpower (USA), General Electric (USA), Gamesa (Spain), Mitsubishi (Japan) and Siemens (Germany). It is not self-evident that the UK market alone could support all these entrants and to encourage too many would run the risk of 'wasting public money' since it must be taken for granted that substantial financial assistance (i.e. subsidy) will be provided in each case. Indeed, Clipper (a subsidiary of United Technologies) received a £5 million assistance package in 2007 to locate a prototype development centre at Blyth. In 2010, Mitsubishi also committed to set up an offshore wind turbine R&D facility in the UK and was promised a £30 million aid package if a manufacturing plant followed. Assuming all these proposed facilities are established, when their capacity exceeds the available domestic market, the extent to which they can offset this through exporting will be decided outside the UK by parents with multiple facilities around the world.

Hopefully, some manufacturing firms in the offshore oil and gas sector will be able to supply components and sub-assemblies to the turbine manufacturers. It is to presumed that DECC's Office of Renewable Deployment (ORED), established in 2009, is much engaged with such issues, since its responsibilities extend to R&D, prototype testing and support for the supply chain. However, OSO's experience in the early 1970s suggests government's main concern will be in meeting its capacity targets rather worrying over-much about where money is spent or the jobs created. All in all, the expansion of offshore wind will offer suppliers and workers some compensation for the decline in offshore oil and gas activity, but even in combination with increased exports, it will probably not be sufficient to prevent an overall contraction of the industry.

What else might be available for the industry in the energy sector? In the immediate future, the best place for it to look is probably in the nuclear generation sector, notwithstanding the fact that all existing capacity will be decommissioned by 2035. A substantial decommissioning market, not confined to power plants, is already is already in existence and has a large export potential. After realising a few years ago that there was no hope of combining its decarbonisation aims with some semblance of security of supply without replacing the UK's substantial nuclear generating capacity, the British government decided that it must create an environment where the necessary private sector investment would be forthcoming.

Bizarrely, at much the same time, it decided to complete the dismemberment of the UK's once world-leading nuclear industry, which had been languishing since the sector fell out of favour 30 or so years ago. This culminated in the sale of British Energy, the privatised owner of most of the UK's existing nuclear generating stations, to the state-controlled Électricité de France (EDF); Centrica (a component of the former BGC) subsequently acquired a minority stake in EDF's new British subsidiary. The significance of the British Energy deal was two-fold. Firstly, EDF acquired most of the sites where future nuclear power stations are likely to be built, since the general presumption is that new nuclear facilities can now only be constructed alongside existing ones. Secondly, all EDF's existing nuclear power stations are to French designs, the latest version of which is seen as one of the two designs currently favoured for new capacity in developed countries. Although EDF is not likely to be the sole builder of new nuclear power plants in the UK, the implications seem obvious.

The other favoured design was designed by Westinghouse Electric, a company of U.S. origin. It was acquired in 1999 by state-owned British Nuclear Fuels (BNFL), which went on in 2000 to acquire the nuclear activities of ASEA Brown Boveri (ABB) and merge them with Westinghouse. In the light of the recognition that a nuclear power revival was probably in the offing, these were sagacious commercial moves, but not safe for a business under political control. In 2005, BNFL was compelled to divest its acquisitions, the purchaser being Toshiba of Japan. Further divestments followed and BNFL itself will soon be wound-up.

Thanks to the UK's early start with nuclear power and its status as a nuclear military power, it already has a well-developed nuclear supply chain, supporting new construction (military only for the moment), operations, maintenance, fuel reprocessing, decommissioning and waste management. It is not proposed to attempt any discussion of UK capabilities as the supply chain (with an emphasis on new builds) has already been analysed in a report for the Department of Business, Enterprise and Regulatory Reform (BERR) by the National Materials Technology Centre (2009). Suffice to say that it found that the nuclear industry and its supply chain employ about 56,000 people, including about 7,000 engaged in the military sector and that the UK was capable of supplying about 70% of the expenditure involved in building a new nuclear power plant, a figure that could be raised to 80% with further investment and training. One of the limitations identified was lack of a capability to produce the very large forgings required for the new reactor design, an area where global capacity is also limited. Sheffield Forgemasters was willing to invest in the equipment requirement but was unable to raise the required funds. In early 2010, the then government offered a loan £80 million, an offer rescinded by its successor a few months later. The opinion of the Office of Nuclear Deployment (OND) on this matter is not known. As in the case of ORED, supply chain issues, though part of its work scope, are not likely to be its top priority.

It is notable that some large suppliers to the nuclear industry are also well known in the offshore oil and gas sector. They include RR, anxious to grow its civil nuclear business and already in 'alliance' with EDF, and AMEC, which not only has a thriving nuclear decommissioning and clean-up business but whose predecessors were heavily involved in previous civil nuclear programmes. With new construction, AMEC will clearly have opportunities as a management contractor, in some areas of engineering design, in procurement and in commissioning. Putting aside the wide range of suppliers of general industrial products such as pumps, valves and controls (of which Weir Group is an example), companies offering robotics and high specification inspection techniques to the oil and gas business, mainly in the subsea sector, should also benefit from an expansion of the nuclear industry. A number of such companies are already involved in the sector.

However, because the nuclear supply chain is already so well established, the benefits to the offshore oil and gas supply chain overall are unlikely to be particularly large, though AMEC and perhaps some of its peers might benefit substantially. Moreover, as yet no company has committed to build a new nuclear power station in the UK. Prospective investors are waiting for government action to improve the economics of their investments. Since direct subsidies are ruled out, it seems likely that there will be regulatory and other administrative alterations to favour nuclear generation, which, it is worth noting, is closer to being competitive with fossil fuel generation than is offshore wind power, as well as being equally capable of reducing carbon emissions.

The idea of industrial policy fell out of favour with the British political classes after it was appreciated that the backward-looking process of intervening in and subsidising large companies in declining markets or with inherent structural problems during the 1960s and 1970s was a wasteful use of public money. OSO survived until the 1990s because it did not fit that model, particularly in that it was inexpensive to operate and (very unusually for the UK) had cross-party support. Almost all other developed countries still maintain industrial policies of some sort or another, whether or not they are so called. Additionally, newly industrialising countries like China and India are intent, like Japan and Korea before them, in building strong technology portfolios (including in the energy sector) under their own control, using whatever method is to hand, not excluding foreign acquisitions. There is little indication that Britain's leaders have an appreciation of the implications of this state of affairs, or indeed much interest in the issue of national control of the civil technologies important for national security.

That said, the purpose of this book was not to formulate new government policies. As stated at the outset, its aims were multiple. It is a contribution to the history of a highly significant phase of UK technical and economic development. It has tried to show how, from an industrial standpoint, the British handled the exploitation of the most significant natural resource they discovered in the twentieth century, which might assist governments and industries faced with future instances of unforeseen, specialist and large-scale new demand to manage their own reactions more effectively. Finally, I have tried to throw some light on how governments can pursue strategic industrial objectives whilst leaving market mechanisms to function with minimal interference, which is something administrations – perhaps even a future British one – may wish to do now or in the future.

To these original aims, I have added a Postscript with the intention of offering some insight into the current UK energy supply chain. I hope the reader will find that I have met my goals, at least in part.

Source Materials

Published Sources

AEA Technology PLC, Professor G. L. Chierici, Compagnie General de Geophysique, Marine & Energy Consulting Gmbh and Smith Rea Energy Associates Limited. (1999), *New Oil and Gas Technology in the Cost Reduction Era*. Brussels: European Commission.

Algar, P. (1980), *Petroleum Review,* January 1980. London: Institute of Petroleum.

Allcock, J. (1999), in *The Development of North Sea Oil and Gas*. Institute of Contemporary British History, seminar, 11 December 1999. http://www.ccbh.ac.uk/witness_northsea_index.php.

Anstey, N. and Hempstead, N. (1995), 'Early Seismic and the Growth of Geophysics' in Moreton, R. (ed.) *Tales from Early UK Oil Exploration 1960–1979*. London: Petroleum Exploration Society of Great Britain.

Archer, R. (1991), *Platform and Module Construction*. Offshore Business, No. 35. Canterbury: Smith Rea Energy.

Arnold, G. (1978), *Britain's Oil*. London: Hamish Hamilton.

Bacon, R. and Eltis, W. (1978), *Britain's Economic Problem – Too Few Producers*. 2nd edition. London: Macmillan.

Bamberg, J. (2000), *British Petroleum and Global Oil 1950 – 1975: The Challenge of Nationalism*. Cambridge: Cambridge University Press.

Bank of England. (1980), 'The North Sea and the United Kingdom economy: some longer-term perspectives and implications'. *Bank of England Quarterly Bulletin*, Vol. 20, No. 4, pp. 449–454. London: Bank of England.

Bank of England. (1982), 'North Sea Oil and Gas: Costs and Benefits'. *Bank of England Quarterly Bulletin*, Vol. 22, No. 1, pp. 56–73. London: Bank of England.

Barnett, C. (1986), *The Audit of War*. London: Macmillan.

Barnett, C. (2001), *The Verdict of Peace*. London: Macmillan.

Bjerrum, L. (1973), 'Geotechnical Problems Involving Foundations of Structures in the North Sea'. *Géotechnique*, (No. 3), pp. 319–359. London: Thomas Telford.

Bolton Committee. (1971), *Small Firms. Report of the Committee of Enquiry on Small Firms*. London: Her Majesty's Stationery Office.

BP. (2010), *Statistical Review of World Energy*. http://www.bp.com.

Broakes, N. (1979), *A Growing Concern*. Weidenfeld and Nicholson: London.

Buchanan Smith, A. (1984a), As reported in Parliamentary Debates (Hansard), Sixth Series, Volume 56, House of Commons Official Report, 12th March 1984 – 23rd March 1984. London: HMSO.

Buchanan Smith, A. (1984b), As quoted in *The Scotsman* 22nd March 1984.

Butler, B. (1999), in *The Development of North Sea Oil and Gas*. Institute of Contemporary British History, seminar, 11 December 1999. http://www.ccbh.ac.uk/witness_northsea_index.php

BVCA. (2003), *Report on Investment Activity*. London: British Venture Capital Association.

Byatt, I. C. R., Hartley, N., Lomax, J. R., Powell, S. and Spencer, P. D. (1982), *North Sea Oil and Structural Adjustment, Government Economic Service. Working Paper No.54*. London: H.M. Treasury.

Cairncross, A. (1992), *The British Economy Since 1945, Economic Policy and Performance, 1945–1990*. Oxford: Blackwell.

Cairns, J. A., Harris, A. H. and Williams, H. C. (1986), *Barriers to Entry in the North Sea Offshore Oil Supply Industry*. University of Aberdeen North Sea Study Occasional Paper, No. 24. Aberdeen: University of Aberdeen.

Cameron Bryce, A. (1999), *Under Sand, Ice and Sea*. Toronto: Erraacht Publishing.

Cameron, P. (1986), *The Oil Supplies Industry: A comparative study of legislative restrictions and their impact*. London: Financial Times Business Information.

Central Statistical Office. (1971), *Financial Statistics*. No.116, Table 77. London: Her Majesty's Stationery Office.

Central Statistical Office. (1977a), *Financial Statistics*. No. 188, Table 9.2. London: Her Majesty's Stationery Office.

Central Statistical Office. (1977b), *Annual Abstract of Statistics*. No. 114, Table 17.9. London: Her Majesty's Stationery Office.

Cook, B. (1999), in *The Development of North Sea Oil and Gas*. London: Institute of Contemporary British History, seminar, 11 December 1999. http://www.ccbh.ac.uk/witness_northsea_index .php.

Cook, L. and Surrey, J. (1983), *Government Policy for the Offshore Industry: Britain compared with Norway and France*. Science Policy Research Unit Occasional Paper Series No. 21. Brighton: University of Sussex.

Corley, T. A. B. (1988), *A History of the Burmah Oil Company. Volume II: 1924–1966*. London: Heinemann.

Crafts, N. (2002), *Britain's Relative Economic Performance 1870–1999*. London: Institute of Economic Affairs.

Cross, M. (1986), 'Barrow-in-Furness – 'A centre of offshore expertise'. *Offshore Commentary*. Vol. 2, No. 2, p. 2. Canterbury: Smith Rea Energy.

Cullen, L. (1990), *Report of the Public Enquiry into the Piper Alpha Disaster*. London: Her Majesty's Stationery Office.

Demsetz, H. (1982), 'Barriers to Entry'. *The American Economic Review*, Vol. 72, No. 1, pp. 47–57.

Denton, A. (1999), *The Development of North Sea Oil and Gas*. Institute of Contemporary British History, seminar, 11 December 1999. http://www.ccbh.ac.uk/witness_northsea_index.php.

Department of Energy. (1974a), *Guidance of the design and construction of offshore installations*. London: Her Majesty's Stationery Office.

Department of Energy. (1974b), *Offshore Installations (Construction and Survey Regulations)*. London: Her Majesty's Stationery Office.

Department of Energy. (1975, 1976a–1978), *Annual 'Blue Books' Offshore Oil and Gas: A summary of orders placed by operators of oil and gas fields on the UK Continental Shelf* or *Offshore (Year): An analysis of orders placed*. London: Her Majesty's Stationery Office.

Department of Energy. (1976b), *The Offshore Energy Technology Board: Strategy for Research and Development*. Energy Paper Number 8. London: Her Majesty's Stationery Office.

Department of Energy/Peat Marwick Mitchell & Co. and Atkins Planning. (1976c), *North Sea Costs Escalation Study*. London: Her Majesty's Stationery Office.

Department of Energy. (1979–1991), *Annual 'Brown Books' Development of the Oil and Gas Resources of the United Kingdom*. London: Her Majesty's Stationery Office.

Department of Energy and Climate Change. (2010), Website. http://www.decc.gov.uk/.

Department of Trade and Industry. (1992–2001), *Annual 'Brown Books' Development of the Oil and Gas Resources of the United Kingdom*. London: Her Majesty's Stationery Office.

Dintenfass, M. (1992), *The Decline of Industrial Britain*. London: Routledge.

Douglas-Westwood Ltd. (2009), *The World Offshore Oil and Gas Production and Spend Forecast 2009–2013*. Canterbury: Douglas-Westwood Ltd.

Douglas-Westwood Ltd. (2010), *UK Offshore Wind: Building Industry*. http://www.renewable-uk. com.

Drury, I. (1986), 'Privatisation Sparks Major BTI Offshore Offensive'. *Offshore Engineer,* May 1986, London: Thomas Telford Ltd, pp. 63–65.

Edgerton, D. (1996), 'Science, Technology and the British Industrial Decline 1870–1970'. in Sanderson, M. (ed.) *New Studies in Economic and Social History*. Cambridge: Cambridge University Press.

Elbaum, B. and Lazonick, W. (1986), 'An Institutional Perspective on British Decline' in Elbaum, B. and Lazonick, W. (eds.) *The Decline of the British Economy*. New York: Oxford University Press, pp 1–17.

Evans, H. and Bevan, J. (1990), *The Divers*. Canterbury Museums: Canterbury.

Expro International Group PLC. (1995), *Listing Document*. London: Robert Fleming & Co.

Financial Times. (1964), Shipping Correspondent: 21st September 1964. London: Financial Times.

Financial Times North Sea Letter. (1980), 30 January 1980, Issue No. 229. London: The Financial Times Business Information Ltd.

Foord, J., Robinson, F., and Sadler, D. (1985), *The Quiet Revolution – Economic and Social Change on Teesside 1965–1985*. A special report for the BBC North East. Newcastle-upon-Tyne: BBC North East.

Forsyth, P. J., and Kay, J. A. (1980), 'The Economic Implications of North Sea Oil Revenues'. *Fiscal Studies*, Vol. 1, No. 3, pp. 1–28. London: Institute of Fiscal Studies.

Franks, K. A. and Lambert, P. F. (1985), *Early California Oil – A Photographic History, 1865–1940*. College Station, Texas: Texas A&M University.

Garrad Hassan Ltd. (2010), *UK Offshore Wind: Staying on Track*. http://www.BWEA.com.

Gibson, P. (1973), As quoted in the *Glasgow Herald*, 26th January 1973.

Goodfellow, R. (1977), *Underwater Engineering*. Tulsa: Petroleum Publishing Company.

Groupement des Entreprises Parapétrolières et Paragazières (GEP). (2003–2010), *The French Oil and Gas Supply and Services Industry*. http://www.gep-france.com.

Guinness, J. (1999), in *The Development of North Sea Oil and Gas*. Institute of Contemporary British History, seminar, 11 December 1999. http://www.ccbh.ac.uk/witness_northsea_index. php.

Hall, W. (1973), 'What British banks are doing about North Sea oil'. *The Banker*, Vol. 123, No. 563, pp. 153–163. London: Financial Times.

Hallwood, P. (1986), *The Offshore Oil Supply Industry in Aberdeen: The Affiliates – Their Characteristics and Importance*. University of Aberdeen North Sea Study Occasional Paper, No. 23. Aberdeen: University of Aberdeen.

Hallwood, C. P. (1990), *Transaction Costs and Trade between Multinational Corporations, a Study of Offshore Oil Production,* World Industry Studies: 9. in Inigo W. (ed.) Cambridge, MA: Unwin Hyman.

Harvie, C. (1994), *Fool's Gold*. London: Hamish Hamilton.

Higgins, G. E. (1996), *A History of Trinidad Oil*. Port-of-Spain: Trinidad Express Newspapers.

Hogben, N. and Standing, R. (1974), 'Wave Loads on Large Structures'. Paper 26 International Symposium on the Dynamics of Marine Vehicles and Structures in Waves. London: Royal Institute of Naval Architects.

Howarth, S. (1997), *A Century in Oil: The "Shell" Transport and Trading Company 1897–1997*. London: George Weidenfeld & Nicholson.

INTSOK. (2010), Website. http://www.intsok.no/.

International Management and Engineering Group of Britain Limited (IMEG). (1972), *Study of Potential Benefits to British Industry from Offshore Oil and Gas Developments*. London: Her Majesty's Stationery Office.

Jamieson, A. G. (2003), *Ebb Tide in the British Maritime Industries: Change and Adaptation 1918–1990*. Exeter: University of Exeter Press.

Jenkin, M. (1981), *British Industry and the North Sea: State Intervention in a Developing Industrial Sector*. London and Basingstoke: The Macmillan Press.

Jennings, J. (1984), 'Opportunities arising from North Sea development. Britain and the Sea: Future Dependence – Future Opportunities', *Greenwich Forum IX*. edited by M. B. F. Ranken. Edinburgh: Scottish Academic Press.

Johnman, J. and Murphy, H. (2002), *British Shipbuilding and the State since 1918*. Exeter: University of Exeter Press.

John Wood Group PLC, (2002), *Listing Document*. London: Cazenove & co. and Credit Suisse First Boston.

Kashani, H. A. W. (2005), 'Regulation and Efficiency: An Empirical Analysis of the United Kingdom Continental Shelf Petroleum Industry'. *Energy Policy*, Vol. 33, (No. 7), pp. 915–925. Amsterdam: Elsevier.

Kassler, P. (1999), in *The Development of North Sea Oil and Gas*. Institute of Contemporary British History, seminar, 11 December 1999. http://www.ccbh.ac.uk/witness_northsea_index.php.

Kemp, A. G. and Stephen, L. (2005), 'Optimising Oil And Gas Depletion In The Maturing North Sea With Growing Import Dependence'. *Oxford Review of Economic Policy*, Vol. 21, No. 1, pp.43–66. Oxford: Oxford University Press.

Kirby, M. W. (1981), *The Decline of British Economic Power since 1870*. London: George Allen and Unwin.

Kirby, M. W. (1991), 'The Economic Record'. in Gourish, T. and O'Day, A. (eds.) *Britain Since 1945*. Basingstoke: Macmillan, pp. 11–38.

Liverman, J., (1999), in *The Development of North Sea Oil and Gas*. Institute of Contemporary British History, seminar, 11 December 1999. http://www.ccbh.ac.uk/witness_northsea_index.php.

Lorenz, E. and Wilkinson, J. (1986), 'The Shipbuilding Industry'. in Elbaum, B. and Lazonick, W. (eds.) *The Decline of the British Economy*. New York: Oxford University Press, pp. 109–134.

Mackay Consultants. (1990), *North Sea Oil & Gas Commentary*, November 1990. Inverness: Mackay Consultants.

Malone, G. (1984), As reported in Parliamentary Debates (Hansard), Sixth Series, Volume 56, House of Commons Official Report, 12th March 1984 – 23rd March 1984. London: HMSO.

Manson, H. (2002), Interview with John Trewhella. University of Aberdeen/British Library Lives in the Oil Industry Oral History Project C963–83 5th February 2002. Aberdeen and London: University of Aberdeen and British Library.

McDonald, R. D. (1974), The Design and Field Testing of the "Triton"Tension-Leg Fixed Platform and its Future Application for Petroleum Product ion and processing in Deep Water. Sixth Offshore Technology Conference (OTC Paper 2104). Houston: Offshore Technology Conference.

McKinstry, S. (1997), 'The Rise and Progress of John Brown Engineering 1966–97: US Technology, Scottish Expertise and English Capital'. *Business History*, Vol. 39, No. 3, pp. 105–133. London: Frank Cass.

McKinstry, S. (1998), 'Transforming John Brown's Shipyard: The drilling rig and Offshore Fabrication Business of Marathon and UIE'. *Scottish Economic & Social History*, Vol. 18, Part 1, pp. 33–59. Edinburgh: Edinburgh University Press.

Mercier, J. (1995), 'A Convincing Case for TLP Technology'. *Offshore Engineer*, pp. 56–58. London: Thomas Telford Ltd.

Ministry of Power, (1967), *Fuel Policy – Cmnd. 3438*. London: Her Majesty's Stationery Office.

Morison, J. R., O'Brien, M. P., Johnson, J. W. and Schaaf, S. A. (1950), 'The Force Exerted by Surface Waves on Piles'. *Petroleum Transactions* Vol. 189, pp. 28–46. New York: American Society of Mechanical Engineers.

Morton, A., (1999), in *The Development of North Sea Oil and Gas*. Institute of Contemporary British History, seminar, 11 December 1999. http://www.ccbh.ac.uk/witness_northsea_index.php.

National Economic Development Office. (1968), *Market – The World*. London: Her Majesty's Stationery Office.

National Economic Development Office. (1970), *Large Industrial Sites*. London: Her Majesty's Stationery Office.

National Economic Development Office. (1971), *What's Wrong on Sites*. London: NEDO.

National Economic Development Office. (1976), *Engineering Construction Performance*. London: Her Majesty's Stationery Office.

National Economic Development Office. (1981), *Preassembly for Process Plant Construction*. London: NEDO.

National Economic Development Office. (1982), *Guidelines for the Management of Major Projects in the Process Industries*. London: NEDO.

North Sea Assets PLC. (1985–1993), Annual Reports and Accounts. Edinburgh: North Sea Assets.

National Materials Technology Centre. (2009), *The Supply Chain for a UK Nuclear New Build Programme*. Department of Business, Enterprise and Regulatory Reform. http://www.decc.gov.uk/.

Ocean Star Offshore Drilling Rig and Museum. (2003–2010), http://www.oceanstaroec.com.

Odell, P. (1999), in *The Development of North Sea Oil and Gas*. Institute of Contemporary British History, seminar, 11 December 1999. http://www.ccbh.ac.uk/witness_northsea_index.php.

Odell, P. R. and Rosing, K. E. (1975), *The North Sea Oil Province*. London: Kogan Page.

Odell, P. R. and Rosing, K. E. (1976), *Optimal Development of the North Sea's Oil Fields*. London: Kogan Page.

Offshore Supplies Office. (1975) Press Release Reference Number 177, 7th August 1975. Glasgow: Department of Energy.

Officer, L. H. (2010), *Purchasing Power of British Pounds from 1264 to Present*. MeasuringWorth. http://www: measuringworth.com/ppoweruk/.

Oil and Gas UK. (2010), *Economic Report 2010*. http://www.oilandgasuk.co.uk.

Oppenheimer, P. M. (1976), 'Employment, Balance of Payments and Oil in the United Kingdom', *The Three Banks Review*, No. 109, pp. 3–25. London: Lloyds Bank.

Paape, A. (1969), *Wave Forces on piles in relation to wave energy spectra*. Delft Hydraulics Laboratory Publication No.69. Delft: Hydraulics Laboratory.

Phelps Brown, E. H. and Browne, M. H. (1968), *A Century of Pay*. London: Macmillan.

Pike, W. (2003), 'Overdue recognition'. *Hart's E&P*, Vol. 76, No. 08, pp. 5 and 17. Houston: Chemical Week Publishing.

Pollard, S. (1984), *The Wasting of the British Economy: British Economic Policy 1945 to the Present*, 2nd edition. London: Croom Helm.

Pollard, S. (1992), *The Development of the British Economy 1914–1990*, 4th edition. London: Edward Arnold.

Pratt, J. A. Priest, T. and Castaneda, C. J. (1997), *Offshore Pioneers: Brown & Root and the History of Offshore Oil and Gas*. Houston: Gulf Publishing.

Reddaway, W. B. (1968), *Effects of U.K. Direct Investment Overseas*. Cambridge: Cambridge University Press.

Reid, M. (1982), *The Secondary Banking Crisis*. London and Basingstoke: The Macmillan Press.

Robinson, C. (1999), in *The Development of North Sea Oil and Gas*. Institute of Contemporary British History, seminar, 11 December 1999. http://www.ccbh.ac.uk/witness_northsea_index.php.

Robinson, C. and Morgan, J. (1976), *Effects of North Sea Oil on the United Kingdom's Balance of Payments*. London: Trade Policy Research Centre.

Robinson, C. and Morgan, J. (1978), *North Sea Oil in the Future*. London: Macmillan for Trade Policy Research Centre.

Rundell, W. Jr. (1977), *Early Texas Oil – A Photographic History, 1866–1936*. College Station, Texas and London: Texas A&M University Press.

Searle, A. (2004), *PLUTO: Pipe-Line Under The Ocean* (second edition). Shanklin: Shanklin Chine.

Schempf, F. J. (2004), 'The History of offshore: developing the E & P infrastructure'. *Offshore Magazine*, pp. 21 and 24. Houston: Penwell Publishing.

Secretary of State for Energy. (1974), *United Kingdom Offshore Oil and Gas Policy. A Report to Parliament* – Cmnd. 5696. London: Department of Energy.

Select Committee on Science and Technology. (1974), Offshore Engineering Session 1974. HC 107. London: Her Majesty's Stationery Office.

Select Committee on Energy. (1984), *North Sea Sun Oil's Decision to Place Abroad a Contract for a Floating Production Vessel*. Session 1983–1984. HC 587. London: Hoer Majesty's Stationery Office.

Select Committee on Energy. (1987), *The Effect of Oil and Gas Prices on Activity in the North Sea*. Session 1986–1987. HC 175. London: Her Majesty's Stationery Office.

Sentance, W. T. C. (1991), *Engineering and Project Management*. Offshore Business, No. 34. Canterbury: Smith Rea Energy.

Shell, U. K. (1982), *The North Sea; a springboard for British Industry*. London: Shell UK.

Smith, N. J. (1978), 'The Offshore Supplies Industry – The British Experience in Its Wider Context', *The Business Economist,* Vol. 10, No. 1 Autumn 1978, pp. 6–20. London: The Society of Business Economists.

Smith, N. (1984), 'What Price the British Service/Supply Sector?' *Offshore Commentary*. Vol. 1, No. 2. (p. 1). Canterbury: Smith Rea Energy.

Smith, N. J. (1985a), 'North Sea Oil Procurement – A UK Perspective'. *Oil and Gas Law and Taxation Review,* No. 10. 1984/85, pp. 267–271. Andover: Sweet and Maxwell.

Smith, N. (1985b), 'Capital Starvation Cripples UK Offshore Companies' *Offshore Commentary*. Vol. 2, No. 1, p. 1. Canterbury: Smith Rea Energy.

Smith, N. (2009), 'Lessons to be Learnt'. *Petroleum Review,* pp. 40–41. 47 London: Energy Institute.

Smith Rea Energy/Hoare Govett. (1983a), *European Industry Perspective*. Offshore business Vol. 1. Canterbury and London: Smith Rea Energy/Hoare Govett Investment Research.

Smith Rea Energy/Hoare Govett. (1983b), *Offshore Contract Drilling*. Offshore business Vol. 2. Canterbury and London: Smith Rea Energy/Hoare Govett Investment Research.

Smith Rea Energy/Hoare Govett. (1985), *Oilfield Mud and Cementing*. Offshore business Vol. 2. Canterbury and London: Smith Rea Energy/Hoare Govett Investment Research.

Smith Rea Energy. (1987), *Subsea Engineering and Construction*. Offshore business, No. 22. Canterbury: Smith Rea Energy.

Smith Rea Energy. (1990a), *Reservoir Engineering and Management*. Offshore business, No. 32, Canterbury: Smith Rea Energy.

Smith Rea Energy. (1990b), *Drilling & Well Servicing*. Offshore business, No. 31. Canterbury: Smith Rea Energy.

Smith Rea Energy. (1990c), *Oilfield Mud and Cementing*. Offshore business, No. 33. Canterbury: Smith Rea Energy.

Swann, C. (2007), *The History of Oilfield Diving*. Santa Barbara: Oceanaut Press.

Tempest, L. P. (1979), 'The Financing of North Sea Oil 1975–1980'. *The Bank of England Quarterly Bulletin,* Vol. 19, No. 1, pp. 31–34. London: Bank of England.

The Economist Intelligence Unit Limited. (1984), *The North Sea and British Industry: The New Opportunities*. London: The Economist Intelligence Unit.

The Monopolies and Mergers Commission. (1982), *Trafalgar House PLC and the Peninsular and Oriental Steam Navigation Company A Report on the Proposed Merger*. London: HMSO.

The Offshore Valuation Group/Boston Consulting Group. (2010), *A Valuation of the UK's Offshore Renewable Energy Resource*. Machynlleth: Public Interest Research Centre.

Thirlwall, A. P. and Gibson, H. D. (1992), *Balance-of-Payments Theory and the United Kingdom Experience*, (4th edition). Basingstoke: Macmillan.

Treasury, (1975), *United Kingdom Balance of Payments*. London: Her Majesty's Stationery Office.

Treasury, (1976), 'The North Sea and the Balance of Payments'. *Economic Progress Report* No. 76. pp. 1–3. London: Her Majesty's Stationery Office.

Treasury, (1977a), 'Interest, Profits and Dividends in the Balance of Payments'. *Economic Trends* No. 284. pp. 101–110. London: Her Majesty's Stationery Office.

Treasury, (1977b), 'The North Sea and the UK Economy'. *Economic Progress Report* No. 89. pp. 3–5. London: Her Majesty's Stationery Office.

Turner, F. R. (1996), *The Maunsell Sea Forts*. Gravesend: F.R. Turner.

Upton, D. (1996), *Waves of Fortune*. Chichester: John Wiley and Sons.

Walmsley, P. (1995), 'Early Days: The Search for Commercial Oil'. in Moreton, R. (ed.) *Tales from Early UK Oil Exploration 1960–1970*. London: Petroleum Exploration Society of Great Britain.

Wass, D. (2008), *Decline to Fall*. Oxford and New York: Oxford University Press.

Wiener, M. J. (1981), *English Culture and the Decline of the Industrial Spirit 1850–1980*. Cambridge: Cambridge University Press.

Williams, G. (1972), Oil and Gas Technology Offshore of the United Kingdom. Second 'Financial Times' North Sea Conference. London: Financial Times.

Williams, G. (1999), in *The Development of North Sea Oil and Gas*. Institute of Contemporary British History, seminar, 11 December 1999. http://www.ccbh.ac.uk/witness_northsea_index. php.

Williamson, S. H. (2010), *Six Ways to Compute the Relative Value of a U.S. Dollar Amount, 1774 to Present*. MeasuringWorth. http://www.measuringworth.com/uscompare/.

Wilson Committee. (1977a), *Progress Report on the Financing of Industry and Trade*. Committee to Review the Functioning of Financial Institutions. London: Her Majesty's Stationery Office.

Wilson Committee. (1977b), *Evidence on the Financing of Trade and Industry*. Vols. 1 and 6. Committee to Review the Functioning of Financial Institutions. London: Her Majesty's Stationery Office.

Wilson Committee. (1978), *The Financing of North Sea Oil*. Research Report No.2. Committee to Review the Functioning of Financial Institutions. London: Her Majesty's Stationery Office.

Woodward, G. H. and Woodward, G. S. (1973), *The Secret of Sherwood Forest*. Oklahoma: University of Oklahoma Press.

Yergin, D. (1991), *The Prize – The Epic Quest for Oil, Power & Money*. New York: Simon & Schuster.

Archive Material

BP Archive files

7201 Report On Offshore Drilling in the Gulf of Mexico, April 1955.

39893: Abu Dhabi Marine Areas. Report on Phase 1 of Offshore Drilling Project, 31st January 1958.

119407 An Outline of the Development of Abu Dhabi Marine Areas, August 1958.

79168: Abu Dhabi Marine Areas: Report on History and Future development of the Project, 31st January 1961.

47677 Abu Dhabi Marine Areas: Reports sent to CFP re Technical Meetings, 8th March 1957 – 27th November 1964.

60267: North Sea (1) UK Offshore 13th November 1961 – 29th December 1964.

11993 Agreement between 1) Offshore Venezuela CA 2) Abu Dhabi Marine Areas Ltd., Arabian Gulf, Das Island 23rd October 1962 – 10th October 1966.

23388: Zakum Development Central Production Platforms, Contract between Abu Dhabi Marine Areas Ltd. and the Join-Venture known as DeLong – Hersent –Wimpey, 16th December 1965 – 19th November 1966.

60284: North Sea (4) Drilling Barges, 3rd December 1964 – 28th December 1967.

6654: North Sea Oil, 24th November 1970 – 31st August 1971.

53556: Forties Field – Exim, 31st December 1970 – 25th July 1974.

11998: Abu Dhabi Marine Areas Technical Advisory Committee 19th – 24th March 1971.

55729: North Sea General [2] Forties 17th January 1972 – 20th July 1972.

34043: Forties Field Development Fixed Platforms, Management Contract between BP Petroleum Development Ltd. and Brown and Root (UK) Ltd., Contract No. NS 9809–85–07–01, 8th August 1972.

6914: The Supply Problem in North Sea Operations, 13th January 1972 – 21st September 1972.

34016: North Sea Forties Field Platforms. Notes of Meetings with Brown & Root Running Sequence 5th August 1971 – 28th September 1972.

37882: Development of Forties Field, 22nd October 1971 – 13th May 1974.

62024: BP/Total Abu Dhabi Marine Areas Ltd., British Petroleum Company Ltd., Total Compagnie Française des Pétroles Zakum Subsea Production Scheme – Wellhead ZK 39 Final Report, 30th September 1972.

66085: North Sea – Forties Field 30th June 1973 – 21st June 1974.

10183 Abu Dhabi and North Sea Exploration and Development 1st November 1973 – 31st August 1975.

File 141261: BP Ventures 2nd February 1974 – 3rd February 1978.

The National Archives (TNA): Public Record Office (PRO) files

CAB 128/37 C (63) 23.5, 4th April 1963

POWE 29/388: North Sea oil and gas exploration, 1962–1964.

CAB 129/117/C.P. (64) 82: CONTINENTAL SHELF BILL: REGULATIONS Memorandum by the Minister of Power, 3rd April 1964.

CAB 128/38 CM (64) 47.7 Continental Shelf: Licenses, 10th September 1964.

PREM 13/925: North Sea oil and gas: allocation of further production licences: May 1965 – July 1966.

CAB 129/122 C. (65) 130: FUEL POLICY Memorandum by the Minister of Power, 11th October 1965.

FO 371/187603: North Sea oil and gas, 1966.

CAB 128/41 CC (66) 5.5 Gas Industry North Sea Gas, 3rd February 1966.

PREM 13/1524: North Sea oil and gas pricing arrangements; part 2. September 1966 – August 1967.

POWE 63/360.The Effects of Devaluation on the Price of North Sea Gas. 29th November 1967.

PET 50/469/01. The Effects of Devaluation on the Price of North Sea Gas: Brief for Steering Group on Energy Policy. 29th November 1967.

CAB 129/153 CP (70) 80: THE ENERGY SCENE Memorandum by the Minister of Technology, 12th October 1970.

CAB 128/47–2 CM 31 (70) 6. THE ENERGY SCENE Memorandum by the Minister of Technology, 19th October 1970.

T 292/178 Department of Trade and Industry: North Sea Oil and Gas, 1st January 1971 – 31st December 1972.

CAB 128/49 1 CM 10 (71) 4, 16th February 1971.

CAB 184/61: North Sea oil and gas, 6th August 1971 – 11th October 1972.

CAB 184/61: Potential benefits to British industry from North Sea oil and gas, 14th November 1972 – 29th October 1973.

PREM 15/1595: Sir David Barran of Shell exposition of study of future energy supply and demand, etc., 21st September 1971 – 14th December 1971.

CAB 128/50 2 CM 29 (72) 2, 6th June 1972.

PREM 13/925 Memo from A.P.L.B. to Prime Minister 8th September 1972.

CAB 184/109: Folio 4. North Sea Oil. 22nd November 1972.

CAB 128/50 2 CM 59 (72) 3, 20th December 1972.

BT 241/2580: GATT Implications on Exploitation of North Sea Oil, February–August 1973.

CAB 128/53 CC 48 (73) 2nd November 1973.

CAB 128/53 CC 53 (73) 8th November 1973.

CAB 128/53 CC 60 (73), 12th December 1973.

CAB 128/53 CC 61 (73) 6, 13th December 1973.

FCO 30/243 – North Sea Oil and Gas, 1st January 1974 – 31st December 1974.

CAB 128/54 CC22 (74). 2nd July 1974.

INF 12/1298 Standing Conference on North Sea Oil – Information Sheet, August 1974.

EG10/64 Memo from T.B. Buyers to A. Blackshaw, 31st January 1975.

EG10/64 Brief entitled The Offshore Supplies Office – Its functions, Powers, Responsibilities and Objectives, 18th July 1975.

UKOOA Records

Council Meetings: Minutes 10th July, 13th July, 1974, 8th January, 1975, 10th July, 12th October, 1977, 9th July, 1980, 8th April 1981.

Executive Officers Meetings: Record Notes, 2nd September 1975, 27th September 1984, 3rd January, 4th February, 27th August 1985.

Full & Fair Opportunity Committee: Annual Reports, 1978, 1981, 1983–1984, 1986–1992. Report to Council, 10th January 1979; File Note, 5th August 1985; Briefing Note to Officers for Meeting with Mr. A. Buchanan-Smith on 10th July 1984, 27th June 1984.

Employment Practices Committee: Annual Report, 1977.

Interest Relief Grant Committee: Annual Report, 1980.

Oceanographic Committee: Annual Report, 1976.

Oil Industry Liaison Committee: Minutes of Oil Industry Liaison Committee Meeting, 13th June 1980; Minutes of Oil Industry Liaison Committee Meeting, 24th November 1982; Record Note of Oil Industry Liaison Committee Meeting, 22nd June 1984; Minutes of Oil Industry Liaison Committee Meeting, 3rd July 1984.

Unclassified: Record of a Meeting between a Delegation of the UKOOA and Mr. John Smith, Parliamentary Under-Secretary of State, Department of Energy, 7th August 1975; Record of UKOOA Meeting with Trade Association Representatives at the CBI on 1st September 1975; Letter from J. Liverman, Department of Energy to G. Williams, UKOOA, 16th September 1975; Draft General History of UKOOA, undated.

Other Unpublished Material

Anglo-Iranian Oil Co. Ltd. (1948), *Our Industry – An Introduction to the Petroleum Industry for the use of the Members of the Company's Staff*, 2nd edition. London Privately circulated.

Association Scientifique et Technique pour Exploitation des Océans. (1986), *Analysis of the European Oil-Related Sector*. Paris. Unpublished.

Baring Brothers & Co. Limited. (1974), *Offshore Oil and Gas in North-West Europe*. London. Privately circulated.

Coopers & Lybrand. (1986), *UK Offshore Support Vessel Industry: Final Report*. London. Unpublished.

Kriedler, T. D. (1997), *The Offshore Petroleum Industry – The Formative Years 1945–1962*. PhD Thesis. Texas Tech University. Unpublished.

Institut Européen des Affaires (INSEAD). (1976), *The Regeneration of an Industry: British Shipping 1969–1973*. Working Paper 4/76. Fontainbleau. Privately circulated.

Lloyds Register. (1998), *Lloyd's Register History 1760–2010*. Unpublished draft.

Mackie, W. (2001), *The Impact of North Sea Oil on the North East of Scotland – A Historical Analysis*. PhD thesis. University of Aberdeen. Unpublished.

Mullen, M. (2002), *Industrial Relations – Teesside 1966–1983: Principal Events/Projects/Factors*. Unpublished memorandum.

Novello, S. and Araujo, L. (2006), *Floating Over Troubled Waters: Knowledge Differentiation and Integration in Offshore Platform Design*. Seminar Paper. Aberdeen University Business School 12th February 2006: privately circulated.

Pike, W. J. (1991), *The Development of the North Sea Oil Industry to 1989, with special reference to Scotland's Contribution*. PhD thesis. University of Aberdeen. Unpublished.

Simmons, M. R. (1979), *Simmons & Company International: Our First Five Years 1975 – 1979*. Houston. Circulated privately.

Smith, N. J. (1980), *OSO 1973–1980 Retrospect and Prospect*. Address to Oil Industries Club 4th March 1980. London. unpublished.

Smith, N. J. (1982), *British Underwater Engineering Limited – Victory Against the Odds or Heroic Defeat?* Seminar Paper. London: Business History Unit London School of Economics and Political Science. Unpublished.

Turner, F. R. (2004), *The Maunsell Sea Forts of the Thames Estuary*. Lecture given to the Canterbury Archaeological Society on 13th November 2004. Canterbury. Unpublished.

Individual Contributors

Alcock, P., Allison, W.E., Benn, A.W., Bevan, J., Billington, C., Borrow. M., Byham, M., Chapman, R., Cotterill, A., Ehret, T., Flemng, M., Fudge, S., Henderson, C., Kemp, A.C., Lang, J., Liverman, J., Luff, R., Mabon, J.D., Morton, A., Mullen, M., Olsen, W., Perren, R., Piggin, R., Pridden, D., Rea, L.C., Robinson, G.H., Smith, N.J., Taylor, P. R., Westwood, J. D., Williams, G., Wilson, R.O., Winchester, R., Wootton, R.

Glossary

Terms, Acronyms and Abbreviations

ABB: ASEA Brown Boveri

ABOI: Association of British Oceanic Industries

ADMA: Abu Dhabi Marine Areas

ADS: atmospheric diving suit/system

AEA/UKAEA: United Kingdom Atomic Energy Authority

AGIP: Azienda Generali Italiana Petroli

AGUT: Advisory Group on Underwater Technology

AIM: Alternative Investment Market

Anticlinal: geological structure where strata slope down from crest

Annex B: field development plan submitted for government approval

AOC: Aker Offshore Contracting

API: American Petroleum Institute

APS: AEA Petroleum Services

ASCo: Aberdeen Service Company

ASME: American Society of Mechanical Engineers

ASTEO: Association Scientifique et Technique pour Exploitation des Océans

B&R: Brown & Root

BAC: British Aircraft Corporation

Barrel: 35 Imperial gallons

Bbl: blue barrel (archaic)

BEFL: Burntisland Engineers and Fabricators

BERR: Department for Business Enterprise and Regulatory Reform

BES: Business Expansion Scheme

BGC: British Gas Corporation

BHRA: British Hydromechanics Research Association

Billion: 1000 million

'BIPST': 'British Institute for Petroleum Science and Technology' (hypothetical only)

BLOC: Ben Line Offshore Contractors

Blow-out: loss of pressure control in an oil or gas well

BMEC: British Marine Equipment Council

BNFL: British Nuclear Fuels Limited

BNOC: British National Oil Corporation

BNSG: Burmah North Sea Group

BOC: British Oxygen Company

BOL: British Oceanics Limited

BOP: blow-out preventer

BOSVA: British Offshore Support Vessels Association

BOTB: British Overseas Trade Board

Bottom-tow: installing offshore pipeline by towing it submerged to site

BP: British Petroleum (formerly Anglo-Persian Oil and Anglo-Iranian Oil)

BPI: British Plasterboard Industries

BRIT: British Indigenous Technology Group

BRP: Bureau de Recherches Pétroliers

BRV: Brown & Root Vickers

BS: British Shipbuilders

BSRA: British Shipbuilding Research Association

BSC: British Steel Corporation

BTG: British Technology Group

BT: British Telecommunications

BUE: British Underwater Engineering

BVCA: British Venture Capital Association

BWEA: British Wind Energy Association; now RenewableUK

Casing crews: installers of well liners

CBMPE: Council of British Manufacturers of Petroleum Equipment

CEPM: Comité d'Études Pétrolières Marine

City: London's financial district, or British financial institutions

CJB: Constructors John Brown

CMPT: Centre for Marine and Petroleum Technology

CNEXO: Centre pour l'Exploitation des Océans

Coiled tubing: means of intervening in well

COMEX: Compagnie Maritime d'Expertises

Conical roller cone bit: hardened toothed wheels revolving in response to bit rotation

Continental shelf: submarine extension of land mass

CoP: Code of Practice

Core: rock sample from drilling; product or service critical to upstream activities

CPF: Compagnie Française du Pétroles (now Total)

CPHS: Community Projects in the Hydrocarbons Sector

CPRS: Central Policy Review Staff (support to Cabinet Office under the Heath government)

CRINE: Cost Reduction Initiative for the New Era

CT: Corporation Tax

DEAL: Digital Energy Atlas & Library

DECC: Department of Energy and Climate Change

DEn: Department of Energy

Derrick: framework to support drilling machinery; type of crane

Deviated drilling: use of directionally slanted wells to drain a reservoir from a single drill site

Development area: area of high unemployment receiving industrial support funds

Development capital: expansion funding for private company

DG: Director General

DP: dynamic positioning/dynamically positioned; automatic anchorless marine station-keeping system

DSV: Diving support vessel

DTI: Department of Trade and Industry (prior to 1970, Board of Trade; between 1974 and 1983, split into separate Departments of Industry and of Trade)

DUCO: Dunlop-Coflexip

DVBO: David Brown Vosper Offshore

E&A: exploration and appraisal

EAE: East Anglian Electronics

EAG: Economic Advisory Group

Early Production System: system for production of oil prior to full field development

ECGD: Export Credits Guarantee Department

EC: European Commission or European Community

EEC: European Economic Community

EDF: Électricité de France

EIC: Energy Industries Council

EMC: European Marine Contractors

ENI: Ente Nationali Idrocarburi

EOR: enhanced oil recovery; tertiary oil production techniques

E&P: exploration and production

EPC: engineering, procurement & commissioning

EPIC: engineering, procurement, installation & commissioning

EPS: early production system or Edinburgh Petroleum Services

ERAP: Entreprise de Recherches et d'Activites Pétrolières

ERC: Energy Resource Consultants

ESP: electrical submersible pump (for raising oil)

ETPM: Entrepose pour les Travaux Pétroliers et Maritimes

EU: European Union

Expat: Expatriate; person living outside their country of origin

Expro: Expro International Group PLC or Shell Exploration & Production

FCI: Finance Corporation for Industry

FCO: Foreign and Commonwealth Office

FFO: Full and Fair Opportunity

Fishing: recovery of objects from oil or gas well

FPV: floating production vessel

FUEL: Furness Underwater Engineering Limited

GATT: General Agreement on Trade and Tariffs

GE: General Electric (of USA)

Geneva Convention: agreement (1958) for settling continental shelf jurisdictions

Geoprosco: Geophysical Prospecting Limited

GERTH: Groupement Européen de Recherches Technologiques pour les Hydrocarbures

GEP: Groupement des Entreprises Parapétrolières et Paragazières

GoM: Gulf of Mexico (USA)

Gravimetric survey: mapping of gravity anomalies

GRC: Geophysical Research Corporation

GSI: Geophysical Services Incorporated

Gusher: see blow-out

GVA: Gotaverken Arendal

Hi-Fab: Highlands Fabricators

HMB: HMB Subwork (after founders Hanley, Messervy and Beveridge)

HMG: Her Majesty's Government

Hyperbaric: at pressure equivalent to surrounding sea-water

ICFC: Industrial and Commercial Finance Corporation (now 3i)

ICI: Imperial Chemical Industries

IDAB: Industrial Development Advisory Board

IDE: Industrial Development Executive

IEP: Infrastructure and Energy Projects Directorate (DTI export promotion unit which absorbed OSO)

IDF: International Drilling Fluids

IDU: Industrial Development Unit

IFP: Institut Français du Pétrole

IHC: Industrieele HandelsCombinatie Holland

IMEG: International Management and Engineering Group of Britain Limited

IMF: International Monetary Fund

IMR: inspection, maintenance and repair

INTSOK: Norwegian Oil and Gas Partners, government/industry export promotion body for oil and gas supply chain

IOR: improved oil recovery; advanced secondary as well as tertiary oil production

IP: Institute of Petroleum (now Energy Institute)

IRC: Industrial Reorganisation Corporation

IRD: International Research & Development

IRG: Interest Relief Grant

ISIS: Internationale de Services Industriels et Scientifiques

ISS: Integrated Subsea Services Limited

ITF: Industry Technology Facilitator

IUOOC: Inter-Union Offshore Oil Committee

Jacket: steel platform sub-structure

Jack-knife mast: open-sided steel drilling tower raised vertically by lifting tackle

Jack-up: seabed supported structure with retractable legs

JBE: John Brown Engineering

JBO: John Brown Offshore

JDR Cable Systems: combination of Jacques Cables (UK) and De Regt Special Cables (Netherlands)

'JIM' suit: one-man atmospheric diving suit (see ADS) named after Jim Jarrett who dived on the wreck of the *Lusitania*

JIP: joint industrial project; research project with multiple funders

Jones Act: US legislation used to exclude foreign vessels from US Continental Shelf

JONSWAP: Joint North Sea Wave Programme

Joule 2 programme: EC R&D programme

KBR: Kellog Brown & Root, design and contracting company 'spun-off' by Halliburton

KCA: Kier Calder Arrow Drilling

LASMO: London and Scottish Marine Oil

LIFT: Licence Initiative for Trading

LNG: liquefied natural gas

'Lock-out': transfer of diver to/from work site under hyperbaric conditions by submersible

LOG: Lyle Offshore Group

Logging: detection of presence of hydrocarbon fluids and water by electrical resistivity

LOGIC: Leading Oil and Gas Competitiveness

Marinising: adaptation of standard product for offshore or marine use

MaTSU: Marine Technology Support Unit (of UKAEA)

MCA: Module Constructors Association

Mentor: Mentor Engineering Consultants

Metocean: meteorological and oceanographic data or setting

MoD: Ministry of Defence

Module: pre-assembled equipment package usually for platform topside

MOIRA: Marine Oil Industry Repair Associates

Monohull: ship–shaped as opposed to semi-submersible vessel

MoS: Minister of State (senior minister without Cabinet status)

MoU: Memorandum of Understanding

MSR: Midland & Scottish Resources

MSV: multi(ple)-function field support vessel

MTD: Marine Technology Directorate

Mud: lubricant, pressure retardant and cuttings transport agent used in drilling

Multiphase: reservoir fluids prior to separation of gaseous from liquid components

Naphtha: oil derived feedstock used for gas production

NCB: National Coal Board

NDT: non-destructive testing

NEB: National Enterprise Board

NEDO: National Economic Development Office

NERC: National Environmental Research Council

Neutrabaric: enclosed subsea environment capable of depressurisation

NORSMEC: North Sea Marine Engineering Construction Company

NRDC: National Research and Development Corporation

NSA: North Sea Assets

NSOC: North Sea Operators Committee (later UKOOA *q.v.*)

NSRI: National Subsea Research Institute, co-ordinating body set up by Subsea UK and a group of universities

NWECS: North West European Continental Shelf

ODECO: Offshore Drilling and Exploration Company

OECD: Organisation for Economic Cooperation and development

OETB: Offshore Energy Technology Board

OGITF: Oil and Gas Industry Task Force

Oil and Gas UK: sucessor to UKOOA (*q.v.*)

OILCO: Offshore Industry Liaison Committee

OND: Office for Nuclear Deployment; unit within DECC

OPA: Oil & Pipeline Agency

ORED: Office for Renewable Energy Deployment; unit within DECC

OPEC: Organisation of Petroleum Exporting Countries

Operator: managing partner in an E&P consortium

OSO: Offshore Supplies Office

OSV: offshore support vessel; most commonly a supply boat or anchor-handling tug

OTO: Oil Taxation Office

OTU: Offshore Technology Unit (of DEn)

OWECo: Oil Well Engineering Company

PAC: Public Accounts Committee (of House of Commons)

PC: Programme Committee (of OETB)

PED: Petroleum Engineering Division (of DEn)

PES: Petroleum Engineering Services

'PISB': 'Petroleum Industry Supply Board' (hypothetical only)

Pluto: pipelines under the ocean

P&O: Peninsular and Orient Steam Navigation Company

PPD: Petroleum Production Division

Private Equity: institutional funding of unlisted companies

PRO: Public Record Office

Project Mohole: deep ocean drilling programme funded by US government

PRT: Petroleum Revenue tax

PSBR: Public Sector Borrowing Requirement

PSL: Progenerative Services

PSN: Production Services Network, 'spin-out' from KBR (*q.v.*)

'PSO': 'Petroleum Supplies Office' (hypothetical only)

PSTI: Petroleum Science and Technology Institute

PUS: Permanent Undersecretary (most UK senior civil rank)

PVT: laboratory analysis of the properties of reservoir fluids

RAP: Régie Autonome des Pétroles

R&D: Research and Development

RD&D: Research, Development and Demonstration

RDL: Redpath Dorman Long

Rent: excess profit arising from natural resource exploitation

Reserves: estimate of oil and gas recoverable with current prices and technology

RGC: Redpath de Groot Caledonian

Riser: pipe bringing reservoir fluids to surface; or housing drill string assembly and mud circulation

RN: Royal Navy

ROV: Remotely operated vehicle (subsea)

RR: Rolls-Royce

Salt dome: structure in sedimentary rocks where a large mass of salt has been forced upwards, often associated with presence of oil

Saturation diving system: system of surface chambers and transport diving bells allowing divers to be pressurised to working depth and to make a number of working excursions prior to depressurisation

SBM: Single Buoy Mooring; marine oil loading device (and trade/company name)

SCDI: Scottish Council for Development and Industry, an economic development organisation with members from the private, public and trade union sectors

SDA: Scottish Development Agency

SEAL: Subsea Equipment Associates Limited

Sealab: experimental subsea habitat funded by US government; or as used in vessel name by Wimpey

SEDCO: South Eastern Drilling Company

Seismic survey: use of sound waves to map subsurface features

Semi-submersible: vessel with column supported deck for wave transparency and achieving motion stabilisation from flooding pontoons

SEn: Scottish Enterprise; succeeded SDA (*q.v.*)

SERC: Science and Engineering Research Council

Slew-ring: mechanism at base of crane pedestal permitting rotary movement

SLP: Sea and Land Pipelines

SMEs: small and medium sized enterprises

SMFC: Ship Mortgage Finance Corporation

SMTRB: Shipbuilding and Marine Technology Research Board

SNP: Scottish National Party

SNPA: Société National de Pétroles d'Aquitaine

Sonar: use of acoustic pulses to detect underwater objects

SoS: Secretary of State (a Cabinet Minister)

SSI: Sub Sea International

SSL: Seismograph Service Limited

SSOS: Subsea Oilfield Services

SSS: R. R. Chapman Sub-Sea Surveys Limited (shortened to Sub-Sea Surveys)

STUC: Scottish Trade Union Congress

Submersible: bottom-supported mobile drilling rig for shallow water; manned mini-submarine

Subsea UK: organisation representing UK-based underwater service and product companies

SWOPS: single well operating production system

TDHS: Technology Development in the Hydrocarbon Sector (an EC scheme for partial funding of R&D)

TFL: through flow-line

TH: Trafalgar House

The Offshore Valuation Group: an organisation of renewable energy companies and government agencies

Therm: unit of gas supply equivalent to 100,000 British thermal units

Thermie: successor EC R&D programme to TDHS (*q.v.*), with greater emphasis on demonstration

TJBO: Trafalgar John Brown Offshore

TLP: tension leg platform (buoyant production platform tension-tethered to seabed)

TNA: The National Archives

Ton: 2240 pounds

Tonne: 1000 kilogrammes

Topside: above platform deck

Tubulars: generic term for steel pipes, tubes and large circular sections

2W/2WT: Wharton Williams/Wharton Williams Taylor

UIE: Union Industrielle et d' Entreprise

UK: United Kingdom of Great Britain and Northern Ireland

UKCS: United Kingdom Continental Shelf

UKOOA: United Kingdom Offshore Operators Association

Umbilical: long protected bundle of electrical and other supply/control lines connecting an ROV or subsea system to its power/control point

UMC: underwater manifold centre (gathering node for subsea production wells)

UMEL: Underwater and Marine Equipment Limited

Upstream: E&P activities (*q.v.*)

US/USA: United States (of America)

USM: Unlisted Securities Market

Value added: net output after deduction of secondary inputs; contribution to economy

Vent stacks: means of emergency gas discharge

Venture capital: start-up or early stage development funding

Venture Management: early OSO function

VI: Vickers Intertech

VOEG: Vickers Offshore Engineering Group

VOL: Vickers Oceanics Limited; later BOL (British Oceanics Limited)

VOPD: Vickers Offshore Projects and Developments

VSEL: Vickers Shipbuilding and Engineering

VUPE: Vickers Underwater Pipeline Engineering

Weather-window: period of calm permitting offshore installation

Whipstock: means of directionally slanting a well

Wildcat well: speculative exploration well

Work-over: subsurface maintenance or reinstatement of an oil or gas producing well

Y-ARD: Yarrow-Admiralty Research Department

Index

Industrial Companies:

Industries and Market Segments:

Oil and Gas Companies:

Oil and Gas Fields:

Political Parties:

Reports:

State and International Agencies: